Physics of Solar Energy

Physics of Solar Energy

C. Julian Chen

Department of Applied Physics and Applied Mathematics
Columbia University

JOHN WILEY & SONS, INC.

This book is printed on acid-free paper. ♾

Published by John Wiley & Sons, Inc., Hoboken, New Jersey

Published simultaneously in Canada

Library of Congress Cataloging-in-Publication Data:
Chen, C. Julian.
Physics of solar energy / C. Julian Chen.
p. cm.
Includes index.
ISBN 978-0-470-64780-6 (acid-free paper); ISBN 978-1-118-04457-5 (ebk); ISBN 978-1-118-04458-2 (ebk); ISBN 978-1-118-04459-9 (ebk); ISBN 978-1-118-04831-3 (ebk); ISBN 978-1-118-04832-0 (ebk) 1. Solar energy. 2. Energy development. 3. Solar radiation. I. Title.
TJ811.C54 2011
621.47--dc23

2011017534

Printed in the United States of America

10 9 8 7 6 5 4 3 2 1

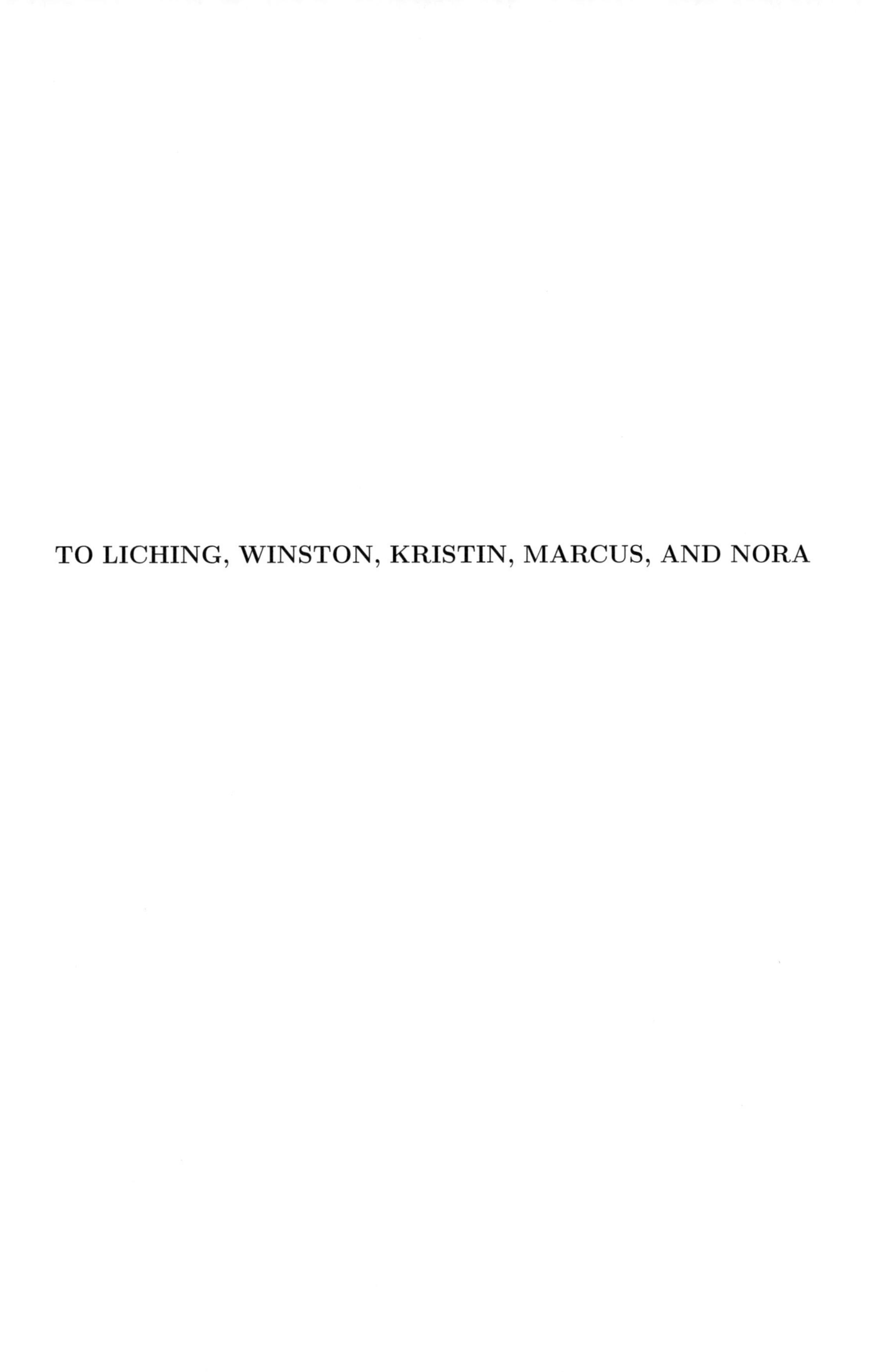

TO LICHING, WINSTON, KRISTIN, MARCUS, AND NORA

Contents

Preface

One of the greatest challenges facing mankind in the twenty-first century is energy. Starting with the industrial revolution in the eighteenth century, fossil fuels such as coal, petroleum, and natural gas have been the main energy resources for everything vital for human society: from steam engines to Otto and diesel engines, from electricity to heating and cooling of buildings, from cooking and hot-water making, from lighting to various electric and electronic gadgets, as well as for most of the transportation means. However, fossil fuel resources as stored solar energy accumulated during hundreds of millions of years are being rapidly depleted by excessive exploration. In addition, the burning of fossil fuels has caused and is causing damage to the environment of Earth.

It is understandable that alternative or renewable energy resources, other than fossil fuels, have been studied and utilized. Hydropower, a derivative of solar energy, currently supplies about 2% of the world's energy consumption. The technology has matured, and the available resources are already heavily explored. Wind energy, also a derivative of solar energy, is being utilized rapidly. The resource of such highly intermittent energy is also limited. Nuclear energy is not renewable. The mineral resource of uranium is limited. The problems of accident prevention and nuclear waste management are still unresolved.

The most abundant energy resource available to human society is solar energy. At 4×10^6 EJ/year, it is ten thousand times the energy consumption of the world in 2007. For example, if 50% of the sunlight shining on the state of New Mexico is converted into useful energy, it can satisfy all the energy needs of the United States.

The utilization of solar energy is as old as human history. However, to date, among various types of renewable energy resources, solar energy is the least utilized. Currently, it only supplies about 0.1% of the world's energy consumption, or 0.00001% of the available solar radiation. Nevertheless, as a result of intensive research and development, the utilization of solar energy, especially solar photovoltaics, is enjoying an amazingly rapid progress. Therefore, it is reasonable to expect that in the latter half of the twenty-first century solar energy will become the main source of energy, surpassing all fossil fuel energy resources.

Similar to other fields of technology, the first step to achieve success in solar energy utilization is to have a good understanding of its basic science. Three years ago, Columbia University launched a master's degree program in solar energy science and engineering. I was asked to give a graduate-level course on the physics of solar energy. In the spring semester of 2009, when the first course was launched, 46 students registered. Columbia's CVN (Columbia Video Network) decided to record the lectures and distribute them to outside students. Because of the high demand, the lectures series for regular students repeated for two more semesters, and the CVN course on the physics of solar energy was repeated for seven consecutive semesters. This book is a compilation of lecture notes.

The basic design of the book is as follows. The first chapter summarizes the energy problem and compares various types of renewable energy resources, including hydropower and wind energy, with solar energy. Chapter 2, "Nature of Solar Radiation," presents the electromagnetic wave theory of Maxwell as well as the photon theory of Einstein. Understanding of blackbody radiation is crucial to the understanding of solar radiation, which is described in detail. Chapter 3, "Origin of Solar Energy," summarizes the astrophysics of solar energy, including the basic parameters and structure of the Sun. The gravitational contraction theory of Lord Kelvin and the nuclear fusion theory of Hans Bethe for the origin of stellar energy are presented. Chapter 4, "Tracking Sunlight," is a self-contained but elementary treatment of the positional astronomy of the Sun for nonastronomy majors. It includes an elementary derivation of the coordinate transformation formulas. It also includes a transparent derivation of the equation of time, the difference of solar time and civil time, as the basis for tracking sunlight based on time as we know it. This chapter is supplemented with a brief summary of spherical trigonometry in Appendix B. The accumulated daily direct solar radiation on various types of surfaces over a year is analyzed with graphics. Chapter 5, "Interaction of Sunlight with Earth," presents both the effect of the atmosphere and the storage of solar energy in the ground, the basis for the so-called shallow geothermal energy. A simplified model for scattered or diffuse sunlight is presented. Chapter 6, "Thermodynamics of Solar Energy," starts with a summary of the basics of thermodynamics followed by several problems of the application of solar energy, including basics of heat pump and refrigeration. Chapters 7–10 deal with basic physics of solar photovoltaics and solar photochemistry. Chapter 7, "Quantum Transition," presents basic concepts of quantum mechanics in Dirac's format, with examples of organic molecules and semiconductors, with a full derivation of the golden rule and the principle of detailed balance. Chapter 8 is dedicated to the essential concept in solar cells, the *pn*-junction. Chapter 9 deals with semiconductor solar cells, including a full derivation of the Shockley–Queisser limit, with descriptions of the detailed structures of crystalline, thin-film, and tandem solar cells. Chapter 10, "Solar Photochemistry," presents an analysis of photosynthesis in plants as well as research in artificial photosynthesis. Various organic solar cells are described, including dye-sensitized solar cells and bilayer organic solar cells. Chapter 11 deals with solar thermal applications, including solar water heaters and solar thermal electricity generators. The vacuum tube collector and the thermal-cipher solar heat collectors are emphasized. Concentration solar energy is also presented, with four types of optical concentrators: through, parabolic dish, heliostat, and especially the compact linear Fresnel concentrator. Chapter 12 deals with energy storage, including sensible and phase-change thermal energy storage systems and rechargeable batteries, especially lithium ion batteries. The last chapter, "Building with Sunshine," introduces architectural principles of solar energy utilization together with civil engineering elements.

Experience in teaching the course has shown me that the student backgrounds are highly diversified, including physics, chemistry, electrical engineering, mechanical engineering, chemical engineering, architecture, civil engineering, environmental science, materials science, aerospace engineering, economy, and finance. Although it is a senior undergraduate and beginning graduate-level course, it must accommodate a broad spectrum of student backgrounds. Therefore, necessary scientific background knowledge is part of the course. The book is designed with this in mind. For example, background knowledge in positional astronomy, thermodynamics, and quantum mechanics is included. For students who have already taken these courses, the background material serves as a quick review and as a reference for the terminology and symbols used in this book. The presentation of the background science is for the purpose of solar energy utilization only, along a "fast track." For example, quantum mechanics is presented using an "empirical" approach, starting from direct perception of quantum states by a scanning tunneling microscope; thus the quantum states are not merely a mathematical tool but a perceptible reality. The scanning tunneling microscope is also an important tool in the research for novel devices in solar energy conversion.

At an insert of the book, a gallery of color graphics and photographs is constructed and compiled. It serves as a visual introduction to the mostly mathematical presentation of the materials, which is useful for intuitive understanding of the concepts.

During the course of giving lectures and writing the lecture notes, I have encountered many unexpected difficulties. Solar energy is a multidisciplinary topic. The subject fields comprise astronomy, thermodynamics, quantum mechanics, solid-state physics, organic chemistry, solid-state electronics, environmental science, mechanical engineering, architecture, and civil engineering. As a unified textbook and reference book, a complete and consistent set of terminology and symbols must be designed which should be as consistent as possible with the established terminology and symbols of the individual fields, but yet be concise and self-consistent. A list of symbols is included toward the end of the book.

I sincerely thank Professors Irving Herman, Richard Osgood, and Vijay Modi for helping me setting up the solar energy course. I am especially grateful to many business executives and researchers in the field of solar energy who provided valuable information: Steve O'Rourke, then Managing Director and Research Analyst of Deutsch Bank, currently Chief Strategy Officer of MEMC Electronics, for detailed analysis of solar photovoltaic industry. John Breckenridge, Managing Director of investment bank Good Energies, for information on renewable energy investment in the world. Robert David de Azevedo, Executive Director of Brazilian American Chamber of Commerce, for information and contacts of renewable energy in Brazil. Loury A. Eldada, Chief Technology Officer of HelioVolt, for manufacture technology of CIGS thin-film solar cells. Ioannis Kymissis, a colleague professor at Columbia University, for two guest lectures in the Solar Energy Course about organic solar cells. Section 10.5 is basically based on literature suggested by him. Vasili Fthenakis, also a colleague professor at Columbia University, for valuable information about economy and environment issues of solar cells. John Perlin, a well-known solar energy historian, for kindly sending me electronic versions of his two books. George Kitzmiller, owner of Miami Pluming and Solar Heating Company, for showing me a number of 80-years-old solar hot water heaters still working in Miami. Margaret O'Donoghue Castillo, President of American Institute of Architects, for introducing me to the geothermal heating and cooling system in AIA, New York City. Mitchell Thomashaw, President of Union College, Maine, for letting me eyewitness the history of solar energy in the United States through brokering the donation of a Carter-era White House solar panel to the Solar Energy Museum in

Dézhōu, China. Academician Hé Zuòxiū, a prominant advocate of renewable energy, for helping me establish contacts in renewable-energy research and industry in China. Lǐ Shēnshēng, Professor Emeritus of Beijing Normal Institute, for kindly gifted me an autographed copy of his out-of-print book Tàiyángnéng Wùlǐxué. Published in 1996, it is probably the first book about the physics of solar energy in any language. Mr. Huáng Míng, founder and CEO of Himin Solar Energy Group and Vice President of International Solar Energy Association, for many inspiring discussions and a visit to Himin Corp, including an impressive production line for vacuum tube solar collectors. Professor Huáng Xuéjié, a long-time researcher of lithium rechargeable batteries and the founder of Phylion Battery Co., for many discussions about electric cars and a tour to the production lines of Phylion. Mire Ma, Vice President of Yingli Green Energy Group, for valuable information and a tour to the entire manufacturing process of solar-grade silicon, solar cells and solar modules. Last but not least, the book could not be written without the patience and support of my wife Liching.

C. Julian Chen

Columbia University
in the City of New York

April 2011

List of Figures

List of Tables

Chapter 1

Introduction

1.1 Solar Energy

According to well-established measurements, the average power density of solar radiation just outside the atmosphere of the Earth is 1366 W/m^2, widely known as the *solar constant.* The definition of the meter is one over 10,000,000 of Earth's meridian, from the North Pole to the equator, see Fig. 1.1. This definition is still pretty accurate according to modern measurements. Therefore, the radius of Earth is $(2/\pi) \times 10^7$ m. The total power of solar radiation reaching Earth is then

$$\text{Solar power} = 1366 \times \frac{4}{\pi} \times 10^{14} \cong 1.73 \times 10^{17}\ \text{W}. \tag{1.1}$$

Each day has 86,400 s, and on average, each year has 365.2422 days. The total energy of solar radiation reaching Earth per year is

$$\text{Annual solar energy} = 1.73 \times 10^{17} \times 86400 \times 365.2422 \cong 5.46 \times 10^{24}\ \text{J}. \tag{1.2}$$

Or 5,460,000 EJ/year. To have an idea of how much energy that is, let us compare it with annual global energy consumption; see Fig. 1.2. In the years 2005–2010, the annual energy consumption of the entire world was about 500 EJ. A mere 0.01% of the annual solar energy reaching Earth can satisfy the energy need of the entire world.

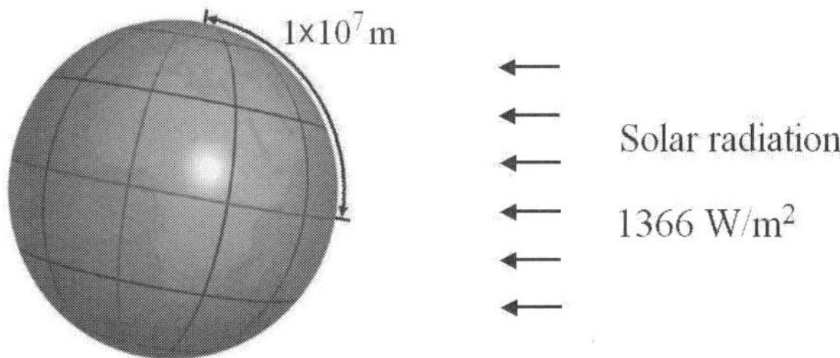

Figure 1.1 Annual solar energy arriving at surface of Earth. The average solar power on the Earth is 1366 W/m^2. The length of the meridian of Earth, according to the definition of the meter, is 10,000,000 m. The total solar energy that arrives at the surface of Earth per year is 5,460,000 EJ.

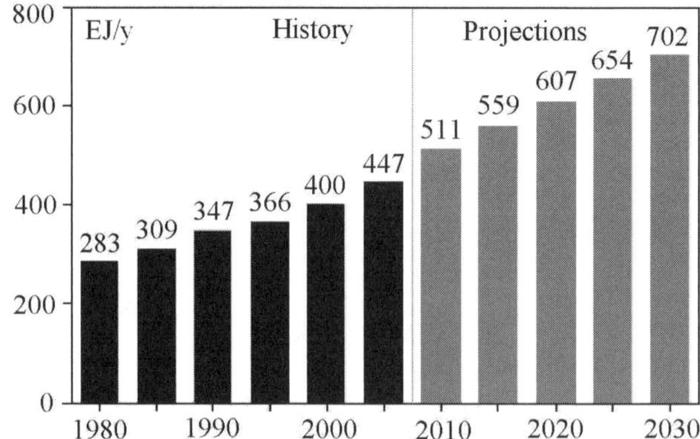

Figure 1.2 World marketed energy consumption, 1980–2030. *Source:* Energy Information Administration (EIA), official energy statistics from U.S. government. History: *International Energy Annual 2004* (May–July 2006), website www.eia.doe.gov/iea. Projections: EIA, *International Information Outlook 2007.*

Not all solar radiation falls on Earth's atmosphere reaches the ground. About 30% of solar radiation is reflected into space. About 20% of solar radiation is absorbed by clouds and molecules in the air; see Chapter 5. About three quarters of the surface of Earth is water. However, even if only 10% of total solar radiation is utilizable, 0.1% of it can power the entire world.

It is interesting to compare the annual solar energy that reaches Earth with the proved total reserve of various fossil fuels; see Table 1.1. The numbers show that the total proved reserves of fossil fuel is approximately 1.4% of the solar energy that reaches the surface of Earth each year. Fossil fuels are solar energy stored as concentrated biomass over many millions of years. Actually, only a small percentage of solar energy was able to be preserved for mankind to explore. The current annual consumption of fossil fuel energy is approximately 300 EJ. If the current level of consumption of fossil

Table 1.1: Proved Resources of Various Fossil Fuels

Item	Quantity	Unit Energy	Energy (EJ)
Crude oil	1.65×10^{11} tons	4.2×10^{10} J/ton	6,930
Natural gas	1.81×10^{14} m^3	3.6×10^7 J/m^3	6,500
High-quality coal	4.9×10^{11} tons	3.1×10^{10} J/ton	15,000
Low-quality coal	4.3×10^{11} tons	1.9×10^{10} J/ton	8,200
Total			36,600

Source: BP Statistical Review of World Energy, June 2007, British Petroleum.

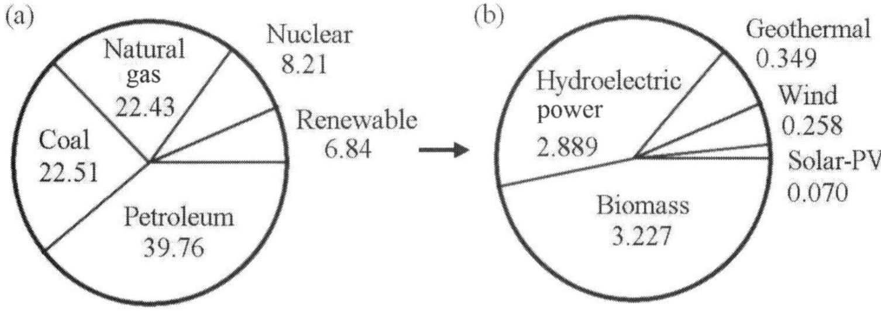

Figure 1.3 U.S. energy consumption, 2006. *Source: Annual Energy Review 2006*, Energy Information Administration (EIA). The unit of energy in the original report is quad, approximately 10^{18}J, or EJ. See Appendix A. In 2006, total U.S. energy consumption was 99.87 quad, almost exactly 100 EJ. Therefore, the energy value in exajoules is almost exactly its percentage. Solar photovoltaic (PV) energy only accounts for 0.07% of total energy consumption in 2006.

fuel continues, the entire fossil energy reserve will be depleted in about 100 years.

Currently, the utilization of renewable energy is still a small percentage of total energy consumption; see Table 1.2. Figure 1.3 shows the percentage of different types of energy in the United States in 2006. The utilization of solar energy through photovoltaic (PV) technology only accounts for 0.07% of total energy consumption. However, globally, solar photovoltaic energy is the fastest growing energy resource. As we will analyze in Section 1.5.4, solar photovoltaics will someday become the dominant source of energy. Figure 1.4 is a prediction by the German Solar Industry Association.

The inevitability that fossil fuel will eventually be replaced by solar energy is simply a geological fact: The total recoverable reserve of crude oil is finite. For example, the United States used to be the largest oil producer in the world. By 1971, about one-half of the recoverable crude oil reserve in the continental United States (the lower 48 states) was depleted. Since then, crude oil production in this area started to decline.

Table 1.2: Renewable Energy Resources

Type	Resource (EJ/year)	Implemented (EJ/year)	Percentage Explored
Solar	2,730,000	0.31	0.0012%
Wind	2,500	4.0	0.16%
Geothermal	1,000	1.2	0.10%
Hydro	52	9.3	18%

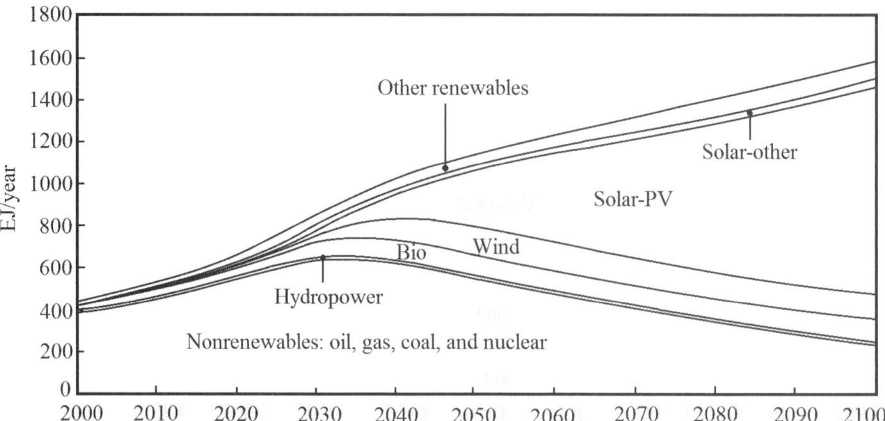

Figure 1.4 Energy industry trend in the twenty-first century. Source of information: German Solar Industry Association, 2007; see www.solarwirtschaft.de. The driving force of twenty-first century energy revolution is economy. Because natural resources of fossil fuels and nuclear materials are finite, the cost of production will increase with time. Solar radiation energy and the raw material to make solar cells, silicon, are inexhaustible. Mass production of solar cells will bring the cost down. At some time, the cost of solar electricity will be lower than that of conventional electricity, to reach *grid parity*. In 2007, it was estimated that grid parity would be reached between 2020 and 2030. After that, an explosive expansion of solar electricity would take place. Recent development indicates that grid parity will be reached around 2015. The rapid implementation of solar electricity will take place sooner than that 2007 prediction. See Section 1.5.4.

Therefore, crude oil production from more difficult geological and environmental conditions must be explored. Not only has the cost of oil drilling increased, but also the energy consumed to generate the crude oil has also increased. To evaluate the merit of an energy production process, the *energy return on energy invested* (EROI), also called *energy balance*, is often used: The definition is

$$\text{EROI} = \frac{\text{energy return}}{\text{energy invested}} = \frac{\text{energy in a volume of fuel}}{\text{energy required to produce it}}. \tag{1.3}$$

In the 1930's, the EROI value to produce crude oil was around 100. In 1970, it was 25. For deep-sea oil drilling, typical value is around 10. Shale oil, shale gas, and tar sands also have low EROI values. If the EROI of an energy production process is decreased to nearly 1, there is no value in pursuing in the process.

On the other hand, although currently the cost of solar electricity is higher than that from fossil fuels, its technology is constantly being improved and the cost is constantly being reduced. As shown in Section 1.5.4, around 2015, the cost of of solar electricity will be lower than conventional electricity, to reach *grid parity*. After that, a rapid growth of solar electrcity will take place; see Fig. 1.4.

1.2 Go beyond Petroleum

Fossil energy resources, especially petroleum, are finite, and depletion will happen sooner or later. The transition to renewable energy is inevitable. This fact was first recognized and quantified by a highly regarded expert in the petroleum industry, Marion King Hubbert (1903–1989). His view is not unique in the oil industry. In 2000, recognizing the eventual depletion of petroleum, former British Petroleum changed its name to "bp beyond petroleum."

In 1956, M. King Hubbert, Chief Consultant of Shell Development Company, presented a widely cited report [40] based on the data available at that time, and predicted that crude oil production in the United States would peak around 1970, then start to decline. His bold and original predictions were scoffed but since then have been proven to be remarkably accurate and overwhelmingly recognized.[1]

His theory started with the discovery that the plots of x, the cumulative production of crude oil Q, versus y, the ratio of production rate P over Q, in the United States follows a straight line; see Fig. 1.5.

The two intersections of the straight line with the coordinate axes are defined as follows. The intersection with the x-axis, Q_0, is the total recoverable crude oil reserve. The value found from Fig. 1.5 is $Q_0 = 228$ billion barrels. The intersection with the y-axis, a, has a dimension of inversed time. The inverse of a is a measure of the duration of crude oil depletion; see below. The value in Fig. 1.5 is $a = 0.0536/\text{year}$. The straight line can be represented by the equation

$$\frac{P}{Q} = a\left(1 - \frac{Q}{Q_0}\right). \tag{1.4}$$

By definition, the relation between P and Q is

$$P = \frac{dQ}{dt}, \tag{1.5}$$

where t is time, usually expressed in years. Using Eq. 1.5, Eq. 1.4 becomes an ordinary differential equation

$$\frac{Q_0\, dQ}{Q(Q_0 - Q)} = a\, dt. \tag{1.6}$$

Equation 1.6 can be easily integrated to

$$\int \frac{Q_0\, dQ}{Q(Q_0 - Q)} = -\ln\left(\frac{Q_0}{Q} - 1\right) = a(t - t_{\text{m}}), \tag{1.7}$$

where t_{m} is a constant of integration to be determined. From Eq. 1.7,

$$Q = \frac{Q_0}{1 + e^{-a(t - t_{\text{m}})}}. \tag{1.8}$$

[1]The mathematics of Hubbert's theory is similar to the equations created by Pierre François Verhurst in 1838 to quantify Malthus's theory on population growth [85].

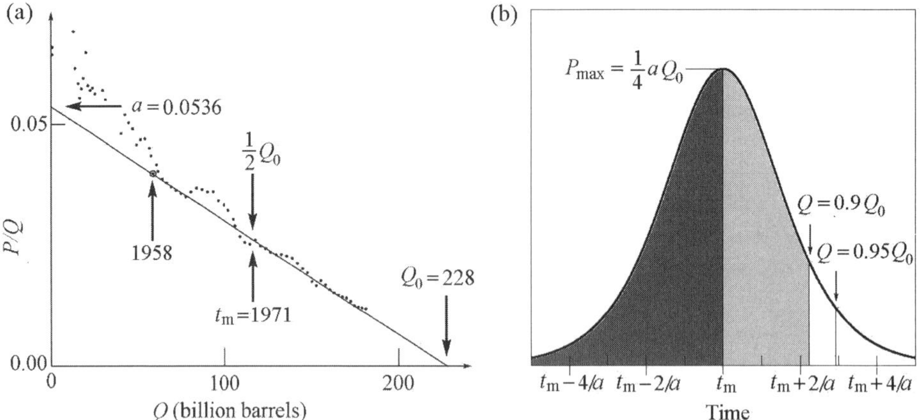

Figure 1.5 Hubbert's curve. (a) In 1956, Merion King Hubbert of Shell Oil studied the data of the cumulative production of crude oil, Q, and the rate of production, P, in the United States. He discovered a linear dependence between P/Q and Q. After Ref. [21]. (b) A curve of Q versus time can be derived from the linear relation, Hubbert's curve (Eq. 1.9). The peak of production occurs at time t_m when one-half of the crude oil is depleted. At time $t_m + 2.197/a$, 90% of the recoverable crude oil is depleted. At time $t_m + 2.944/a$, 95% of the recoverable crude oil is depleted.

The initial and final conditions, $Q = 0$ at $t = -\infty$ and $Q = Q_0$ at $t = +\infty$, are satisfied. The time at which one half of the crude oil is depleted, $Q = Q_0/2$ at $t = t_m$, can be determined from historical data.

The rate of production P can be obtained using Eqs. 1.5 and 1.8,

$$P = \frac{dQ}{dt} = \frac{1}{4} a Q_0 \operatorname{sech}^2 \frac{a(t - t_m)}{2}. \tag{1.9}$$

Equation 1.9 represents a bell-shaped curve[2] symmetric with respect to t at $t = t_m$, see Fig. 1.5(b). Therefore, $t = t_m$ is also the time (year) of maximum production rate, $P_0 = aQ_0/4$. The quantity a is a measure of the rate of oil field depletion. Actually, the time when 90% of crude oil is depleted can be determined by Eq. 1.8,

$$\frac{Q_0}{1 + e^{-a(t_{0.9} - t_m)}} = 0.9\, Q_0, \tag{1.10}$$

which yields $t_{0.9} = t_m + 2.197/a$. Defining depletion time as the time when 95% of crude oil is depleted yields $t_{0.95} = t_m + 2.944/a$.

Figure 1.6 shows the crude oil production rate in the United States from 1920 to 2010. The solid curve is a least-squares fitting with a Hubbert curve; Eq. 1.9. The peak reached in 1971 represents the crude oil production of the lower 48 states (excluding Alaska and Hawaii). There is another peak at around 1989. In 1977, the U.S. Congress passed a law to start drilling for crude oil in Alaska. Because Alaska did not produce

[2] By definition, $\operatorname{sech} x = 1/\cosh x = 2/(e^x + e^{-x})$.

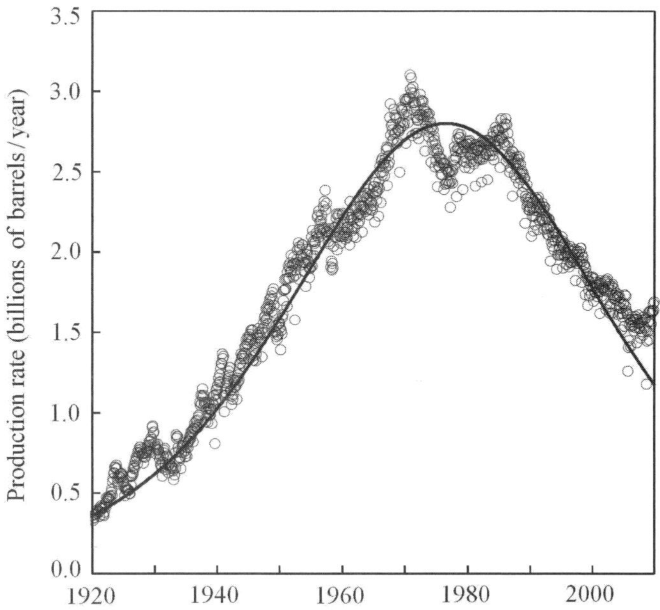

Figure 1.6 Rate of U.S. production of crude oil. Circles: actual U.S. production rate of crude oil. *Source:* EIA (U.S. Energy Information Administration). Solid curve: Hubbert function, using a least-squares fit to actual data. The sudden increase of production in 1977 and the second peak in 1989 are due to oil production in Alaska; see Fig. 1.7.

any crude oil before the 1970s, according to the theory of Hubbert, it should be treated as a stand-alone case independent of the lower 48 states. Plotting the data of crude oil production in Alaska published by the EIA, except for the earlier years, the ratio P/Q shows a rather accurate linear dependence on the accumulative production Q; see Fig. 1.7. From the plot $Q_0 = 17.3$ billion barrels, $a = 0.1646$, and $t_{\mathrm{m}} = 1989.38$, at about May 1989. Using those parameters, a Hubbert curve is constructed; see Fig 1.7(b). As shown, except in the early years, the production data follow the Hubbert curve rather accurately.

The date of depletion can be estimated from the parameters. For the entire United States, $a = 0.0536$. The date of 95% depletion is

$$t_{0.95} = 1971 + \frac{2.944}{0.0536} \approx 2026. \tag{1.11}$$

For Alaska, the date of 95% depletion is

$$t_{0.95} = 1989 + \frac{2.944}{0.1646} \approx 2007. \tag{1.12}$$

The depletion date of Alaska crude oil is sooner than the depletion date for the entire United States. Although crude oil in Alaska started to produce much later than the

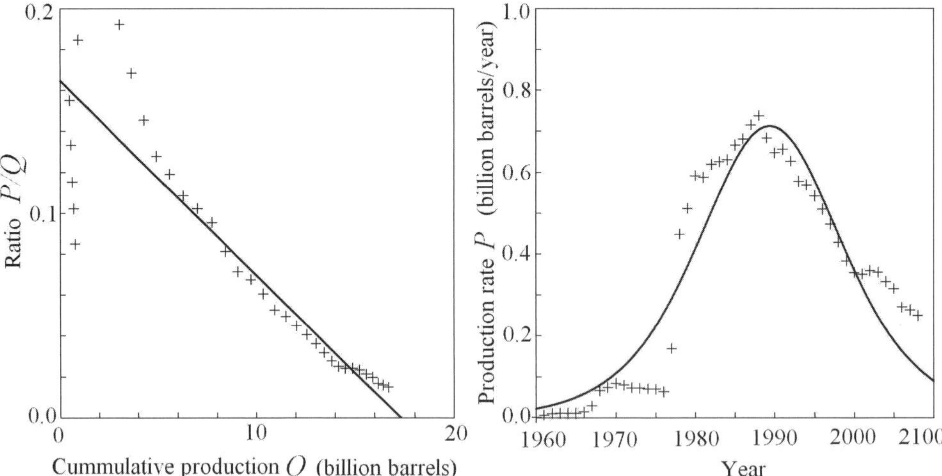

Figure 1.7 Production of crude oil in Alaska. (a) The ratio of P/Q versus Q for Alaska. *Source:* EIA. (b) A Hubbert curve constructed accordingly. As shown, except in the early years, the production data follows the Hubbert curve rather accurately.

lower 48 states, it is extracted much more aggressively there than in the other U.S. states.

Because different countries started crude oil production at different times, a better way of applying the Hubbert theory is to look at each country individually. Hubbert made an estimate based on the data available in the 1950s for the entire world and predicted a peak of crude oil production in the 2000s. Estimates based on more recent data came up with similar results. Recent data show that this peak actually already occurred. The process of discovery and depletion of other nonrenewable energy resources, such as natural gas and coal, follows a similar pattern. As resources are dwindling, the engineering and environmental costs of fossil fuel exploration are increasing rapidly. The Deepwater Horizon oil spill was a wakeup call that in the twenty-first century we must find and utilize renewable energy resources to gradually replace fossil energy resources.

1.3 Other Renewable Energy Resources

Because of the limited reserve of fossil fuel and the cost, from the beginning of the industrial age, renewable energy resources have been explored. Although solar energy is by far the largest resource of renewable energy, other renewable energy resources, including hydropower, wind power, and shallow and deep geothermal energy, have been extensively utilized. Except for deep geothermal energy, all of them are derived from solar energy.

1.3.1 Hydroelectric Power

Hydroelectric power is a well-established technology. Since the late nineteenth century, it has been producing substantial amounts of energy at competitive prices. Currently, it produces about one sixth of the world's electric output, which is over 90% of all renewable energy. As shown in Fig. 1.8, for many countries, hydropower accounts for a large percentage of total electricity. For example, Norway generates more than 98% of all its electricity from hydropower; in Brazil, Iceland, and Colombia, more than 80% of electricity is generated by hydropower. Table 1.3 lists the utilization of hydropower in various regions on the world.

The physics of hydropower is straightforward. A hydropower system is characterized by the *effective head*, the height H of the water fall, in meters; and the *flow rate*, the rate of water flowing through the turbine, Q, in cubic meters per second. The power carried by the water mass is given as

$$P(\mathrm{kW}) = g \times Q \times H, \tag{1.13}$$

where g, 9.81 m/s^2, is the gravitational acceleration. Because a 2% error is insignificant, in the engineering community, it always takes $g \approx 10\,\mathrm{m/s}^2$. Thus, in terms of kilowatts,

$$P(\mathrm{kW}) = 10 \times Q \times H. \tag{1.14}$$

The standard equipment is the Francis turbine, invented by American engineer James B. Francis in 1848. With this machine, the efficiency η of converting water power to mechanical power is very high. Under optimum conditions, the overall efficiency of converting water power into electricity is greater than 90%, which makes it one of the most efficient machines. The electric power generated by the hydroelectric system is

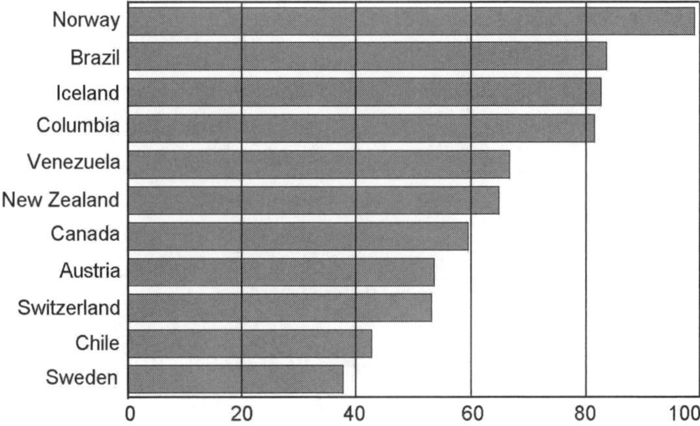

Figure 1.8 Percentage of electricity generation from hydropower in various countries. Norway generates virtually all its electricity from hydropower; in Brazil, Iceland, and Colombia, more than 80% of electricity is generated by hydropower. .

Table 1.3: Regional Hydropotential and Output

Region	Output (EJ/year)	Resource (EJ/year)	Percentage Explored
Europe	2.62	9.74	27%
North America	2.39	6.02	40%
Asia	2.06	18.35	11%
Africa	0.29	6.80	4.2%
South America	1.83	10.05	18%
Oceania	0.14	0.84	17%
World	9.33	51.76	18%

Source: Ref. [14], Chapter 5.

$$P(\text{kW}) = 10\,\eta\,QH. \tag{1.15}$$

A significant advantage over other renewable energy resources is that hydropower provides an energy storage mechanism of very high round-trip efficiency. The energy loss in the storage process is negligible. Therefore, the hydropower station together with the reservoir makes a highly efficient and economic energy storage system. Figure 1.9 is a photo of one of the world's largest hydropower station, the Itaipu hydropower station, which supplies about 20% of Brazil's electricity.

Figure 1.9 Itaipu hydropower station at border of Brazil and Paraguay. With a capacity of 14.0 GW, the Itaipu hydropower station is one of the world's largest and generates about 20% of Brazil's electricity.

1.3.2 Wind Power

The kinetic energy in a volume of air with mass m and velocity v is

$$\text{Kinetic energy} = \frac{1}{2}mv^2. \tag{1.16}$$

If the density of air is ρ, the mass of air passing through a surface of area A perpendicular to the velocity of wind per unit time is

$$m = \rho v A. \tag{1.17}$$

The wind power P_0, or the kinetic energy of air moving through an area A per unit time, is then

$$P_0 = \rho v A \times \frac{1}{2}v^2 = \frac{1}{2}\rho v^3 A. \tag{1.18}$$

Under standard conditions (1 atm pressure and 18°C), the density of air is 1.225 kg/m^3. If the wind speed is 10 m/s, the wind power density is

$$P_0 \approx 610 \text{ W/m}^2. \tag{1.19}$$

It is of the same order of magnitude as the solar power density.

However, the efficiency of a wind turbine is not as high as that of hydropower. Because the air velocity before the rotor, v_1, and the air velocity after the rotor, v_2, are different, see Figure 1.10, the air mass flowing through area A per unit time is determined by the *average wind speed* at the rotor,

$$m = \rho A \, \frac{v_1 + v_2}{2}. \tag{1.20}$$

Thus, the kinetic energy picked up by the rotor is

$$\text{Kinetic energy difference} = \frac{1}{2}mv_1^2 - \frac{1}{2}mv_2^2. \tag{1.21}$$

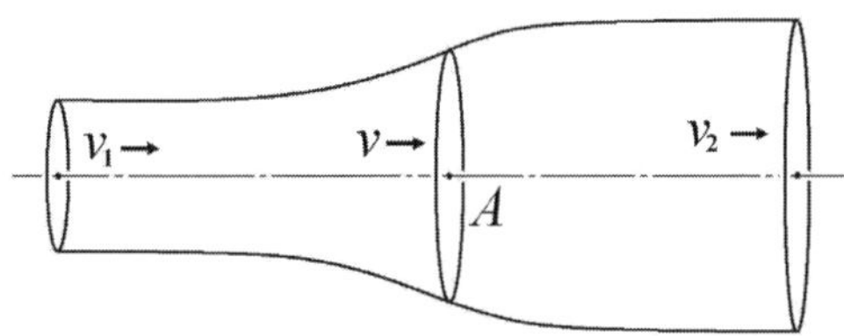

Figure 1.10 Derivation of Betz theorem of wind turbine. Wind velocity before the turbine rotor is v_1, and wind velocity after the turbine rotor is v_2. The velocity at the rotor is the average velocity, and the power generated by the rotor is related to the difference in kinetic energy.

Combining Eqs. 1.20 and 1.21, we obtain an expression of the wind power P picked up by the rotor,

$$P = \frac{1}{4}\rho A(v_1 + v_2)\left[v_1^2 - v_2^2\right]. \tag{1.22}$$

Rearranging Eq. 1.22, we can define the fraction C of wind power picked up by the rotor, or the *rotor efficiency*, as

$$P = \frac{1}{2}\rho v_1^3 A \left[\frac{1}{2}\left(1 + \frac{v_2}{v_1}\right)\left(1 - \frac{v_2^2}{v_1^2}\right)\right] = P_0\, C. \tag{1.23}$$

Hence,

$$C = \frac{1}{2}\left(1 + \frac{v_2}{v_1}\right)\left(1 - \frac{v_2^2}{v_1^2}\right). \tag{1.24}$$

Let $x = v_2/v_1$ be the ratio of the wind speed after the rotor and the wind speed before the rotor. Then we have

$$C = \frac{1}{2}(1 + x)\left(1 - x^2\right). \tag{1.25}$$

The dependence of rotor efficiency C with speed ratio x is shown in Fig. 1.11. It is straightforward to show that the maximum occurs at $x = 1/3$ where $c = 16/27 = 59.3\%$. This result was first derived by Albert Betz in 1919 and is widely known as Betz's theorem or the Betz limit.

The estimate of worldwide available wind power varies. A conservative estimate shows that the total available wind power, 75 TW, is more than five times the world's total energy consumption. In contrast to hydropower, currently, only a small fraction of wind power has been utilized. However, it is growing very fast. From 2000 to 2009,

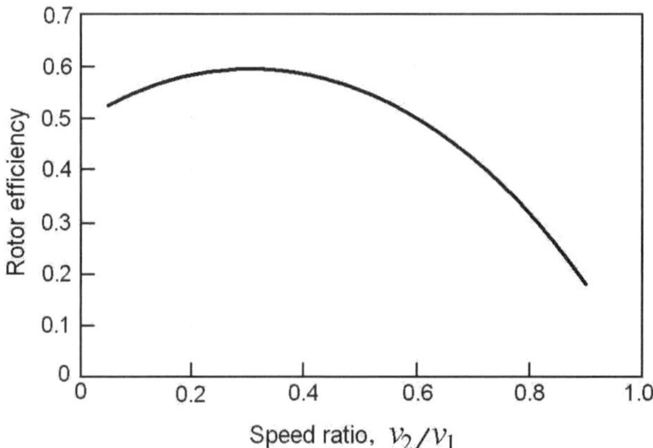

Figure 1.11 Efficiency of wind turbine. See Eq. 1.26. As shown, the maximum efficiency is 16/27, which occurs at a speed ratio of $v_2/v_1 = 1/3$.

Figure 1.12 Wind turbines in Copenhagen. A photo taken by the author in Copenhagen, Denmark, 2006. The statue of the little mermaid, a national symbol of Denmark, is staring at a dense array of wind turbines rather than the Prince.

total capacity grew nine fold to 158.5 GW. The Global Wind Energy Council expects that by 2014, total wind power capacity will reach 409 GW.

Because of a shortage of conventional energy resources, in the late nineteenth century, Denmark began to developed wind power and accelerated production after the 1970 energy crisis. Denmark is still the largest manufacturer of wind turbines, led by Vestas Cooperation, and it has about 20% of wind power in its electricity blend. Figure 1.12 is a photo the author took in Copenhagen. The little mermaid is staring at a dense array of wind turbines instead of the Prince.

However, Denmark's success in wind energy could not be achieved without its neighbors: Norway, Sweden, and Germany [45]. Because wind power is intermittent and irregular, a stable supply of electricity must be accomplished with a fast-responding power generation system with energy storage. Fortunately, almost 100% of the electricity in Norway is generated by hydropower, and the grids of the two countries share a 1000-MW interconnection. In periods of heavy wind, the excess power generated in Denmark is fed into the grid in Norway. By using the reversible turbine, the surplus electrical energy is stored as potential energy of water in the reservoirs. In 2005, the author visited the Tonstad Hydropower Station in Norway on a Sunday afternoon. I asked a Norwegian engineer why the largest turbine was sitting idle. He explained that one of the missions of that power station is to supply power to Denmark. On Monday morning, when the Danes brew their coffee and start to work, that turbine would run full speed.

1.3.3 Biomass and Bioenergy

Over the many thousands of years of human history, until the industrial revolution when fossil fuels began to be used, the direct use of biomass was the main source of energy. Wood, straw, and animal waste were used for space heating and cooking. Candle (made of whale fat) and vegetable oil were used for light. The mechanical power of the horse was energized by feeding biomass. In less developed countries of the world, this situation remains the norm. Even in well-developed countries, direct use of biomass is still very common: for example, firewood for fireplaces and wood-burning stoves.

Biomass is created by photosynthesis from sunlight. For details, see Section 10.1. Although the efficiency of photosynthesis is only about 5% and land coverage by leaves is only a few percent, the total energy currently stored in terrestrial biomass is estimated to be 25,000 EJ, roughly equal to the energy content of the known fossil fuel reserve of the world, see Table 1.1. The energy content of the annual production of land biomass is about six times the total energy consumption of the world; see Table 1.4.

Currently there is a well-established industry to generate liquid fuel using biomass for transportation. Two approaches are widely used: produce ethanol from sugar and produce biodiesel from vegetable oil or animal oil.

Ethanol from Sugar Fermentation

The art of producing wine and liquor from sugar by fermentation has been known for thousands of years. Under the action of the enzymes in certain yeasts, sugar is converted into ethanol and CO_2:

$$C_6H_{12}O_6 \longrightarrow 2(C_2H_5OH) + 2(CO_2). \tag{1.26}$$

At the end of the reaction, the concentration of ethanol can reach 10–15% in the mixture, using specially cultured yeast, it can be up to 21%. Ethanol is then extracted by distillation.

One of the most successful examples is the production of ethanol from sugarcane in Brazil. An important number in the energy industry is *energy balance*, or EROI,

Table 1.4: Basic Data of Bioenergy

Item	in EJ/year	in TW
Rate of energy storage by land biomass	3000 EJ/year	95 TW
Total worldwide energy consumption	500 EJ/year	15 TW
Worldwide biomass consumption	56 EJ/year	1.6 TW
Worldwide food mass consumption	16 EJ/year	0.5 TW

Source: Ref. [14], page 107.

Figure 1.13 Costa Pinto Production Plant of sugar ethanol. The foreground shows the receiving operation of the sugarcane harvest; on the right in the background is the distillation facility where ethanol is produced. Courtesy of Mariordo.

the ratio of energy returned over energy invested; see Eq. 1.3. According to various studies, the energy balance in Brazil for sugar ethanol is over 8, which means that, in order to produce 1 J equivalent of ethanol, about 0.125 J of input energy is required. Also, the cost to produce 1 gal of ethanol in Brazil is about \$0.83, much less than the cost of 1 gal of gasoline. This is at least partially due to the climate and topography of São Paulo, the south-east state of Brazil, a flat subtropical region with plenty of rainfall and sunshine. Since the advance of the flex-fuel automobiles in 2003 which can efficiently use an mixture of gasoline and ethanol of any proportion, the consumption of ethanol dramatically increased because of its low price. In 2008, as a world first, ethanol overtook gasoline as Brazil's most used motor fuel [49]. Figure 1.13 is a panoramic view of the Costa Pinto Production Plant for producing ethanol located in Piracicaba, São Paulo state. The foreground shows the receiving operation of the sugarcane harvest. On the right side, in the background, is the distillation facility where ethanol is produced. This plant produces all the electricity it needs from the bagasse of sugarcane left over by the milling process, and it sells the surplus electricity to public utilities.

Although the Brazil government makes no direct subsidy for the use of ethanol, there is continuous government-supported research to improve the efficiency of production and mechanization of the process. It is an important factor for the success of the Brazilian sugar-ethanol project. From 1975 to 2003, yield has grown from 2 to 6 m^3/ha. Recently, in the state of São Paulo, it has reached 9 m^3/ha. Figure 1.14 shows the annual production of fuel-grade ethanol in Brazil from 1975 to 2010. Although Brazil currently produces more than 50% of the fuel for domestic automobiles and about 30% of the world's traded ethanol, it only uses 1.5% of its arable land.

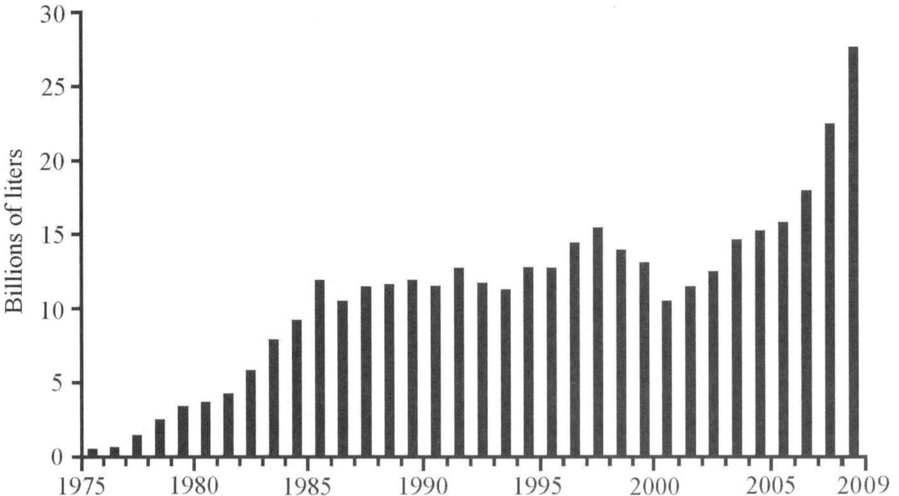

Figure 1.14 Annual production of ethanol in Brazil. *Source: Anuário Estatístico da Agroenergia* 2009, Ministério da Agricultura, Pecuária e Abastecimento, Brazil.

Biodiesel from Vegetable Oil or Animal Fat

Another example of using biomass for liquid fuel is the production of biodiesel from vegetable oil or animal fat. The chemical structures of vegetable oil and animal fat molecules are identical: triglyceride formed from a single molecule of glycerol and three molecules of fatty acid; see Fig. 1.15. A fatty acid is a carboxylic acid (characterized by a —COOH group) with a long unbranched carbon hydride chain. With different types of fatty acids, different types of triglycerides are formed. Although vegetable oil can be used in diesel engines directly, the large molecule size and the resulting high viscosity as well as the tendency of incomplete combustion could damage the engine. Commercial biodiesel is made from reacting triglyceride with alcohol, typically methanol or ethanol. Using sodium hydroxide or potassium hydroxide as catalyst, the triglyceride is transesterified to form three small esters and a free glycerin; see Fig. 1.15. The ester is immiscible with glycerin, and its specific gravity (typically 0.86–0.9 g/cm^3) is much lower than that of glycerin (1.15 g/cm^3). Therefore, the biodiesel can be easily separated from the mixture of glycerin and residuals.

The biodiesel thus produced has a much smaller molecule size than the triglycerides, which provides better lubrication to the engine parts. It was reported that the property of biodiesel is even better than petroleum-derived diesel oil in terms of lubricating properties and cetane ratings, although the calorific value is about 9% lower. Another advantage of biodiesel is the absence of sulfur, a severe environmental hazard of petroleum-derived diesel oil.

The cost and productivity of biodiesel depend critically on the yield and cost of the feedstock. Recycled grease, for example, used oil in making French fries and grease re-

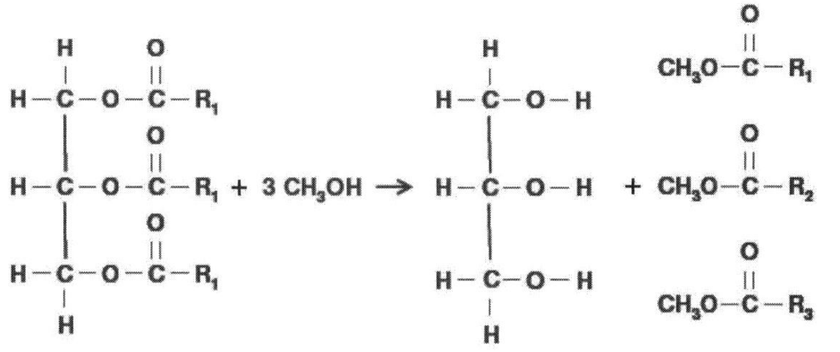

Fat/oil (triglyceride) + Methanol → Glycerol + Fatty acid methyl esters

Figure 1.15 Production process of biodiesel. By mixing triglyceride with alcohol, using a catalyst, the triglyceride is transesterified to form three esters and a free glycerin. The ester, or the biodiesel, has a much smaller molecule size, which provides better lubrication to the engine parts.

covered from restaurant waste, is a primary source of the raw materials. Byproducts of the food industry, such as lard and chicken fat, often considered unhealthy for humans, are also frequently used. However, the availability of those handy resources is limited. Virgin oil is thus the bulk of the feedstock of biodiesel. The yield and cost of virgin oil vary considerably from crop to crop; see Table 1.5.

In Table 1.5, several crops for producing biofuels are listed, including those for producing ethanol. Two of them are sugar-rich roots (sugar beet and sweet sorghum). Harvesting these roots takes much more energy and is more labor intensive than harvesting sugarcane. Therefore, the energy balance (the ratio of energy produced versus the energy required to produce it) is often around 2, much lower than the case of sugarcane, which is higher than 8. The energy balance of corn is also lower (around 2), because the first step is to convert corn starch into sugar, which requires energy and labor. Palm oil, originally from Africa, has the highest yield per unit area of land.

Table 1.5: Yield of Biofuel from Different Crops

For Ethanol	m^3/ha	For Biodiesel	m^3/ha
Sugar beet (France)	6.67	Palm oil	4.75
Sugarcane (Brazil)	6.19	Coconut	2.15
Sweet sorghum (India)	3.50	Rapeseed	0.95
Corn (U.S.)	3.31	Peanut	0.84

Source: Ref. [14], Chapter 4.

Figure 1.16 Oil palm fruit. The size and structure of oil palm fruit are similar to a peach or a plum. However, the soft tissue of the fruit contains about 50% palm oil. The yield of palm oil per unit plantation area is much higher than any other source of edible oil. The kernel is also rich in oil but of a different type. The oil palm kernel oil is a critical ingredient of soap.

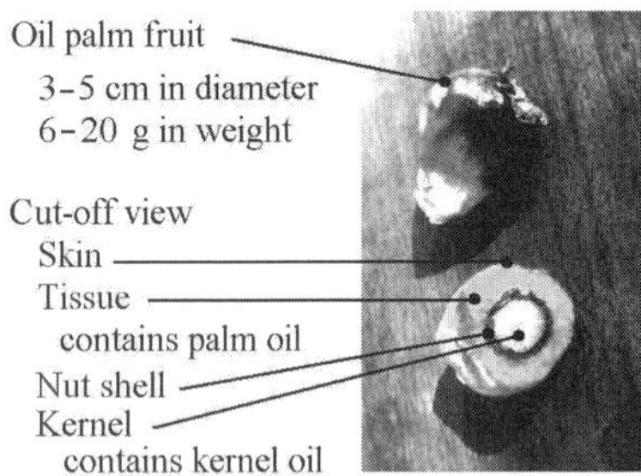

A photograph of the oil palm fruit is shown in Fig. 1.16. The fruit is typically 3–5 cm in diameter. The soft tissue of the fruit contains about 50% palm oil. The kernel contains another type of oil, the palm kernel oil, a critical ingredient for soap. Under favorable conditions, the yield of palm oil could easily reach 5 tonnes per hectare per year, far outstriping any other source of edible oil. Because it contains no cholesterol, it is also a healthy food oil. Currently, palm oil is the number one vegetable oil on the world market (48 million tonnes, or 30% of the world market share), with Malaysia and Indonesia the largest producers. Unlike other types of oil-producing plants (such as soybean and rapeseeds), which are annual, oil palms are huge trees; see Fig. 1.17. Once planted, an oil palm can produce oil for several decades.

Figure 1.17 Wild oil palms in Africa. Oil palms are native trees in Africa which have supplied palm oil for centuries. Shown is a photo of wild oil palms taken by Marco Schmidt on the slopes of Mt. Cameroon, Cameroon, Africa.

1.3.4 Shallow Geothermal Energy

By definition, geothermal energy is the extraction of energy stored in Earth. However, there are two distinct types of geothermal energy depending on its origin: shallow and deep geothermal energy. Shallow geothermal energy is the solar energy stored in Earth, the origin of which will be described in Section 5.4. The temperature is typically some 10°C off that of the surface. The major application of shallow geothermal energy is to enhance the efficiency of the electrical heater and cooler (air conditioner) by using a vapor compression heat pump or refrigerator. Deep geothermal energy is the heat stored in the core and mantel of Earth. The temperature could be hundreds of degrees Celsius. It can be used for generating electricity and large-scale space heating. In this section, we will concentrate on shallow geothermal energy. Deep geothermal energy is presented in the following section.

The general behavior of the underground temperature distribution is shown in Fig. 1.18. At a great depth, for example, 20–30 m underground, the temperature is the annual average temperature of the surface, for example, $\overline{T} = 10°\text{C}$. At the surface, the temperature varies with the seasons. In January, the temperature is the lowest, for example, $\overline{T} - \Delta T = 0°\text{C}$. In July, the temperature is the highest, for example, $\overline{T} + \Delta T = 20°\text{C}$. There are diurnal variations, but the penetration depth is very small. Because of the finite speed of heat conduction, at certain depth, typically -5 to -10 meters below the surface, the temperature profile is *inverted.* In other words, in the summer, the temperature several meters underground is *lower* than the annual average; and in the winter, the temperature several meters underground is *higher* than the annual average.

The solar energy stored in Earth is universal and of very large quantity. In much of the temperate zone, it can be used directly for space cooling. By placing heat exchange structures underground and guiding the cool air through ducts to the living space, a virtually free air-conditioning system can be built. In areas with average temperature

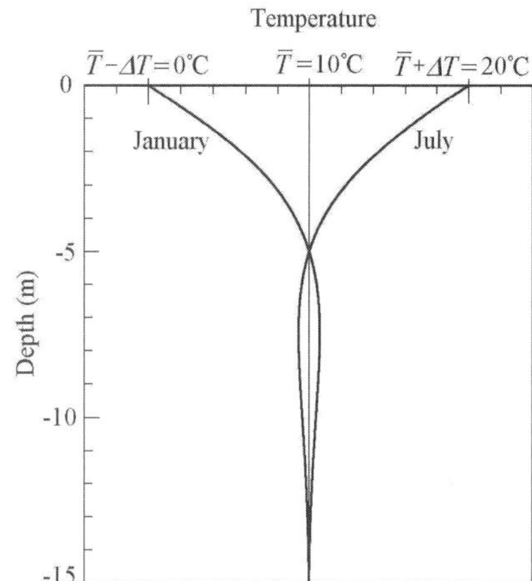

Figure 1.18 Shallow geothermal energy. Seasonal variation of underground temperature. On the surface, the summer temperature is much higher than the winter temperature. Deeply underground, e.g., minus 20 m, the temperature is the annual average temperature of the surface. In the Summer, the temperature several meters underground is *lower* than the annual average; in the winter, the temperature several meters underground is *higher* than the annual average. The energy stored in Earth can be used for space heating and cooling, to make substantial energy savings.

close to or slightly below 0°C, underground caves can be used as refrigerators, also virtually free of energy cost.

The major application of the shallow geothermal energy is the space heating and cooling systems using vapor-compression heat pump or refrigerator, taking the underground mass as a heat reservoir. Details will be presented in Chapter 6.

1.3.5 Deep Geothermal Energy

The various types of renewable energies presented in the previous sections are derivatives of solar energy. Deep geothermal energy, on the other hand, is the only major energy source not derived from solar energy. At the time Earth was formed from hot gas, the heat and gravitational energy made the core of Earth red hot. After Earth was formed, the radioactive elements continuously supplied energy to keep the core of Earth hot. Figure 1.19 is a schematic cross section of Earth. The crest of Earth, a relatively cold layer of rocks with a relatively low density (2–3 g/cm^3), is divided into several *tectonic plates*. The thickness varies from place to place, from 0 to some 30 km. Underneath the crest is the *mantle*, a relatively hot layer of partially molten rocks with relatively high density (3–5.5 g/cm^3). It is the reservoir of magna for volcanic activities. From about 3000 km and down is the core of Earth, which is believed to be molten iron and nickel, with the highest density (10–13 g/cm^3).

The heat content of the mantle and the core is enormous. In principle, by drilling a deep well to the hot part of Earth, injecting water, superheated steam can be produced to drive turbines to generate electricity. In general, such operation is prohibitively expensive and difficult.

Most current geothermal power stations are located either in the vicinity of edges of tectonic plates or in regions with active volcanoes, where the thickness of Earth's crest is less than a few kilometers and drilling to hot rocks is practical. Figure 1.20 shows the regions on Earth where deep geothermal energy can be extracted.

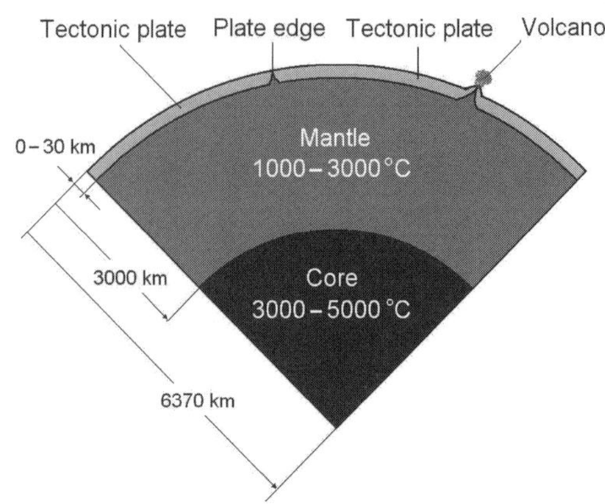

Figure 1.19 Deep geothermal energy. The origin of deep geothermal energy is the core of Earth. First, during the formation of Earth, gravitational contraction generated heat. Then, nuclear reactions in Earth continuously supplied energy. Because of the thickness of the tectonic plates, deep geothermal energy is economical only at the edges of the plate or near the volcanoes.

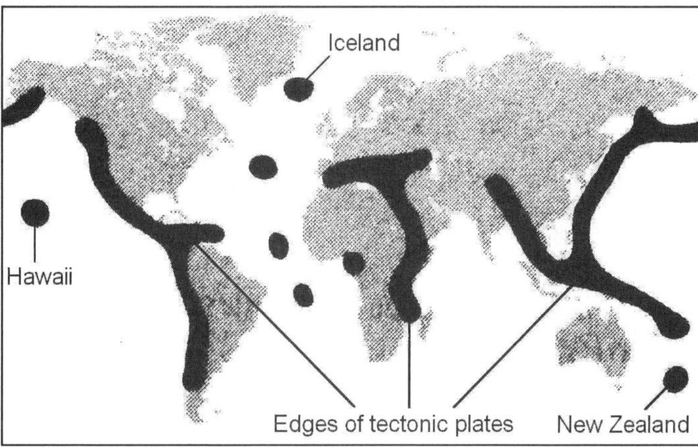

Figure 1.20 Regions for deep geothermal energy extraction. At the edges of tectonic plates and regions with active volcanoes, deep geothermal energy can be extracted economically.

Being rich in active volcanoes, Iceland has an unusual advantage in the utilization of deep geothermal energy. In 2008, about 24% of Iceland's electricity was geothermal and 87% of the buildings were heated by geothermal energy. Figure 1.21 is a photograph of the Nesjavellir Geothermal Power Station, the second largest in Iceland, with a capacity of 120 MW.

Figure 1.21 Nesjavellir geothermal power station, Iceland. Due the high concentration of volcanoes, Iceland has an unusual advantage of utilizing geothermal energy. Shown here is Nesjavellir Geothermal Power Station with a capacity of 120 MW.

1.4 Solar Photovoltaics Primer

It is clear that in the first half of the twenty-first century, fossil fuel will be depleted to an extent that it cannot support the energy demand of human society. There are various types of renewable energy resources. Many of them have limitations, including hydropower, wind energy, and geothermal energy. Solar thermal applications such as solar water heaters can fill only a small part of the total energy demand. Solar photovoltaics is the single most promising substitute for fossil energy. In this section, we will present an elementary conceptual overview of photovoltaics. Details will be presented in Chapters 2–4 and 7–10.

1.4.1 Birth of Modern Solar Cells

In 1953, Bell Labs set up a research project for devices to provide energy source to remote parts of the world where no grid power was available. The leading scientist, Darryl Chapin, suggested using solar cells, and his proposal was approved by his supervisors.

At that time, the photovoltaic effect in selenium, discovered in the 1870s, was already commercialized as a device for the measurement of light intensity in photography. Figure 1.22(a) is a schematic. A layer of Se is applied on a copper substrate, then covered by a semitransparent film of gold. When the device is illuminated by visible light, a voltage is generated, which in turn generates a current. The intensity of electric current depends on the intensity of light. It has been a standard instrument in the first half of the twentieth century for photographers to measure light conditions. This device is much more rugged and convenient than photoresistors because there are no moving parts and no battery is required.

Chapin started his experiment with selenium photocells. He found that the efficiency, 0.5%, is too low to generate sufficient power for telephony applications. Then, a stroke of unbelievable luck, two Bell Lab scientists involved in the pioneering effort to develop silicon transistors, Calvin Fuller and Gerald Pearson, joined Chapin in using

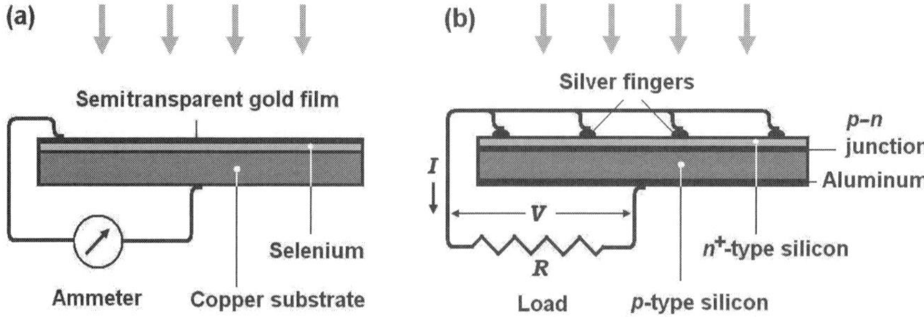

Figure 1.22 Selenium solar cell and silicon solar cell. (a) The selenium photovoltaic cell was discovered in the middle of the nineteenth century and was used for measuring light intensity in photography. (b) The silicon photovoltaic cell was invented at Bell Labs in 1954 using the technology for silicon transistors.

Figure 1.23 Inventors of silicon solar cells. Left to right: Gerald Pearson (1905–1987), Darryl Chapin (1906–1995), and Calvin Fuller (1902–1994). In 1953 Bell Labs set up a research project to provide energy sources for remote parts of the world where no grid power was available. Utilizing the nascent technology to make silicon transistors, in 1954, they designed and demonstrated the first silicon solar cells. The efficiency achieved, 5.7%, makes the solar cell a useful power source. Courtesy of AT&T Bell Labs.

the nascent silicon technology for solar cells; see Fig 1.23. In 1954, a solar cell with 5.7% efficiency was demonstrated [67]. A schematic is shown in Fig. 1.22(b).

The silicon solar cell was made from a single crystal of silicon. By judicially controlling the doping profile, a *p–n* junction is formed. The *n*-side of the junction is very thin and highly doped to allow light to come to the *p–n* junction with very little attenuation, but the lateral electric conduction is high enough to collect the current to the front contact through an array of silver fingers. The back side of the silicon is

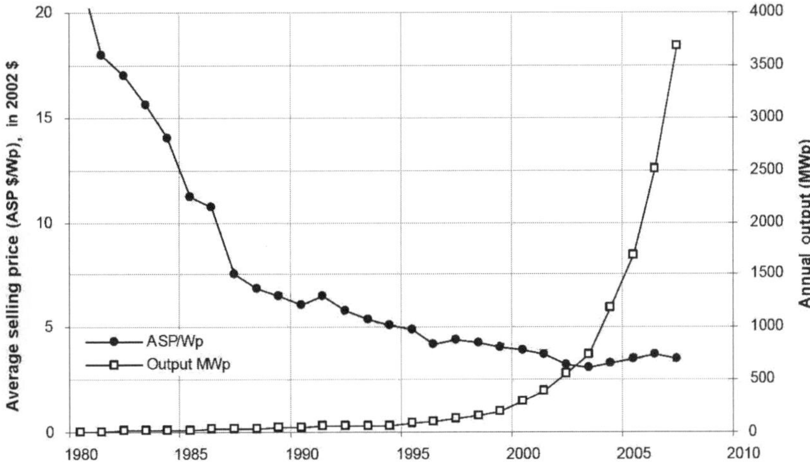

Figure 1.24 Average price and installation of solar cells: 1980–2007. The average price of solar cells dropped threefold from more than $20 per peak watt in 1980 to $6.5 per peak watt in 1990. The installation of solar cells steadily increased during that period. *Source: Solar Photovoltaic Industry, 2008 Global Outlook*, Deutsch Bank.

covered with a metal film, typically aluminum. The basic structure of the silicon solar cell has remained almost unchanged until now.

The initial demonstration of the solar cell to the public in New York City was a fanfare. And the cost of building such solar cells was very high. From the mid 1950s to the early 1970s, photovoltaics research and development were directed primarily toward space applications and satellite power. In 1976, the U.S. Department of Energy (DOE) was established. A Photovoltaics Program was created. The DOE, as well as many other international organizations, began funding research in photovoltaics at appreciable levels. A terrestrial solar cell industry was quickly established. Economies of scale and progress in technology reduced the price of solar cells dramatically. Figure 1.24 shows the evolution of price and annual PV installation from 1980 to 2007.

1.4.2 Some Concepts on Solar Cells

Following are a list of key terms and concepts regarding solar cells.

Standard Illumination Conditions

The efficiency and power output of a solar module (or a solar cell) are tested under the following standard conditions: 1000 W/m^2 intensity, 25°C ambient temperature, and a spectrum that relates to sunlight that has passed through the atmosphere when the sun is at 42° elevation from the horizon [defined as air mass (AM) 1.5, see Plate 1].

Fill Factor

The *open-circuit voltage* $V_{\rm op}$ is the voltage between the terminals of a solar cell under standard illumination conditions when the load has infinite resistance that is open. In this situation, the current is zero. The *short-circuit current* $I_{\rm sc}$ is the current of a solar cell under standard illumination conditions when the load has zero resistance. In this case, the voltage is zero. By using a resistive load R, the voltage V will be smaller than $V_{\rm op}$, and the current I is smaller than $I_{\rm sc}$. The power $P = IV$. The maximum power output is determined by the condition

$$dP = d(IV) = IdV + VdI = 0. \tag{1.27}$$

Figure 1.25 shows the relation among these quantities. Denoting the point of maximum power by $I_{\rm mp}$ and $V_{\rm mp}$, we have $P_{\rm max} = I_{\rm mp}V_{\rm mp}$.

The fill factor of a solar cell FF is defined as

$$\mathrm{FF} = \frac{P_{\rm max}}{I_{\rm sc}V_{\rm oc}} = \frac{I_{\rm mp}V_{\rm mp}}{I_{\rm sc}V_{\rm oc}}. \tag{1.28}$$

The typical value of the fill factor is between 0.8 and 0.9.

Efficiency

The efficiency of a solar cell is defined as the ratio of the output electric power over the input solar radiation power under standard illumination conditions at the maximum power point. Efficiencies for various solar cells are shown in Plate 5.

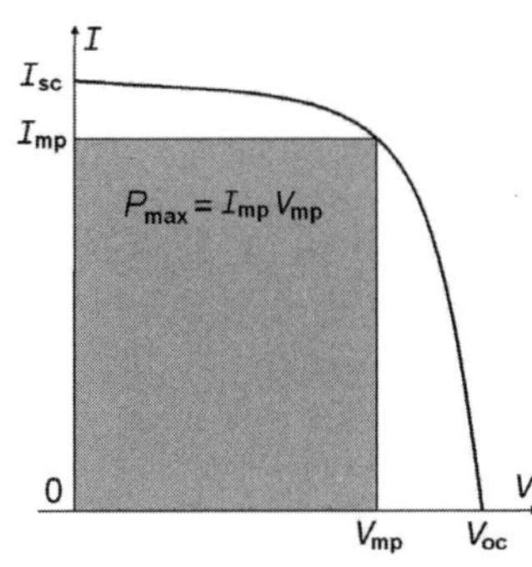

Figure 1.25 Maximum power and fill factor. By connecting a load resistor to the two terminals of a solar cell, the solar cell supplies power to the load. The maximum power point occurs when $P = IV$ reaches maximum. At that point, $P_{max} = I_{mp}V_{mp}$. Obviously, there is always $I_{mp} < I_{sc}$ and $V_{mp} < V_{oc}$. The fill factor of a solar cell is defined as $FF = P_{max}/I_{sc}V_{oc} = I_{mp}V_{mp}/I_{sc}V_{oc}$.

Peak Watt

The "peak watt" (Wp) rating of a solar module is the power (in watts) produced by the solar module under standard illumination conditions at the maximum power point. The actual power output of a solar cell obviously depends on the actual illumination conditions. For a discussion of solar illumination, see Chapter 4.

1.4.3 Types of Solar Cells

The crystalline silicon solar cell was the first practical solar cell invented in 1954. The efficiency of such solar cells as mass produced is 14–20%, which is still the highest in single-junction solar cells. It also has a long life and a readiness for mass production. To date, it still accounts for more than 80% of the solar cell market. There are two versions of the crystalline silicon solar cell: monocrystalline and polycrystalline. Amorphous silicon thin-film silicon solar cells are much less expensive than the crystalline ones. But the efficiency is only 6–10%. In between are CIGS (copper indium gallium selenide) and CdTe–CdS thin film solar cells, with a typical efficiency of around 10% and account for about 15% of the market. Because of the very high absorption coefficient, the amount of materials required is small, and the production process is simpler; thus the unit price per peak watt is lower than crystalline silicon solar cells. To date, organic solar cells still have low efficiency and a short lifetime, and the market share is insignificant, see Plate 5. Table 1.6 summarizes several significant types of solar cells.

1.4.4 Energy Balance

It takes energy to produce solar cells. Therefore, a study of the EROI is important. Here we discuss the energy balance for the most expensive case, crystalline silicon solar cells. The energy investment includes that for producing silicon feedstock, ingot and wafers, cell production, module assembly, and installation. A standard benchmark number to evaluate the energy balance of photovoltaics is *payback time*. By setting the solar cells in a given solar illumination condition, the solar cells will generate energy in the form of electricity. Payback time is the number of years it takes for the electricity

Table 1.6: Types of Solar Cells

Type	Efficiency (%)	Cost ($/Wp)	Market share (%)
Monocrystalline Si	17–20	3.0	30
Polycrystalline Si	15–18	2.0	40
Amouphous Si	5–10	1.0	5
CIGS	11–13	1.5	5
CdTe-CdS	9–11	1.5	10

generated by the solar cell to compensate for the energy invested in the production and installation process.

Figure 1.26 represents a conservative estimate of payback time for crystalline silicon solar cells based on European insolation conditions [4]. For Central Europe, with annual insolation of 1000 W/m^2, the payback time is 3.6 years. Its lifetime is typically 25 years, which results in an EROI of 7. In Southern Europe and most places in the United States, the EROI is above 10. The EROI for thin-film solar cells is even better. However, because of lower efficiency, these require more space to generate the same power.

1.5 Above Physics

Good physics does not always lead to successful industrial implementation and to conferring a great benefit to mankind. Solar energy is no exception. Economics and politics play a significant role. In this section, we will review some important historical lessons and analyze the economical and political context for which the utilization of solar energy can become a success and thus benefit mankind.

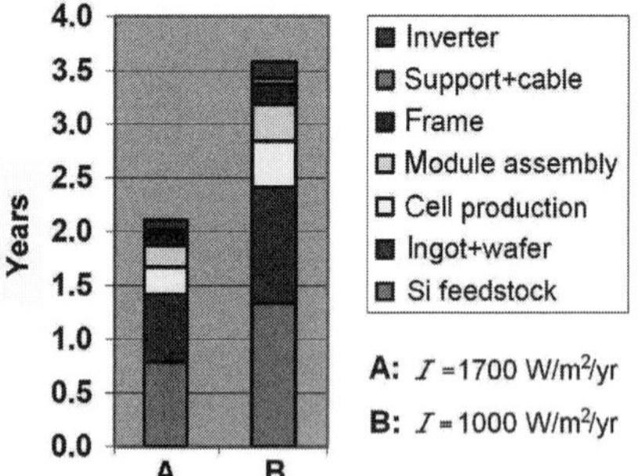

Figure 1.26 Payback time for crystalline silicon solar cells. The energy invested for a solar cell includes the energy for producing silicon feedstock, ingot and wafers, cell production, and installation. Even under unfavorable insolation conditions, such as central Europe, the payback time is less than 15% of its lifetime, which results in an EROI of 7 (see Ref. [4]).

1.5.1 Economics of Solar Energy

The early history of solar water heaters in the United States vividly illustrates the interplay of physics, engineering and economics [17]. In the nineteenth century, before the invention of modern hot-water systems, making hot water for a bath was expensive and difficult. Water had to be heated in a large pot over fire, then scooped into the bathtub. It was especially expensive in California, where fuel such as coal had to be imported and wood was precious. Artificial gas and electricity were very expensive. However, sunlight is plenty there, and the weather is mild.

In 1891, Clarence Kemp patented an effective and usable solar water heater, named Climax (U.S. Patent 451,384). First marketed in Maryland, Kemp's business was not very successful. Then he sold the exclusive right to two Pasadena businessmen and made a great commercial success in California. By 1900, 600 units were sold in southern California alone. However, the Climax water heater had a drawback in that it took a few hours of sunlight to heat up the water and after sunset the water temperature would drop quickly. Therefore, it could only be used in the afternoon of a sunny day.

In 1910, William J. Bailey invented and patented the Day-and-Night solar water heater (U.S. Patent 966,070), which resolved the major problems and became the prototype of later solar water heaters; see Fig. 1.27. First, the heat collector A is made of a parallel grid of copper pipes welded on a flat piece of copper plate. Second, it uses a water tank C placed above the heat collector, heavily insulated by cork, D. Such an arrangement enables water circulation by *natural convection* and effective *energy storage.* When water is heated by sunlight, the specific gravity decreases. It flows automatically upward through pipe B into water tank C. The colder water then flows automatically downwards through pipe E back into the heat collector A. If the insulation is sufficient, the water can stay hot overnight. Therefore, it works in the day as well as in the night. Although the Day-and-Night system cost about $180 at that time, much higher than a Climax system, it quickly conquered the consumers. Climax was forced out of business. By the end of World War I, more than 4000 Day-and-Night solar water heaters were

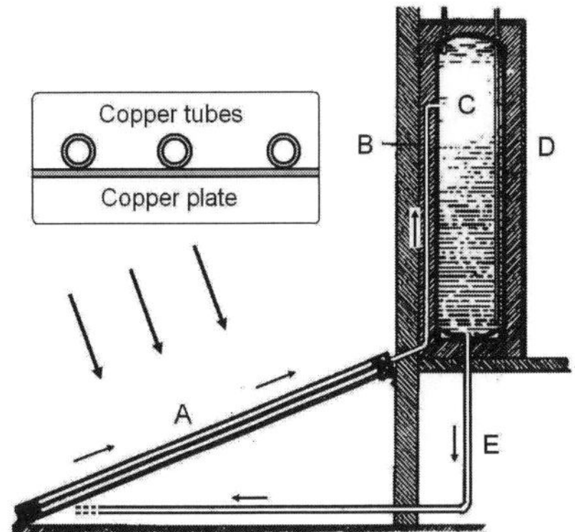

Figure 1.27 Day-and-Night solar water heater. The inset shows the copper tubes and copper plate of the solar heat collector A. The system works under natural convection: the water heated by sunlight in collector A rises to the insulated tank C. The cold water flows down from tank C through pipe E back to the solar heat collector A. *Source*: U.S. Patent 966,070, 1911.

sold.

In the early 1920s, abundant natural gas was discovered in the Los Angeles Basin. The price of natural gas in 1927 was only a quarter of that in 1900 for town gas. The gas-operated water heater, much cheaper in initial investment than the solar heater and more convenient to use, gradually replaced the once popular solar water heaters. Bailey's company, being quite experienced in water heater systems, quickly adapted into a gas heater business. Still keeping Day-and-Night as the company name, it soon became one of the largest producers of gas water heaters in the nation.

The downfall of the solar water heater business in California was not the end of it. Florida, with a real estate boom in the 1920s through the 1940s and no natural gas available, became the sweet spot of solar water heaters; see Fig. 1.28. It is estimated that from 25,000 to 60,000 solar water heaters were installed in Miami during 1920–1941. During World War II, price of copper skyrocketed. After the war, the price of electricity plummeted. The result was the gradual replacement of solar water heaters by electric water heaters. In the United States, the solar water heater had lost its glory.

However, after World War II, elsewhere in the world, solar water heaters gained momentum, especially in Israel. A desert area without energy resources similar to California in the late nineteenth century, solar water heaters could bring sizeable economic benefit. A significant advance in solar thermal technology, selective absorption coating, was invented in Israel in the 1950s, which greatly improved the efficiency of solar water heaters. Later, Israel became the first country to require that all new buildings must have solar water heaters.

Figure 1.28 A Day-and-Night solar water heater in Florida. From 1920 to 1941, more than 25,000 solar water heaters were manufactured and installed in Florida. After 80 years, thousands of them are still working. The photo, taken by the author in Miami in August 2010, is a solar water heater installed in 1937. The insulated water tank is disguised as a chimney. Even with a broken pane, it is still working properly.

1.5.2 Moral Equivalence of War

Government policies on energy have a major effect on renewable energy development. In the United States, the new energy policies during the Carter administration in the 1970s created a golden period for renewable energy research and development.

After World War II, the United States enjoyed cheap crude oil, staying below $20 per barrel (inflation adjusted in January 2008 dollars) for three decades. In 1973, an oil embargo triggered the first energy crisis. The price of crude oil jumped dramatically; see Fig. 1.29.

Coincidentally, the timing of the energy crisis matched the prediction of M. King Hubbert in 1956 that shortly after 1970 the production of crude oil in the United States would peak and start to decline; see Section 1.2. The coincidence is not accidental. As crude oil production in the United States started to decline, consumption was still growing. In 1971, the United States paid $3.7 billion for importing crude oil; in 1977, it increased 10-fold to $37 billion (in 1977 dollars). Obviously, excessive dependence on foreign oil poses severe economic and security threats.

On April 18, 1977, then President Jimmy Carter delivered a televised speech about his new energy policy. He called the struggle for greater energy independence the *moral equivalence of war* — one that "will test the character of the American people." He said:

> Tonight I want to have an unpleasant talk with you about a problem unprecedented in our history. With the exception of preventing war, this is the greatest challenge our country will face during our lifetimes. The energy

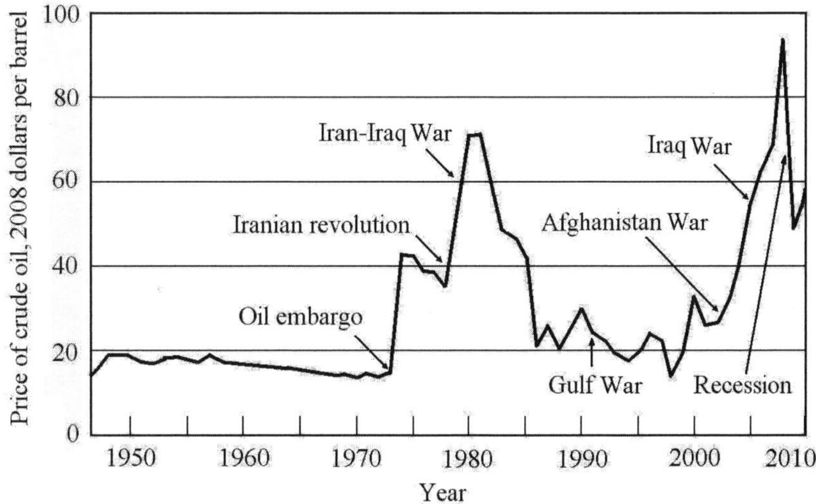

Figure 1.29 History of crude oil price, 2008 dollars. *Source:* U.S. Energy Information Administration.

> crisis has not yet overwhelmed us, but it will if we do not act quickly.
>
> It is a problem we will not solve in the next few years, and it is likely to get progressively worse through the rest of this century.
>
> We must not be selfish or timid if we hope to have a decent world for our children and grandchildren.
>
> We simply must balance our demand for energy with our rapidly shrinking resources. By acting now, we can control our future instead of letting the future control us.

The major points of Carter's energy policy included energy conservation, increasing domestic traditional energy exploration, and developing renewable energy resources. In his words, "we must start now to develop the new, unconventional sources of energy we will rely on in the next century." A few days later President Carter signed the Department of Energy Organization Act and on August 4, 1997, formed the U.S. Department of Energy. Then, the National Energy Act (NEA) was established in 1978 with tax incentives for renewable energy projects, especially solar energy. This legislation initiated a significant boost to the research, development, and installation of solar water heaters, solar cells, and solar-operated buildings.

To lead the public by example, on June 20, 1979, Carter installed a solar water heater with 32 panels on the roof of the White House; see Fig. 1.30. At the ceremony, Carter reflected upon his own idealism:

> A generation from now, this solar heater can either be a curiosity, a museum piece, an example of a road not taken, or it can be a small part of one of the greatest and most exciting adventures ever undertaken by the American people . . . to harness the power of the sun to enrich our lives as we move away from our crippling dependence on foreign oil.

In 1978, the Carter administration enacted the first National Energy Act (NEA) to promote fuel efficiency and renewable energy. The research and development funding for renewable energy is greatly increased. Part of the 1978 NEA is an Energy Tax Act that gave an income tax credit to private residents who use solar, wind, or geothermal sources of energy. The 1978 Energy Tax Act was expired in 1986. However, many other countries followed the example of the United States and provided government financial support for renewable energy utilization.

As anticipated by Jimmy Carter, during the rest of the twentieth century, several factors made the energy problem "progressively worse": Due to a steady decline and an increasing consumption, crude oil import into the United States increased from 1.8 billion barrels in 1980 to 5.0 billion barrels in 2000s. The price of crude oil (in 2008 dollars) increased from about \$20 to more than \$100 a barrel in late 2000s; see Fig. 1.29. The petroleum crisis in the 1970s reappeared, but with an even more gruesome context: According to Hubbert, in the early 2000s, the world's crude oil production peaked and started to decline; see Section 1.2. The world's two most populous countries, India and China, are experiencing rapid economic development, which consume a growing

Figure 1.30 Jimmy Carter dedicates solar water heater on top of the White House, June 20, 1979. He hoped that this would be "a small part of one of the greatest and most exciting adventures ever undertaken by the American people, ... to harness the power of the sun to enrich our lives as we move away from our crippling dependence on foreign oil". Courtesy of Jimmy Carter Library, Atlanta, GA.

proportion of the dwindling production of the world's crude oil. Both India and China have very limited crude oil resource and therefore have an even more severe energy problem.

1.5.3 Solar Water Heaters over the World

As we have presented, the solar water heater was invented in the United States and it was quite popular in the first half of the twentieth century. However, despite the energy crisis and strong government incentive in the 1970s, the installation volume in the United States is still very low. Nevertheless, in recent decades, the solar water heater has enjoyed an explosive growth globally, especially in China. As shown in Fig. 1.31, in 2007, China installed 80% of the new solar water heaters with 16 GW capacity; the total installation capacity of solar water heaters is 84 GW, accounting for two-thirds of the world's total.

A Huge Virgin Market

The market for water heaters in China is quite similar to that in California in the late nineteenth century. Up to the 1980s, nearly one billion people in China had no running

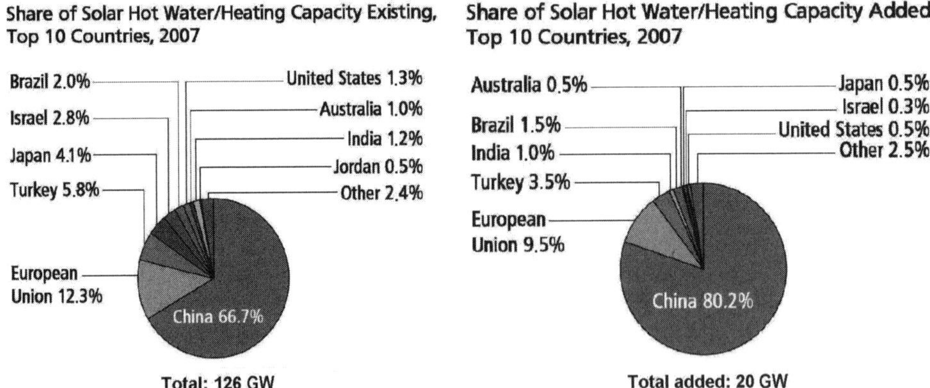

Figure 1.31 Global installations of solar water heater. Globally, solar water heater installation is increasing rapidly. The most growth is in China. *Source:* Renewables Global Status Report, 2009 Update, (http://www.ren21.net).

hot water. Improvement in the standard of living has made hot water a necessity of life. However, natural gas and heating oil are expensive and not generally available. Electricity is very expensive. The cost of equipment for making hot water using fossil fuel is comparable with that using sunlight. China represents perfect soil in which the solar water heater industry can grow.

Advances in Technology and Economy of Scale

Until recently, most solar water heaters in the Western world used flat-plate heat collectors similar to the Day-and-Night system (Fig 1.28) and the White House solar panels (Fig 1.30). However, the structure is rather inconvenient for mass production. It uses a lot of copper, a major factor of its demise during and after World War II. The heat loss due to conduction through the glass window and the back plate is significant. A lot of metal parts exposed to the elements limit its lifetime. The vacuum tube heat collector, invented by American engineer William L. R. Emmet in 1911 (U.S. Patent 980,505) has superb properties. However, for several decades, it was expensive and complicated to produce. In early 1980s, vacuum tube solar heat collectors were improved. An extremely simple and effective solar water heater was invented and perfected; see Fig 1.32.

Figure 1.32(a) shows the design of evacuated-tube solar water heater. Each heat collector is a double-walled glass tube. The space between the outer tube and the inner tube is evacuated to a high vacuum, similar to a Dewar flask. On the outer surface of the inner tube, a *selective absorption film* is applied. For sunlight, which is mostly visible and near infrared, the absorption coefficient is around 95%. For far-infrared radiation from the hot water (80–100°C), the emission coefficient is around 5%. The vacuum sleeve perfectly blocks thermal conduction. The entire system works automatically under the principle of natural convection. The water heated by sunlight, having a lower specific gravity, flows upward to the insulated tank. Similarly, the cold water flows downward from the tank into the heat collector tubes. A photo is shown in Fig 1.32(b). Typically, 10–40 evacuated tubes are used in one system. The water tank is insulated by foam polyurethane. The temperature only drops a few degrees overnight. Therefore, it can provide hot water day and night.

The design also facilitates shipping and storage. The tubes, the tank, and the parts of the frame are shipped in three separate rectangular cardboard boxes. The system is then assembled at the installation site.

Because the evacuated tubes are made of borosilicon glass (Pyrex), the selective absorption film is under high vacuum, and the tank is usually made of high-grade stainless steel. Unless it is broken by brute force, the system could work for decades. In addition, these parts are suitable for automatic mass-production. Currently, some 200 million evacuated tubes and some 10 million insulated water tanks are being manufactured every year.

The huge market enables cost advantages by the economies of scale. The producer's average cost per unit falls as scale is increased. Due to relentless effort of automation, the manufacturing cost of evacuated tubes is reduced to a few dollars per piece, unimaginable a few decades ago. The reduced price of solar water heaters further increases the size of the market. The expansion of the market further provides grounds to improve the manufacturing process. As a result, the solar water heater business in China is sustainable *without government financial incentives.*

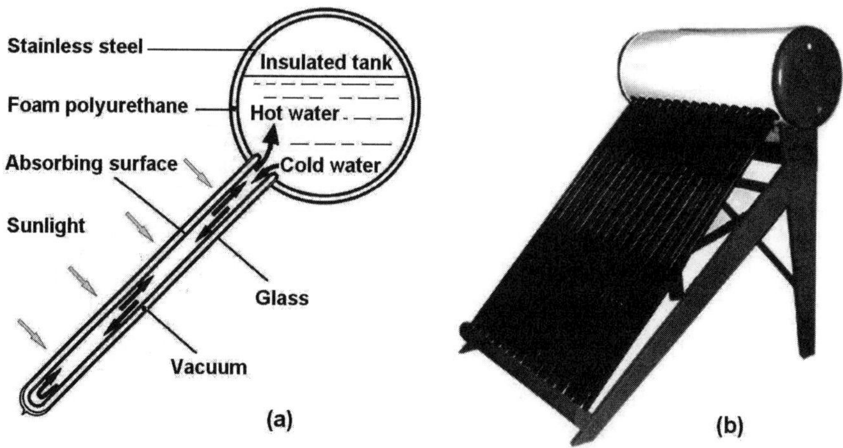

Figure 1.32 Evacuated-tube solar water heater. (a) Schematic of evacuated-tube solar water heater. Each heat collector is a double-walled glass tube. The system works automatically under the principle of natural convection. (b) Photograph of system.

Himin Model of Solar Energy Business

It is instructive to learn how the colossal solar water heater industry has been able to grow over such a short period of time. For that purpose, I visited the world's largest manufacturer of solar water heaters, Himin Solar Energy Group, and met the founder and CEO, Mr. Huang Ming in Beijing, when he attended the Plenary Meeting of People's Congress. A number of internal documents of this company were then collected.

In 1978, when Huang Ming was a student at East China Petroleum Institute, he learned that the crude oil reserve in the world would be depleted within 50 years. China's crude oil would be depleted even sooner. Several years after graduation, he became a highly regarded research engineer specializing in oil well drilling. However, his personal experience reinforced his pessimism about the future of petroleum.

In 1987, by chance, he found Duffie and Beckman's book *Solar Energy Thermal Processes*. Reading it from cover to cover and doing hands-on experiments using funds from the sale of a patent, he became convinced that sunlight is the ultimate solution to the energy problem. Solar energy became his lifetime passion. Since then, he spends 8 h at the Petroleum Institute, 8 h working at home on solar energy projects, and 8 h sleeping and eating. He gave his hand-made solar water heater to friends and relatives as gifts and installed a system at a children's entertainment center.

In 1995, he quitted his job in petroleum industry, and started his own business. He chose the Chinese name of the company, Huangming, as a homophone of his personal name, to symbolize his dedication. Within ten years, his company has grown into one of the largest solar water heater manufacturers on the world without government subsidy. In 2009, Himin produced more than two million square meters of solar heat collectors, equivalent to 2 GW of peak solar energy utilization. In May 2006, Huang Ming was

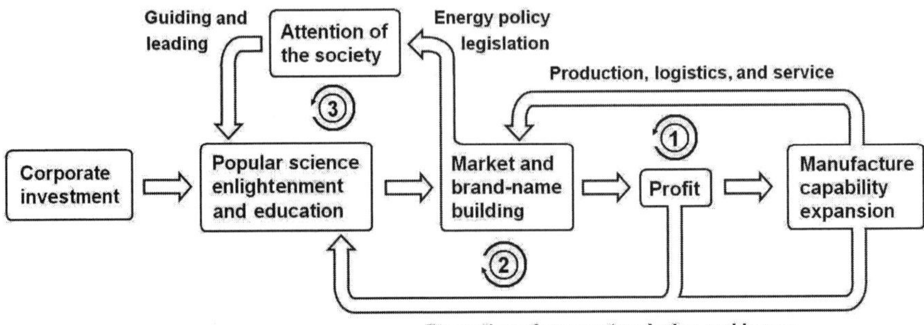

Figure 1.33 Himin model of solar energy business. The Himin model consists of three loops. Loop 1 is similar to a conventional business cycle. Strict quality control and good after sale service build up the brand name. Large-scale mass production reduces unit cost. Loop 2 is to invest heavily in popular science education as the major marketing method. Loop 3 is to push energy policy legislation to favor renewable energy and to promote public awareness. *Source:* Minutes of the 14th Conference of United Nations Commission on Sustainable Development, New York, 2006.

invited by United Nations to present Himin's business mode at the 14th Meeting of United Nations Commission on Sustainable Development, 2006. It was later known as the "Himin model of solar energy business".

Figure 1.33 is a sketch of the Himin model. A central point is to create a new market through popular science enlightenment and education. In 1996, right after the birth of Himin, the weekly newsletter *Popular Solar Energy* was established. It became Himin's continuing marketing tool, with accumulated distribution of 300 million copies in 2010. Himin organized numerous popular science tours, traveling 80 million kilometers over China. Because for an average Chinese family a solar water heater is still a major capital investment, the decision to purchase a set must be based on careful thinking. Such educated people were the first customers and then became volunteer marketers for the product. The Himin model is represented by three loops; see Fig. 1.33.

Loop 1 is the main production cycle. Himin's strategy is to pursue the highest standard of excellence. The retail price of their product is among the most expensive on the market. However, because of their extensive research and quality control, the products have less trouble and last a long time. The quality ensures their reputation in the marketplace.

Loop 2 emphasizes the importance of popular science enlightenment and education. In addition to corporate investment, a substantial portion of the profit is invested in science education to ensure that customers understand how the system works and how to choose a good product or part.

Loop 3 emphasizes the importance of pushing for energy policy legislation and raising public attention to renewable energy. This is also a significant factor of its success. In 2003, local city folks, especially Himin employees, elected Huang Ming to become a member of the People's Congress. He then mobilized 60 fellow congressmen to propose a Renewable Energy Act, which was passed in the spring of 2005. The legislation has motivated central and local governments to set up renewable-energy projects and raises public support to solar energy technology and products.

In 2008, the International Solar Energy Society (ISES) chose Dezhou, the location of Himin, to host the 2010 International Solar City Congress. Around that time, Huang Ming was elected as the vice president of the ISES. The venue of the congress, the Dezhou Apollo Temple, an 800,000-ft^2 museum, ballroom, and hotel with 65% of its energy supplied by solar, was completed in 2009; see Plate 17.

1.5.4 Photovoltaics: Toward Grid Parity

Because water heaters consume less than 10% of total energy, the bulk of energy needed, especially electricity, can only be supplied through photovoltaics or other means of solar electricity generation. Compared with solar water heaters, the total power of solar PV installed in the world is much smaller. In 2008, 6.08 GWp of PV was installed in the world. The accumulative installation in 2008 is 15 GWp. Figure 1.34 shows the yearly installation and growth rate of the entire world from 1990 to 2008.

As shown, the installation in terms of peak watts of PV is only about one tenth of solar water heaters. The limiting factor is simply economics. As shown in Section 1.4.1,

Table 1.7: Cost Per Kilowatt-Hour of Solar Electricity for Various Cases

Insolation	By Cost of Installed PV per Peak Watt				
kWh/m^2/day	\$ 2	\$ 4	\$ 6	\$ 8	\$ 10
3	\$ 0.073	\$ 0.146	\$ 0.219	\$ 0.292	\$ 0.365
4	\$ 0.054	\$ 0.109	\$ 0.164	\$ 0.219	\$ 0.273
5	\$ 0.043	\$ 0.087	\$ 0.131	\$ 0.175	\$ 0.219
6	\$ 0.036	\$ 0.073	\$ 0.109	\$ 0.146	\$ 0.182

although practically usable solar cells were invented in 1954, the manufacturing cost was very high. The major applications were space and military fields. In the late 1970s and early 1980s, stimulated by the National Energy Act, dramatic improvement in efficiency and reduction in cost were achieved; see Fig. 1.24. The gradual reduction in the manufacturing cost of solar cells continues in the 1990s and early 2000s. In 2003, the price of solar cell per peak watt has dropped to \$6.50. However, solar electricity is still much more expensive than the electricity generated by traditional energy resources, especially by coal and hydropower, which is \$0.05 to \$0.10 per kWh. Including the supporting structure, inverters and necessary instruments, the cost of installed solar panels per peak watt in 2003 was about \$10. Table 1.7 shows the cost of generating 1 kWh of electric energy by solar cells in regions with different insolation.

To jump start the utilization of solar energy, in the 1990s, many European countries established the *feed-in tariff (FIT) law*, which guarantees a solar photovoltaic system

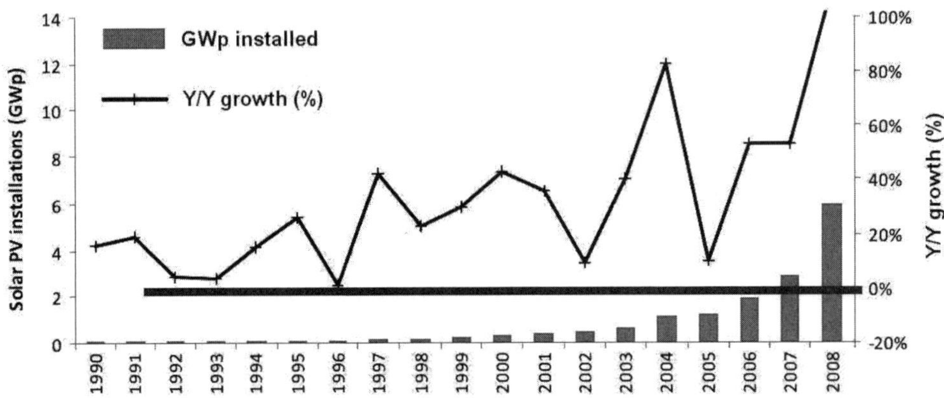

Figure 1.34 Installation of solar photovoltaics: 1990 to 2008. Bar graph: world annual installation of solar photovoltaic panels. Solid curve: growth rate by year. *Source*: After *Solar Photovoltaic Industry, 2008 Global Outlook*, Deutsch Bank.

access to the grid with the purchase price of thus generated electricity based on cost. The purchase guarantees could be extended to 20 or 25 years, but the rate could decline based on expected cost reductions. The law significantly expended the market for solar photovoltaics. However, it also caused unexpected gyrations. For example, in 2007, Spain raised the date for large PV systems from €0.18/kWp to €0.42/kWp. Immediately, the installation surged from 61 mWp in 2006 to 591 mWp in 2007, then to 2700 mWp in 2008. The Spanish government suddenly found that the rate was not sustainable and reduced it to €0.32/kWp, to take effect in 2009. In 2009, the installation is reduced to less than 200 MWp.

The dramatic increase of demand in 2007–2008 nevertheless gave a thrust to an unprecedented boom in the solar cell industry, especially in the United States and Asia. The economy of scale, in the form of vertical integration, again works. First Solar in the Unites States is vertically integrated in a sense that it designs and manufactures equipment for solar cell production by themselves, thus its production capability is quickly expended. Yingli Solar in Baoding, China, and Renewable Energy Corporation in Norway, both manufacturers of polycrystalline silicon solar cells, are so vertically integrated that they start with producing pure silicon from silica mines and end up with solar panel installation. In 2009, 49% of the world's solar cells were produced in China and Taiwan, mostly crystalline silicon solar cells with high efficiency. Figure 1.35 shows the statistics from two manufacturers. As shown, in 2009, the manufacturing

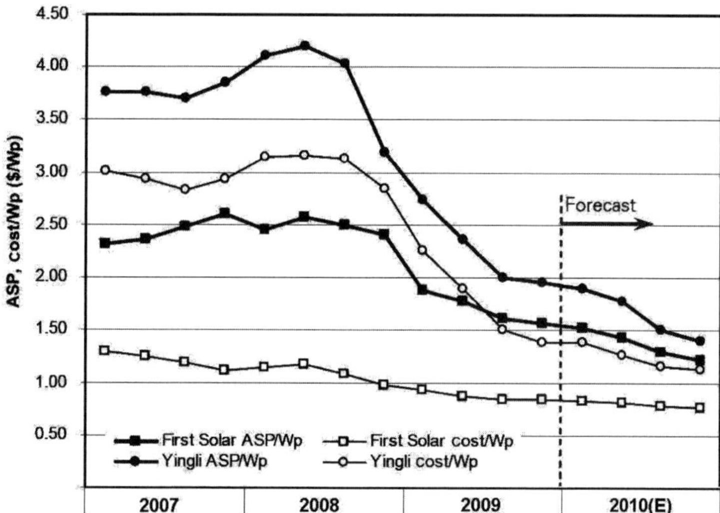

Figure 1.35 Price of solar cells from two major suppliers: 2007 to 2010. Existing data and forecast for two representative solar cell manufacturers. First Solar is the world's largest manufacturer of CdTe-CdS thin film solar cells. Yingli is the world's largest manufacturer of polycrystalline solar cells, which is also one of the world's most vertically integrated solar cell manufacturers. After *Solar Photovoltaic Industry, 2008 global outlook*, Deutsch Bank.

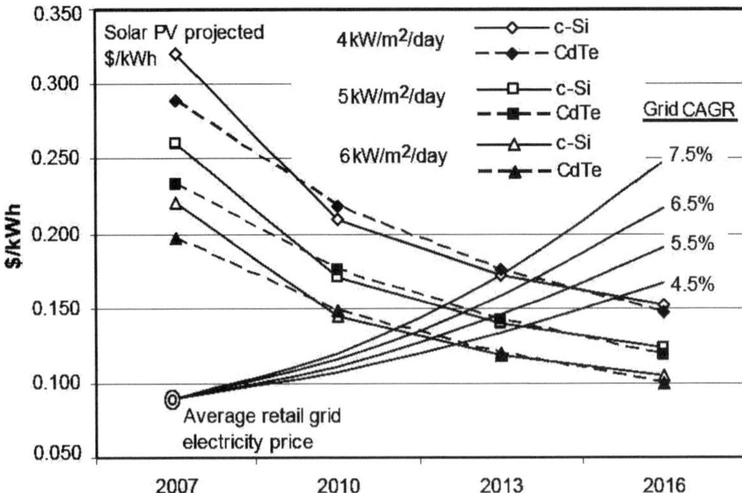

Figure 1.36 Prediction of grid parity. According to a forecast by the Deutsch Bank, the two most promising solar cells, crystalline silicon solar cells and CdTe thin-film solar cells, will arrive at grid parity around year 2013. Beyond 2013, the cost of solar electricity could be lower than that of traditional sources of electricity for many places in the world. *Source*: After *Solar Photovoltaic Industry, 2008 Global Outlook*, Deutsch Bank.

cost of First Solar's CdTe thin-film solar cells dropped to $0.80 per peak watt, and the retail price is $1.50 per peak watt. Yingli's polycrystalline silicon solar cells, with an efficiency close to 20%, is manufactured with $1.50 per peak watt, and a and the retail price of $2 per peak watt. The cost and price is expected to decrease continuously in the years to come. (Recently, Yingli announced a selling price of polycrystalline silicon solar panels of less than $1 per peak watt.)

According to statistics and analysis by Deutsch Bank, because of the combined effect of explosive expansion of solar cell production capability and the recession, in 2009, the price of solar cells on the international market dropped significantly. A brutal shake-up of the industry, suppliers of both pure silicon and solar cells, took place. After a brief period of oversupply, the price of solar cells will be reduced to a level where the cost of solar electricity will be comparable to electricity generated by, for example, coal-burning power stations, that is, will reaches *grid parity*; see Fig. 1.36. The pace of grid parity depends on the local situation and thus varies from place to place. For places with high electricity cost, such as Hawaii, Connecticut, California, and New York, especially those places with high insolation, grid parity will take place earlier. For areas with low electricity prices, for example, West Virginia and central-west China, especially places with low insolation, grid parity will be reached later. However, the trend that the cost of fossil fuel electricity will increase and the cost of solar electricity will decrease is inevitable. Solar electricity will gradually replace fossil fuel electricity.

Figure 1.37 shows the average annual growth rates of renewables from 2002 to

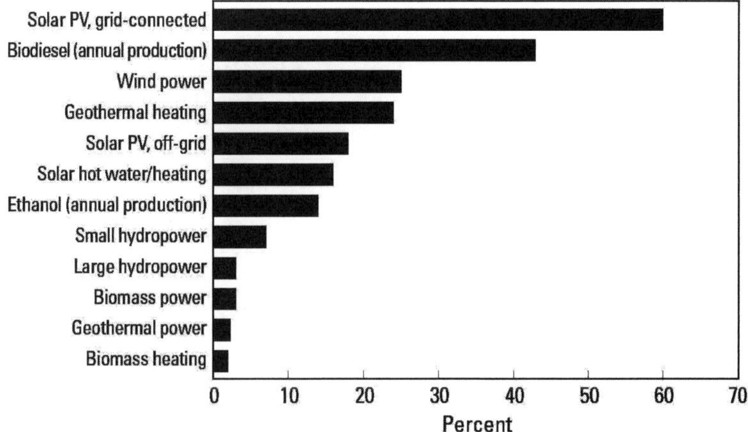

Figure 1.37 Average annual growth rates of renewables: 2002–2006. The average growth rate for grid-connected solar photovoltaics far exceeds other renewable energy capacities, even more for traditional energy resources. If in the future the rate of growth of photovoltaics can maintain 40%, in 20 years, or by 2030, solar photovoltaics will account for 50% of the energy supply and become the dominant energy source. *Source*: After *Renewables 2007 Global Status Report*, REN21 [1].

2006, as reported by REN21 [1]. As shown, the average growth rate for grid-connected solar photovoltaics, more than 60%, far exceeds other energy resources. In 2006, the percentage of solar photovoltaics was only a miserable 0.07%. However, if in the future the rate of growth of photovoltaics can maintain 40%, in 20 years, or by 2030, solar photovoltaics will supply more than 50% of the energy consumption.

Problems

1.1. In the United States, the British thermal unit (Btu) is defined as the energy to raise the temperature of one pound water by one degree Fahrenheit. Show that to a good approximation, 1 Btu equals 1 kJ.

1.2. Approximately (to ±5%), how much energy is in one billion barrels of petroleum in gigajoules (GJ) and megawatt-hours (MWh)?

1.3. The area of New Mexico is 121,666 square miles. The average annual insolation (hours of equivalent full sunlight on a horizontal surface) is 2200 h. If one-half of the area of New Mexico is covered with solar panels of 10% efficiency, how much electricity can be generated per year? What percentage of US energy needs can be satisfied? (The total energy consumption of the United States in 2007 was 100 EJ).

1.4. The area of Tibet is 1,230,000 km^2. The average annual insolation (hours of equivalent full sunlight on a horizontal surface) is 3000 h. If one-half the area of Tibet is covered with solar panels of 10% efficiency, how much electricity per year can be generated? What percentage of the world's energy needs can be satisfied? (The total energy consumption of the world in 2007 was 500 EJ.)

1.5. A solar oven has a concentration mirror of 1 m^2 with a solar tracking mechanism. If the efficiency is 75%, on a sunny day, how long will it take to melt one kg of ice at 0°C at the same temperature? How long will it take to heat it to the boiling point? How long will it take to evaporate it at 100°C?

1.6. For wind speeds of 20, 30, 40, ..., up to 100 mph, calculate the wind power density (in watts per square meter).

1.7. The distance from the equator to the North Pole along the surface of Earth is 1.00×10^7 m. If the average solar radiation power density on Earth is one sun, how much energy is falling on Earth annually? If the annual energy consumption of the entire world in 2040 is 800 EJ, what percentage of solar energy is required to supply the world's energy need in 2040? (*Hint*: One day equals 24×60×60=86,400 seconds.)

1.8. Using a solar photovoltaic field of 1 square mile (2.59 km^2) with efficiency of 15%, how many kilowatt-hours will this field generate annually at locations of average daily insolation (on flat ground) of 3 h (Alaska), 4 h (New York), 5 h (Georgia), and 6 h (Arizona)? An average household consumes 1000 kWh per month. How many households can this field support in the four states, respectively?

Chapter 2

Nature of Solar Radiation

Solar energy comes to Earth in the form of radiation, or sunlight, with spectral components mostly in the visible, near infrared, and near ultraviolet. To study the properties of sunlight, we need to consider and understand it from two points of view: as an electromagnetic wave and as a flow of photons. The first point of view is essential for all solar thermal applications and the antireflection coatings for solar cells. The second point of view is essential with regard to solar cells and solar photochemistry. The unification of the two points of view is represented by quantum electrodynamics, one of the most fruitful and matured fields in modern physics. Here, for simplicity, we will present an elementary treatment of these two points of view separately.

2.1 Light as Electromagnetic Waves

Up to the middle of the nineteenth century, electromagnetic phenomena and light have been considered as totally independent entities. In 1865, in a monumental paper *A Dynamic Theory of the Electromagnetic Field*, James Clerk Maxwell (Fig 2.1) proposed that light is an electromagnetic wave [58]. In that paper, he developed a complete set of

Figure 2.1 James Clerk Maxwell. Scottish physicist (1831–1879), one of the most influential physicists along with Isaac Newton and Albert Einstein. He developed a set of equations describing electromagnetism, known as the *Maxwell's equations*. In 1865, based on those equations, he predicted the existence of electromagnetic waves and proposed that light is an electromagnetic wave [58]. He also pioneered the kinetic theory of gases, and created a science fiction character *Maxwell's demon*. Portrait courtesy of Smithsonian Museum.

equations explaining electromagnetic phenomena, now known as *Maxwell's equations*. Based on those equations, he predicted the existence of electromagnetic waves, propagating in free space with a speed that equals exactly the speed of light, which was then verified experimentally by Heinrich Hertz. Maxwell's bold postulation that light is an electromagnetic wave has since become one of the cornerstones of physics.

2.1.1 Maxwell's Equations

In vacuum, or free space, Maxwell's equations are

$$\nabla \cdot \mathbf{E} = \frac{\rho}{\varepsilon_0}, \tag{2.1}$$

$$\nabla \cdot \mathbf{B} = \mathbf{0}, \tag{2.2}$$

$$\nabla \times \mathbf{E} = -\frac{\partial \mathbf{B}}{\partial t}, \tag{2.3}$$

$$\nabla \times \mathbf{B} = \varepsilon_0 \mu_0 \frac{\partial \mathbf{E}}{\partial t} + \mu_0 \mathbf{J}. \tag{2.4}$$

Electric current cannot exist in free space. For linear, uniform, isotropic materials, the current density $\mathbf{J}$ is determined by the electric field intensity $\mathbf{E}$ through Ohm's law,

$$\mathbf{J} = \sigma \mathbf{E}. \tag{2.5}$$

The names, meanings, and units of the physical quantities in these equations are listed in Table 2.1. For example, the electric constant has an intuitive meaning as follows. A capacitor made of two parallel conducting plates with area A and distance d has a capacitance $C = \varepsilon_0 A/d$ in farads. Similarly, the electric constant has an intuitive meaning as follows. An inductor made of a long solenoid of N loops with cross-sectional area A and length l has an inductance $L = \mu_0 N^2 A/l$ in henrys.

Table 2.1: Quantities in Maxwell's Equations

Symbol	Name	Unit	Meaning or Value
$\mathbf{E}$	Electric field intensity	V/m	
$\mathbf{B}$	Magnetic field intensity	T (tesla)	N/A·m
ρ	Electric charge density	C/m^3	
$\mathbf{J}$	Electric current density	A/m^2	
ε_0	Electric constant (permittivity of free space)	F/m	8.85×10^{-12} F/m
μ_0	Magnetic constant (permeability of free space)	H/m	$4\pi \times 10^{-7}$ H/m
σ	Conductivity	$(\Omega \cdot m)^{-1}$	

2.1.2 Vector Potential

To treat the electromagnetic field in space, a convenient method is to use the *vector potential*. From Eq. 2.2, it is possible to construct a vector field $\mathbf{A}$ which satisfies

$$\mathbf{B} = \nabla \times \mathbf{A}. \tag{2.6}$$

Then, Eq. 2.2 is automatically satisfied. Substituting Eq. 2.6 into Eq. 2.3, one obtains

$$\nabla \times \mathbf{E} = -\frac{\partial}{\partial t}\nabla \times \mathbf{B}. \tag{2.7}$$

For any function $\phi(\mathbf{r})$, $\nabla \times [\nabla\phi(\mathbf{r})] = \mathbf{0}$, it is possible to set up the vector potential $\mathbf{A}$ such that

$$\mathbf{E} = -\frac{\partial \mathbf{A}}{\partial t} - \nabla\phi, \tag{2.8}$$

where ϕ is the electrostatic potential arising from the charges. The choice of the vector potential is not unique. By adding a gradient of an arbitrary function to it, values of the electric field and magnetic field do not change. This is called the *gauge invariance* of the vector potential. It is possible to define a vector potential which satisfies the condition

$$\nabla \cdot \mathbf{A} = 0. \tag{2.9}$$

Equation 2.9 is called the *Coulomb gauge*, which is the most convenient gauge to treat nonrelativistic problems of an atomic system and an independent electromagnetic wave. In fact, using Eq. 2.9 and the first Maxwell equation Eq. 2.1, one obtains

$$\nabla^2\phi = -\frac{\rho}{\varepsilon_0}, \tag{2.10}$$

which means that the scalar potential is generated by the static charges only. It is thus convenient for treating the problems of interactions between the radiation field and atomic systems. For details of the gauge problem, see, for example, *The Quantum Theory of Radiation* by Walter Heitler [37].

2.1.3 Electromagnetic Waves

In this section, we study the electromagnetic waves in free space, that is, where the electric charge ρ and current $\mathbf{J}$ are zero. Substituting Eqs 2.6 and 2.8 into Eq. 2.4, we have

$$\nabla \times \nabla \times \mathbf{A} + \varepsilon_0\mu_0\frac{\partial^2 \mathbf{A}}{\partial t^2} = 0. \tag{2.11}$$

Using the identity

$$\nabla \times \nabla \times \mathbf{A} \equiv \nabla(\nabla \cdot \mathbf{A}) - \nabla^2\mathbf{A} \tag{2.12}$$

and Eq. 2.9, Eq. 2.11 becomes

$$\nabla^2\mathbf{A} - \varepsilon_0\mu_0\frac{\partial^2 \mathbf{A}}{\partial t^2} = 0. \tag{2.13}$$

Introducing

$$c = \frac{1}{\sqrt{\varepsilon_0 \mu_0}}, \tag{2.14}$$

Eq. 2.13 becomes

$$\nabla^2 \mathbf{A} - \frac{\mathbf{1}}{\mathrm{c}^2} \frac{\partial^2 \mathbf{A}}{\partial \mathbf{t}^2} = \mathbf{0}, \tag{2.15}$$

which is a wave equation with velocity c. Because of Eqs. 2.6 and 2.8, the electric field intensity and the magnetic field intensity also satisfy the same wave equation,

$$\nabla^2 \mathbf{E} - \frac{1}{\mathrm{c}^2} \frac{\partial^2 \mathbf{E}}{\partial t^2} = 0 \tag{2.16}$$

and

$$\nabla^2 \mathbf{B} - \frac{1}{\mathrm{c}^2} \frac{\partial^2 \mathbf{B}}{\partial t^2} = 0, \tag{2.17}$$

According to values of ε_0 and μ_0 coming from electromagnetic measurements in 1860s, the velocity of electromagnetic waves should be 3.1×10^8 m/s. On the other hand, experimental values of the speed of light at that time were 2.98×10^8–3.15×10^8 m/s. The difference was within experimental error. Maxwell proposed thusly [58]:

> The agreement of the results seems to show that light and magnetism are affections of the same substance, and that light is an electromagnetic disturbance propagated through the field according to electromagnetic laws.

Maxwell's theory of electromagnetic waves was experimentally verified by Heinrich Hertz in 1865. From recent electrical measurements, one finds $1/\sqrt{\varepsilon_0 \mu_0} = 2.998 \times 10^8$m/s, which is exactly the speed of light in a vacuum, c.

2.1.4 Plane Waves

An electromagnetic wave with circular frequency ω in space is defined as

$$\mathbf{A}(x, y, z, t) = \mathbf{A}(x, y, z)\, e^{-i\omega t}. \tag{2.18}$$

To study the properties of electromagnetic waves, we consider the case that the wave propagates in one direction, say z. In this case, the field intensities only depend on z. Equation 2.15 becomes

$$\frac{d^2 \mathbf{A}}{dz^2} + \frac{\omega^2}{c^2} \mathbf{A} = \mathbf{0}. \tag{2.19}$$

The general solution is

$$\mathbf{A} = \mathbf{A_0} e^{i(k_z z - \omega t)}. \tag{2.20}$$

where $\mathbf{A}_0$ is a constant, and the z-component of the *wavevector* k_z is defined as

$$k_z = \frac{\omega}{c}. \tag{2.21}$$

2.1.5 Polarization of Light

Although in general the vector potential could have x, y, z-components, because of Eq. 2.9, the z-component of the vector potential must be zero,

$$\nabla \cdot \mathbf{A} = \frac{\partial A_x}{\partial x} + \frac{\partial A_y}{\partial y} + \frac{\partial A_z}{\partial z} = ik_z A_z = 0. \tag{2.22}$$

This means that A_z must be a constant over the entire space. Because we are interested in electromagnetic waves, or the variation of electromagnetic fields, we can simply set $A_z = 0$. The waves are *transverse*. In other words, the intensity vectors are perpendicular to the direction of propagation.

The direction of the vector potential could be either x or y or a linear combination of x- and y-components. For the x-component of $\mathbf{A}$, we have

$$A_x = A_{x0} e^{i(k_z z - \omega t)}, \quad A_y = 0, \quad A_z = 0. \tag{2.23}$$

The electric field intensity, according to Eq. 2.8, is

$$E_x = i\omega A_{x0}\, e^{i(k_z z - \omega t)}, \quad E_y = 0, \quad E_z = 0. \tag{2.24}$$

And the magnetic field intensity, according to Eq. 2.6, is

$$B_x = 0, \quad B_y = ik_z A_{x0}\, e^{i(k_z z - \omega t)}, \quad B_z = 0. \tag{2.25}$$

Therefore, the only nonvanishing components of the electric field intensity and the magnetic field intensity are E_x and B_y. According to Eq. 2.21, they are in phase and proportional,

$$E_x = c\, B_y. \tag{2.26}$$

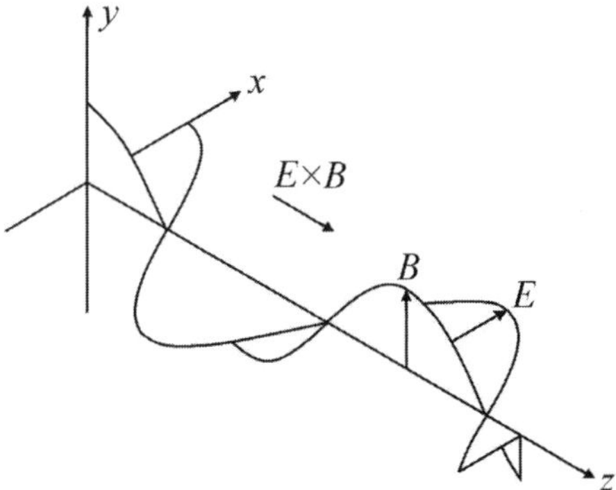

Figure 2.2 Electromagnetic wave. The electromagnetic wave is *transverse*, where the intensity vectors $\mathbf{E}$ and $\mathbf{B}$ are perpendicular to the direction of propagation. The electric field intensity $\mathbf{E}$ is perpendicular to the magnetic field intensity $\mathbf{B}$. The energy flux vector $\mathbf{S} = \mu_0^{-1}\mathbf{E} \times \mathbf{B}$ is formed from $\mathbf{E}$ and $\mathbf{B}$ by a right-hand rule.

In summary, according to the electromagnetic theory of light, the electric field intensity vector is perpendicular to the direction of the propagation of light. The magnetic field intensity is perpendicular to both the direction of the electric field intensity vector and the direction of the propagation of light, and its magnitude is proportional to the electric field intensity. See Fig. 2.2.

2.1.6 Motion of an Electron in Electric and Magnetic Fields

In this section, the interaction of the radiation field—electric and magnetic fields varying with time—with the electrons is studied using classical mechanics as a preparation for a quantum-mechanical treatment.

The standard method of developing the quantum mechanics of a dynamic system is first to cast the classical equation of motion into Hamiltonian format. The Hamiltonian $H(\mathbf{p}, \mathbf{r})$ of a dynamic system is a function of its coordinate $\mathbf{r}$ and corresponding momentum $\mathbf{p}$, representing the total energy. For example, for an electron with charge q moving in an electric field with potential $\phi(\mathbf{r})$, the Hamiltonian is

$$H = \frac{1}{2m_e}\mathbf{p}^2 + q\phi(\mathbf{r}). \tag{2.27}$$

The equations of motion in the Hamiltonian format are a pair of first-order ordinary differential equations:

$$\dot{p}_x = -\frac{\partial H}{\partial x}, \tag{2.28}$$

$$\dot{x} = \frac{\partial H}{\partial p_x}. \tag{2.29}$$

There are similar equations for y and z. Using Eqs 2.29 and 2.27, the expression of momentum is found to be identical to the usual definition,

$$\mathbf{p} = m_e\mathbf{v} = m_e\dot{\mathbf{r}}, \tag{2.30}$$

where a dot means taking a derivative with respect to time t. Applying Eqs. 2.28 and 2.29 to Eq. 2.27, one finds

$$m_e\ddot{\mathbf{r}} = -q\nabla\phi(\mathbf{r}) = q\mathbf{E}, \tag{2.31}$$

which is Newton's equation of motion, where $\mathbf{E}$ is electric field intensity.

For the motion of an electron in both electric and magnetic fields, the standard method is to insert the vector potential $\mathbf{A}$ into the expression of momentum by simply substituting $\mathbf{p}$ with $\mathbf{p} - q\mathbf{A}$ in the Hamiltonian,

$$H = \frac{1}{2m_e}(\mathbf{p} - q\mathbf{A})^2 + q\phi(\mathbf{r}), \tag{2.32}$$

or

$$H = \frac{1}{2m_e}\left[(p_x - qA_x)^2 + (p_y - qA_y)^2 + (p_z - qA_z)^2\right] + q\phi(x, y, z). \tag{2.33}$$

Applying Eq. 2.29 to Eq. 2.32, one obtains

$$p_x = m_e\dot{x} + qA_x \tag{2.34}$$

and so on. In vector form, it is

$$\mathbf{p} = m_e\dot{\mathbf{r}} + q\mathbf{A}, \tag{2.35}$$

which is the definition of momentum in a magnetic field. Applying Eq. 2.28 to Eq. 2.32 and using Eq. 2.35, for the x-component, yield

$$\frac{dp_x}{dt} = q\left[\frac{\partial A_x}{\partial x}\dot{x} + \frac{\partial A_y}{\partial x}\dot{y} + \frac{\partial A_z}{\partial x}\dot{z}\right] - q\frac{\partial \phi}{\partial x}. \tag{2.36}$$

The familiar Newton equation of motion, similar to Eq. 2.31, can be obtained from Eqs. 2.33 and 2.36. The x-component is given as

$$m_e\ddot{x} = \frac{dp_x}{dt} - q\frac{dA_x}{dt}. \tag{2.37}$$

Note that

$$\frac{dA_x}{dt} = \frac{\partial A_x}{\partial t} + \frac{\partial A_x}{\partial x}\dot{x} + \frac{\partial A_x}{\partial y}\dot{y} + \frac{\partial A_x}{\partial z}\dot{z}, \tag{2.38}$$

for example, for the x-component, one obtains,

$$m_e\ddot{x} = q\left[\dot{y}\left(\frac{\partial A_y}{\partial x} - \frac{\partial A_x}{\partial y}\right) + \dot{z}\left(\frac{\partial A_z}{\partial x} - \frac{\partial A_x}{\partial z}\right) - \frac{\partial A_x}{\partial t} - \frac{\partial \phi}{\partial x}\right]. \tag{2.39}$$

Using Eqs. 2.6 and 2.8, in vector form, the equation of motion is

$$m_e\ddot{\mathbf{r}} = q\mathbf{E} + q\dot{\mathbf{r}} \times (\nabla \times \mathbf{A}) = q\mathbf{E} + q\dot{\mathbf{r}} \times \mathbf{B}. \tag{2.40}$$

which is Newton's equation of motion including the magnetic force. Therefore, the correctness of the Hamiltonian, Eq. 2.32, is verified. We will use the Hamiltonian in the quantum-mechanical treatment of the interaction of radiation with atomic systems.

2.2 Optics of Thin Films

Maxwell's theory of light plays a critical role in the understanding of selective absorption films for solar thermal applications and antireflection films in photovoltaics. The general theory with an arbitrary incident angle is rather complicated. However, for applications related to solar energy, it suffices to study the case of normal incidence, which demonstrates most of the related physics. First, let us extend Maxwell's equations to dielectrics.

2.2.1 Relative Dielectric Constant and Refractive Index

Maxwell's equations, Eqs. 2.1–2.4 are used for the case of a vacuum. To describe electromagnetic phenomena in a nonmagnetic medium, the electric constant ε_0 is replaced by the electric constant of the medium, ε. Maxwell's equations are

$$\nabla \cdot \mathbf{E} = \frac{\rho}{\varepsilon}, \tag{2.41}$$

$$\nabla \cdot \mathbf{B} = \mathbf{0}, \tag{2.42}$$

$$\nabla \times \mathbf{E} = -\frac{\partial \mathbf{B}}{\partial t}, \tag{2.43}$$

$$\nabla \times \mathbf{B} = \varepsilon \mu_0 \frac{\partial \mathbf{E}}{\partial t} + \mu_0 \mathbf{J}. \tag{2.44}$$

Following the procedures in Section 2.1.1, we found the wave equations for the electric field intensity and the magnetic field intensity:

$$\nabla^2 \mathbf{E} - \frac{1}{v^2} \frac{\partial^2 \mathbf{E}}{\partial t^2} = 0, \tag{2.45}$$

$$\nabla^2 \mathbf{B} - \frac{1}{v^2} \frac{\partial^2 \mathbf{B}}{\partial t^2} = 0, \tag{2.46}$$

where the velocity v is given as

$$v = \frac{1}{\sqrt{\varepsilon \mu_0}}. \tag{2.47}$$

Table 2.2: Dielectric Constant and Refractive Index of Selected Materials

Material	Wavelength	ε_r	n
Silicon	1.39 μm	12.2	3.49
Germanium	2.1 μm	16.8	4.10
TiO_2	2.0 μm	5.76	2.4
SiO_2	Visible	2.40	1.55
Window glass	Visible	2.40	1.55
ZnS	Visible	5.43	2.33
CeO_2	Visible	3.81	1.953
CaF_2	Visible	2.06	1.435
MgF_2	Visible	1.91	1.383

Source: American Institute of Physics Handbook, 3rd Ed, McGraw-Hill, New York, 1982.

Comparing with Eq. 2.14, the relation of v with c is

$$\frac{c}{v} = \sqrt{\frac{\varepsilon}{\varepsilon_0}}. \tag{2.48}$$

Defining the relative dielectric constant of the medium as

$$\varepsilon_r \equiv \frac{\varepsilon}{\varepsilon_0}, \tag{2.49}$$

the ratio of the speed of light in a vacuum and the speed of light in the medium, defined as the *refractive index* n, is

$$n \equiv \frac{c}{v} = \sqrt{\varepsilon_r}. \tag{2.50}$$

In general, the relative dielectric constant and the refractive index depend on the frequency or wavelength of the electromagnetic wave. For application in solar energy devices, the most relevant case is solar radiation in the visible or infrared. Table 2.2 shows the relative dielectric constant and refractive index of several materials often used in solar energy devices.

For electromagnetic waves propagating in the z direction with wavevector k and electric field intensity in x, similar to Eqs. 2.24–2.26, the nonzero components are

$$E_x = E_0\, e^{i(kz-\omega t)} \tag{2.51}$$

$$B_y = \frac{k}{\omega} E_0\, e^{i(kz-\omega t)}. \tag{2.52}$$

The wavevector k is given as

$$k = \frac{\omega}{v} = \frac{\omega n}{c}. \tag{2.53}$$

And, according to Eq. 2.50, the electric and magnetic fields are in phase and proportional,

$$B_y = \frac{1}{v} E_x = \frac{n}{c} E_x. \tag{2.54}$$

2.2.2 Energy Balance and Poynting Vector

Let us study the energy balance in an electromagnetic field by considering a unit volume with relatively uniform fields. If the current density is $\mathbf{J}$ and the electric field intensity is $\mathbf{E}$, the ohmic energy loss per unit time per unit volume is $\mathbf{J} \cdot \mathbf{E}$. Using Eq. 2.44, the expression of energy loss becomes

$$\mathbf{J} \cdot \mathbf{E} = -\frac{1}{\mu_0} \mathbf{E} \cdot (\nabla \times \mathbf{B}) + \varepsilon \mathbf{E} \cdot \frac{\partial \mathbf{E}}{\partial t}. \tag{2.55}$$

Using the mathematical identity

$$\mathbf{E} \cdot (\nabla \times \mathbf{B}) = -\nabla \cdot (\mathbf{E} \times \mathbf{B}) + \mathbf{B} \cdot (\nabla \times \mathbf{E}), \tag{2.56}$$

Eq. 2.55 becomes

$$\mathbf{J}\cdot\mathbf{E} = \nabla\cdot\left(\frac{1}{\mu_0}\mathbf{E}\times\mathbf{B}\right) + \frac{1}{\mu_0}\mathbf{B}\cdot(\nabla\times\mathbf{E}) - \varepsilon\mathbf{E}\cdot\frac{\partial\mathbf{E}}{\partial t}. \tag{2.57}$$

Using Eq. 2.43, Eq. 2.57 becomes

$$\mathbf{J}\cdot\mathbf{E} = -\nabla\cdot\left(\frac{1}{\mu_0}\mathbf{E}\times\mathbf{B}\right) - \frac{\partial}{\partial t}\left(\frac{\varepsilon}{2}E^2 + \frac{1}{2\mu_0}B^2\right). \tag{2.58}$$

The right-hand side of Eq. 2.58 has a straightforward explanation. The energy density of the electromagnetic fields is

$$W = \frac{\varepsilon}{2}E^2 + \frac{1}{2\mu_0}B^2, \tag{2.59}$$

and the power density of the electromagnetic field per unit area is

$$\mathbf{S} = \frac{1}{\mu_0}\mathbf{E}\times\mathbf{B}. \tag{2.60}$$

The vector $\mathbf{S}$ is called the *Poynting vector* after its discoverer.

For an electromagnetic wave, according to Eq. 2.54, $cB_y = nE_x$. The magnitude of the Poynting vector along the direction of propagation is

$$S_z = \frac{n}{\mu_0 c}E_x^2. \tag{2.61}$$

2.2.3 Fresnel Formulas

Consider two media of refractive indices n_1 and n_2 with an interface at $z = 0$, as shown in Fig. 2.3. The incident light is moving in the z direction with wavevector k_{I},

$$k_{\mathrm{I}} = \frac{\omega n_1}{c}. \tag{2.62}$$

The field intensities of the incident light are

$$E_{\mathrm{I}} = I\,e^{i(k_{\mathrm{I}}z-\omega t)}, \tag{2.63}$$

$$B_{\mathrm{I}} = \frac{n_1}{c}I\,e^{i(k_{\mathrm{I}}z-\omega t)}, \tag{2.64}$$

where I is a constant characterizing the intensity of incident light.

For transmitted light, the wavevector is determined by the refractive index of medium 2,

$$k_{\mathrm{T}} = \frac{\omega n_2}{c}. \tag{2.65}$$

The field intensities of the transmitted light are

$$E_{\mathrm{T}} = T\,e^{i(k_{\mathrm{T}}z-\omega t)}, \tag{2.66}$$

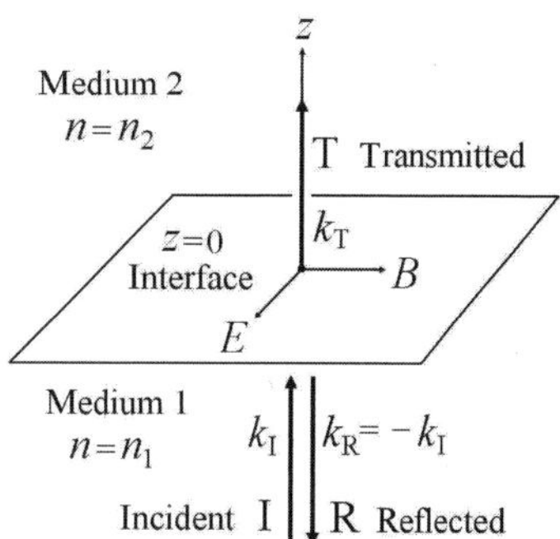

Figure 2.3 Derivation of Fresnel formulas. Two media with indices of refraction n_1 and n_2 share an interface at $z = 0$. The incident light has a wavevector k_I. The wavevector of transmitted light is k_T. The wavevector of reflected light is identical to that of incident light but with opposite sign. By applying Maxwell's equations at the interface, the relations between the three components of light can be derived.

$$B_\mathrm{T} = \frac{n_2}{c} T\, e^{i(k_\mathrm{T} z - \omega t)}, \tag{2.67}$$

The constant T characterizing the intensity of the transmitted light is to be determined by the boundary conditions required by Maxwell's equations.

For reflected light, because it is in the same medium as the incident light, the absolute value of the wavevector is identical to that of the incident light. However, the direction of z is reversed. By using the same notation k_I, the field intensities of the reflected light are

$$E_\mathrm{R} = R\, e^{i(-k_\mathrm{I} z - \omega t)}, \tag{2.68}$$

$$B_\mathrm{R} = -\frac{n_1}{c} R\, e^{i(-k_\mathrm{I} z - \omega t)}. \tag{2.69}$$

Notice the negative sign of the magnetic field intensity B_R. Again, the constant R characterizes the intensity of the reflected light.

On the interface, $z = 0$, following Eqs. 2.1 and 2.2, both electric field intensity and magnetic field intensity should be continuous. In other words,

$$E_\mathrm{I} + E_\mathrm{R} = E_\mathrm{T}, \tag{2.70}$$

$$B_\mathrm{I} + B_\mathrm{R} = B_\mathrm{T}. \tag{2.71}$$

Using Eqs. 2.63–2.71, we find

$$I - R = T, \tag{2.72}$$

$$n_1(I + R) = n_2 T. \tag{2.73}$$

The solutions of Eqs.2.72 and 2.73 are

$$R = \frac{n_2 - n_1}{n_2 + n_1} I, \tag{2.74}$$

$$T = \frac{2n_1}{n_2 + n_1} I. \tag{2.75}$$

Equations 2.74 and 2.75 are the *Fresnel formulas* for the case of normal incidence. Obviously, if $n_1 = n_2$, there is no reflected light, and 100% of incident light is transmitted through the interface.

The power densities of the incident, transmitted, and reflected light can be evaluated using Eqs. 2.74 and 2.75 and the expression of the Poynting vector, Eq. 2.60. For incident light, the magnitude is

$$S_{\rm I} = \frac{1}{\mu_0} E_{\rm I} B_{\rm I} = \frac{n_1}{\mu_0 c} I^2. \tag{2.76}$$

For transmitted light,

$$S_{\rm T} = \frac{1}{\mu_0} E_{\rm T} B_{\rm T} = \frac{n_2}{\mu_0 c} T^2. \tag{2.77}$$

Using Eq. 2.74,

$$S_{\rm T} = \frac{4n_1 n_2}{(n_1+n_2)^2} \frac{n_1}{\mu_0 c} I^2 = \frac{4n_1 n_2}{(n_1+n_2)^2} S_{\rm I}. \tag{2.78}$$

A dimensionless coefficient of transmission is defined as

$$\mathcal{T} \equiv \frac{S_{\rm T}}{S_{\rm I}} = \frac{4n_1 n_2}{(n_1+n_2)^2}. \tag{2.79}$$

Following Eqs. 2.77 and 2.78, the intensity of reflected light can be determined, and a dimensionless coefficient of reflection is defined as

$$\mathcal{R} \equiv \frac{S_{\rm R}}{S_{\rm I}} = \left(\frac{n_1 - n_2}{n_1 + n_2}\right)^2. \tag{2.80}$$

For semiconductors, the reflection loss can be significant. For example, for silicon, $n = 3.49$. The reflection coefficient is

$$\mathcal{R} = \frac{(1-3.49)^2}{(1+3.49)^2} \approx 0.3076. \tag{2.81}$$

More than 30% of light is lost by reflection. To build high-efficiency solar cells, an *antireflection coating* is essential. We will discuss this in Section 9.4.

2.3 Blackbody Radiation

It was known for centuries that a hot body emits radiation. At around 700°C, a body becomes red hot. At even higher temperatures, a body emits much more radiation, and the color changes to orange, yellow, white, and even blue. In the late nineteenth century, in order to understand phenomena related to industry technology such as steel making and incandescent light bulbs, heat radiation became a hot subject for physicists.

Although all hot bodies emit radiation, blackbodies emit the maximum amount of radiation at a given temperature. At equilibrium, radiation emitted must equal radiation absorbed. Therefore, the body that emits the maximum amount also absorbs the maximum amount—which should look black. Practically, a blackbody is constructed by opening a small hole on a large cavity, as shown in Fig. 2.4. Any light ray passing through the hole with area A experiences multiple reflections on the internal surface of the cavity. If the material is not absolutely shiny, after several impingements, the light will eventually be completely absorbed by the cavity. Therefore, the small hole on the large cavity always looks black, which is a good example of a blackbody.

2.3.1 Rayleigh–Jeans Law

The energy density of radiation as a function of its frequency was studied in the late nineteenth century by Lord Rayleigh and then by Sir James Jeans using classical statistical physics. They treated standing electromagnetic waves in a cavity as individual modes, and the modes follow the equal-partition law of Maxwell–Boltzmann statistics.

Consider a closed cubic cavity with reflective inner surfaces of side L. A sinusoidal electromagnetic wave with frequency ν satisfies the following equation:

$$\nabla^2 \mathbf{A} + \frac{4\pi^2\nu^2}{c^2}\mathbf{A} = 0. \tag{2.82}$$

Assuming that the cavity is made of metal. On the walls of the cavity, electrical field intensity vanishes. Therefore, the vector potential vanishes. The general solution of Eq. 2.82 satisfying that condition is

$$\mathbf{A} = \mathbf{A}_0 \sin(k_x x)\, \sin(k_y y)\, \sin(k_z z). \tag{2.83}$$

The wavevectors are defined by

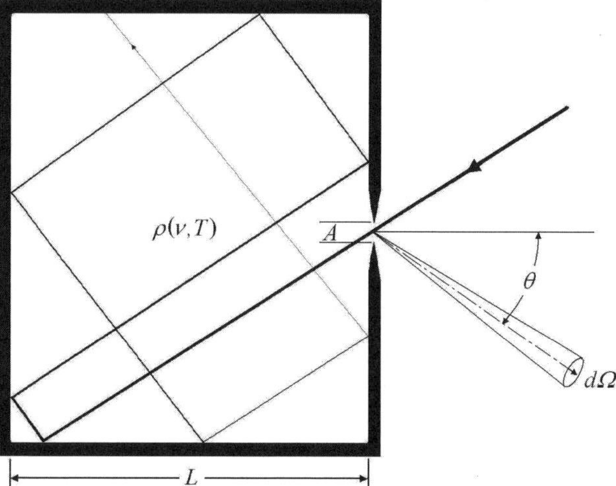

Figure 2.4 Blackbody radiation. A large cavity with a small hole is a good blackbody. The light enters the hole will experience multiple reflections, and all be absorbed and thus looks black. A blackbody emits maximum amount of radiation when heated.

$$k_x = \frac{\pi\, n_x}{L}, \quad k_y = \frac{\pi\, n_y}{L}, \quad k_z = \frac{\pi\, n_z}{L}, \tag{2.84}$$

where n_x, n_y, and n_z are positive integers. By direct substitution one finds that the solution, Eq. 2.83, satisfies differential equation 2.82 and the boundary conditions at the walls. Each set of the integers, n_x, n_y, n_z, represents a pattern of electromagnetic wave in the cavity. Inserting Eq. 2.84 into Eq. 2.82 yields

$$k_x^2 + k_y^2 + k_z^2 = \frac{4\pi^2\nu^2}{c^2}, \tag{2.85}$$

and in terms of the numbers n_x, n_y, and n_z, Eq. 2.85 becomes

$$n_x^2 + n_y^2 + n_z^2 = \frac{4\nu^2 L^2}{c^2}. \tag{2.86}$$

Now, we count the number of standing waves with frequencies ν by considering a sphere of radius $\sqrt{n_x^2 + n_y^2 + n_z^2} = 2\nu L/c$. The number N of modes with positive n_x, n_y, and n_z up to ν is

$$N = \frac{1}{8}\frac{4}{3}\pi\left(\frac{2\nu L}{c}\right)^3 = \frac{4\pi\nu^3 L^3}{3c^3}, \tag{2.87}$$

For each type of standing wave, there are two polarizations. Therefore, the number of modes of standing electromagnetic waves is

$$N = \frac{8\pi\nu^3 L^3}{3c^3}, \tag{2.88}$$

where L^3 is the volume, and the *density of states* at frequency ν is

$$\frac{d}{d\nu}\left(\frac{N}{L^3}\right) = \frac{8\pi\nu^2}{c^3}. \tag{2.89}$$

According to Maxwell–Boltzmann statistics, at absolute temperature T, each degree of freedom contributes energy $k_\mathrm{B}T$, where k_B is the Boltzmann constant, and the energy density is

$$\rho(\nu, T) = \frac{d}{d\nu}\left(\frac{N}{L^3}\right) k_\mathrm{B}T = \frac{8\pi\nu^2}{c^3}\, k_\mathrm{B}T. \tag{2.90}$$

Equation 2.90 is the energy density of radiation per unit frequency interval in a cavity of temperature T. It is not directly observable. The directly observable quantity is the spectral radiance $u(\nu, T)$, that is, the energy radiating from a unit area of the hole per unit frequency range. To calculate $u(\nu, T)$ from $\rho(\nu, T)$, first we consider a simplified situation: If the field has a well-defined direction of radiation with velocity c, we have

$$u(\nu, T) = c\,\rho(\nu, T). \tag{2.91}$$

Because the hole is small, the radiation field in a cavity is isotropic. As the radiation only comes through a hole of well-defined direction, $u(\nu, T)$ should be a fraction of $c\rho(\nu, T)$. The value of the fraction can be determined using the following argument. Consider a sphere of radius R. The surface area of the sphere is $4\pi R^2$. If the radiation inside the sphere is allowed to emit over all directions, the area is $4\pi R^2$. If the radiation is allowed to emit in only one direction, the area is a disc with radius R, that is, πR^2. Consequently, the factor is 1/4. Equation 2.91 becomes

$$u(\nu, T) = \frac{1}{4} c\rho(\nu, T). \tag{2.92}$$

Following is a more detailed proof of the factor 1/4. Consider the radiation from a small hole of area A on the cavity; see Fig. 2.4. Because the electromagnetic wave is isotropic and the speed of light is c, the energy radiated through a solid angle $d\Omega$ at an angle θ is

$$\frac{dE}{dt\, d\Omega} = \frac{c}{4\pi} \rho(\nu, T)\, A \cos\theta \tag{2.93}$$

because the area of the hole observed from an angle θ is $A \cos\theta$. Integrating over the hemisphere, the total irradiation per unit area is

$$u(\nu, T) = \frac{c}{4\pi} \int_0^{\pi/2} 2\pi \cos\theta \sin\theta \, d\theta \, \rho(\nu, T) = \frac{c}{4} \rho(\nu, T), \tag{2.94}$$

confirming Eq. 2.92. Using Eq. 2.90, we finally obtain the Rayleigh–Jeans distribution of blackbody radiation,

$$u(\nu, T) = \frac{2\pi\nu^2}{c^2} k_B T. \tag{2.95}$$

The Rayleigh–Jeans distribution fits the low-frequency behavior of the experimental energy density very well. However, as the frequency increases, the spectral irradiance increases, the total irradiation energy is infinite. This contradicts the experimental fact that the total blackbody radiation is finite, and the spectral density has a maximum; see Fig. 2.5.

2.3.2 Planck Formula and Stefan–Boltzmann's Law

In 1900, Max Planck found an empirical formula that fits accurately the experimental data,

$$u(\nu, T) = \frac{2\pi\nu^2}{c^2} \frac{h\nu}{e^{h\nu/k_B T} - 1}. \tag{2.96}$$

The constant h in the formula, Planck's constant, was initially obtained by fitting with experimental blackbody radiation data. Later, Planck found a mathematical explanation of his formula by assuming that the energy of radiation can only take discrete values. Specifically, he assumed that the energy of radiation with frequency ν can only take integer multiples of a basic value $h\nu$, the *energy quantum*,

$$\epsilon = 0,\;\; h\nu,\;\; 2h\nu,\;\; 3h\nu,\;\; \dots . \tag{2.97}$$

According to Maxwell–Boltzmann statistics, the probability of finding a state with energy $nh\nu$ is $\exp(-nh\nu/k_B T)$. The average value of energy of a given component of radiation with frequency ν is

$$\bar{\epsilon} = \frac{\sum_{n=0}^{\infty} nh\nu\, e^{-nh\nu/k_B T}}{\sum_{n=0}^{\infty} e^{-nh\nu/k_B T}} = \frac{h\nu}{e^{h\nu/k_B T} - 1}. \tag{2.98}$$

instead of $k_B T$. By replacing the expression $k_B T$ in Eq. 2.90 with Eq. 2.98, we recovered Eq. 2.96.

Initially, Max Planck believed that the quantization of energy is only a mathematical trick to reconcile his empirically obtained formula with the knowledge of physics known at that time. The profound significance of the concept of quantization of radiation and the meaning of Planck's constant were discovered by Albert Einstein in his interpretation of the photoelectric effect, which is the conceptual foundation of solar cells.

By integrating the spectral radiance over frequency, the total radiation is found to be

$$\begin{aligned} U(T) &= \int_0^\infty \frac{2\pi h\nu^3}{c^2} \frac{d\nu}{e^{h\nu/k_B T} - 1} \\ &= \frac{2\pi h}{c^2} \left(\frac{k_B T}{h}\right)^4 \int_0^\infty \frac{x^3\, dx}{e^x - 1} \\ &= \frac{2}{15} \frac{\pi^5 k_B^4}{c^2\, h^3}\, T^4. \end{aligned} \tag{2.99}$$

Here a mathematical identity is applied,

$$\int_0^\infty \frac{x^3\, dx}{e^x - 1} = \frac{\pi^4}{15}. \tag{2.100}$$

Equation 2.99 is *Stefan–Boltzmann's law*, discovered experimentally before the Planck formula and backed by an argument using thermodynamics. The constant in Eq. 2.99,

$$\sigma \equiv \frac{2}{15} \frac{\pi^5 k_B^4}{c^2 h^3} = \frac{\pi^2 k_B^4}{60\, c^2 \hbar^3} = 5.67 \times 10^{-8} \frac{\mathrm{W}}{\mathrm{m}^2 \cdot \mathrm{K}^4}, \tag{2.101}$$

is called the *Stefan–Boltzmann's constant.* It can be memorized using the mnemonic: 45678. The total radiance is proportional to the fourth power of absolute temperature, and the coefficient is 5.67 times the inverse eighth power of 10.

For applications in solar cells, the electron volt is the most convenient unit of photon energy; see Fig. 2.5. The Planck formula for blackbody spectral irradiance in terms of photon energy ϵ in units of electron volts is

$$u(\epsilon, T) = \frac{2\pi q^4}{c^2\, h^3} \frac{\epsilon^3}{e^{\epsilon/\epsilon_T} - 1} = 1.587 \times 10^8 \frac{\epsilon^3}{e^{\epsilon/\epsilon_T} - 1} \frac{\mathrm{W}}{\mathrm{m}^2 \cdot \mathrm{eV}}, \tag{2.102}$$

where $\epsilon_T = k_B T/q$ is the value of $k_B T$ in electron volts. Numerically, it equals $\epsilon_T = T/11,600$. For the Sun, $T_\odot$=5800 K; thus $\epsilon_\odot$ =0.5 eV. At the location of Earth, the radiation is diluted by the distance from the Sun to Earth, the astronomical constant $A_\odot = 1.5 \times 10^{11}$ m. Introducing a geometric factor f representing the solid angle of the Sun with radius $r_\odot = 6.96 \times 10^8$m as observed from Earth

$$f = \left(\frac{r_\odot}{A_\odot}\right)^2 = \frac{\left[6.96 \times 10^8\right]^2}{\left[1.5 \times 10^{11}\right]^2} = 2.15 \times 10^{-5}, \tag{2.103}$$

the spectrum of the AM0 solar radiation (outside the atmosphere at the location of Earth) is

$$u_\oplus(\epsilon, T) = f u_\odot(\epsilon, T) = 3.41 \times 10^3 \frac{\epsilon^3}{e^{\epsilon/\epsilon_\odot} - 1} \frac{\mathrm{W}}{\mathrm{m}^2 \cdot \mathrm{eV}}. \tag{2.104}$$

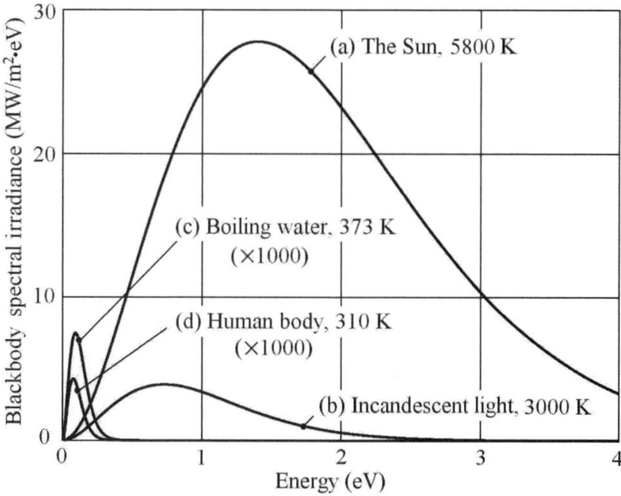

Figure 2.5 Blackbody spectral irradiance. The blackbody spectral irradiance, or the radiation power emitted per square meter per unit energy interval (here in electron volts) at an energy value (also in electron volts) at four different temperatures ise shown. The maximum of solar irradiance is at 1.4 eV, with a value of 27.77 MW/m^2·eV. The temperature of the filament of an incandescent light is about 3000 K. The radiation power density at the filament surface is only about 7% that on the Sun. The spectral irradiance from a blackbody at the boiling points of water and the human body are also shown, in units of kW/m^2·eV.

Table 2.3: Blackbody Radiation at Different Temperatures

Radiator	Temperature (K)	Power (W/m^2)	Peak ϵ (eV)	Peak λ (μm)	Peak u ($W/m^2{\cdot}eV$)
The Sun	5800	6.31×10^7	1.410	0.88	2.81×10^7
Light bulb	3000	4.59×10^6	0.728	1.70	3.88×10^6
Boiling water	373	1.10×10^3	0.091	13.6	7.46×10^3
Human body	310	5.24×10^2	0.075	16.5	4.28×10^3

The position of the peak in blackbody spectral irradiance can be of a transcendental equation

$$\frac{d}{dx}\left[3\log x - \log\left(e^x - 1\right)\right] = 0, \tag{2.105}$$

and can be obtained by numerical computation,

$$x = 2.82. \tag{2.106}$$

In other words, the peak of blackbody spectral irradiance is at

$$\epsilon_{\text{MAX}} = 2.82\,\epsilon_T = 2.43 \times 10^{-4}\,T\,(\text{eV}). \tag{2.107}$$

The peak value for the function $x^3/(e^x - 1)$ is 1.42. Therefore, the peak value of the spectral irradiance is

$$u_{\text{MAX}} = 1.42\,\frac{2\pi q k_{\text{B}}^3}{c^2\,h^3}\,T^3 \cong 1.44 \times 10^{-4}\,T^3\;\frac{\text{W}}{\text{m}^2 \cdot \text{eV}}. \tag{2.108}$$

Table 2.3 lists the data for some frequently encountered cases.

2.4 Photoelectric Effect and Concept of Photons

The photoelectric effect was discovered accidentally by Heinrich Hertz in 1887 during experiments to generate electromagnetic waves. Since then, a number of studies have been conducted in an attempt to understand the phenomena. Around 1900, Phillip Lenard did a series of critical studies on the relation of the kinetic energy of ejected electrons with the intensity and wavelength of the impinging light [50]. His results were in direct conflict with the wave theory of light and inspired Albert Einstein to develop his theory of photons.

Figure 2.6 shows schematically the experimental apparatus of Phillip Lenard. The entire setup was enclosed in a vacuum chamber. An electric arc lamp, using carbon rods or zinc rods as the electrodes, generates strong UV light. A quartz window allows such UV light to shine on a target made of different metals. The target and a counter electrode are connected to an adjustable power supply. An ammeter is used to measure the electric current generated by the UV light, the *photocurrent*, especially when the voltages between the two electrodes are very small. By gradually increasing the voltage, which tends to reflect the electrons back to the target, the photocurrent is reduced. The voltage with which the photocurrent becomes zero is recorded as the *stopping voltage*.

The stopping voltage is apparently related to the kinetic energy of the electrons ejected from the target:

$$qV = \frac{1}{2}mv^2. \tag{2.109}$$

Understandably, the photocurrent varies with the intensity of light. By changing the magnitude of the current that drives the arc or the distance from the arc lamp to the target, the photocurrent could change by two orders of magnitude: for example, from 4.1 to 276 pA. An unexpected and dramatic effect Lenard observed was that no matter how strong or how weak the light is, and no matter how large or how small the photocurrent is, the stopping voltage does not change; see Table 2.4. The stopping voltage changes only when the material for the electric arc lamp changes. However, for a given type of arc, the stopping voltage stays unchanged.

The effect Lenard observed has no explanation in the framework of the wave theory of light. According to the wave theory of light, the more intense the light is, the more kinetic energy the electrons acquire.

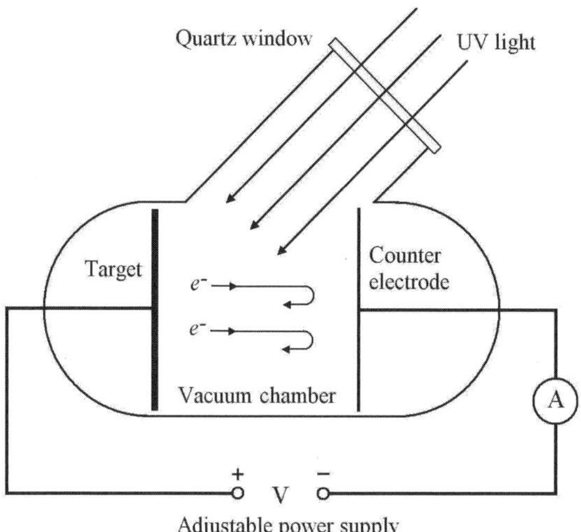

Figure 2.6 Lenard's apparatus for studying photoelectric effect. A quartz window allows the UV light from an electric arc lamp to shine on a target. The voltage between the target and the counter electrode is controlled by an adjustable power supply. An ammeter is used to measure the electric current generated by the UV light, the *photocurrent*. By gradually increasing the voltage (with the polarity as shown), the photocurrent is reduced. The voltage with which the photocurrent becomes zero is recorded as the *stopping voltage* [50].

Table 2.4: Stopping Voltage for Photocurrent

Rod material	Driving current (A)	Distance to target (cm)	Photocurrent (pA)	Stopping voltage (V)
Carbon	28	33.6	276	-1.07
Carbon	20	33.6	174	-1.12
Carbon	28	68	31.7	-1.10
Carbon	8	33.6	4.1	-1.06
Zinc	27	33.6	2180	-0.85
Zinc	27	87.9	319	-0.86

Source: P. Lenard, *Annalen der Physik*, **8**, 167 (1902) [50].

2.4.1 Einstein's Theory of Photons

In 1905, while employed as a patent examiner at the Swiss Patent Office, Albert Einstein wrote five papers, published in *Annalen der Physik*, that initiated the twentieth century revolution in science. For general public, Einstein is mostly known for his theory of relativity. Therefore, when the Swedish Academy announced in 1922 that Einstein had won the Nobel Prize "for services to theoretical physics and especially for the discovery of the law of the photoelectric effect," referring to his paper *On a Heuristic Viewpoint Concerning the Production and Transformation of Light* [27], the public was surprised. In hindsight, the Nobel Committee was correct: His paper on photoelectric effect is considered the boldest, the most revolutionary, and the most original. Although its predictions were fully verified by experiments, for many years, several prominent physicists did not accept Einstein's concept of photons. Here is a quote from Einstein [27]:

> According to the assumption considered here, when a light ray starting from a point is propagated, the energy is not continuously distributed over an ever increasing volume, but it consists of a finite number of energy quanta, localized in space, which move without being divided and which can be absorbed or emitted only as a whole.

According to Einstein, light, when it interacts with matter, appears as a flow of individual and indivisible particles. When a photon interacts with an electron, either it is absorbed or there is no interaction. The energy value of a photon, ϵ, depends on its frequency,

$$\epsilon = h\nu, \tag{2.110}$$

where $h = 6.63 \times 10^{-34}\,\mathrm{J \cdot s}$ is Planck's constant and ν is the frequency of light. For example, for green light, $\lambda = 0.53\,\mu\mathrm{m}$, and the frequency is $5.6 \times 10^{14}\,\mathrm{s}^{-1}$. The energy of the photon is $3.7 \times 10^{-19}\mathrm{J}$, or 2.3 eV.

When a photon interacts with an electron in the metal, it transfers the entire energy to the electron. The electron could escape from the metal by overcoming the *work function* ϕ of the metal, typically a few electron volts. If the energy of the photon is smaller than the work function of the metal, the electron would stay in the metal. If the energy of the photon is greater than the work function of the metal, then the electron can escape from the metal surface *with an excess kinetic energy,*

$$\frac{1}{2}mv^2 = h\nu - \phi. \tag{2.111}$$

The kinetic energy of an escaping electron can be measured by an external voltage, or electric field, to turn it back onto the target. Voltage that just is enough to cancel the kinetic energy is called the *stopping voltage*,

$$q\,V_{\text{stop}} = \frac{1}{2}mv^2 = h\nu - \phi, \tag{2.112}$$

where q is the electron charge, 1.60×10^{-19} C. According to Einstein's quantum theory of light, the stopping voltage is linearly dependent on the frequency of the photon and *independent of the intensity of light.* The slope should be a universal constant, which provides a direct method to determine the value of Planck's constant,

$$\frac{\Delta V_{\text{stop}}}{\Delta \nu} = \frac{h}{q}. \tag{2.113}$$

2.4.2 Millikan's Experimental Verification

Einstein's theory of photons was rejected by a number of prominent physicists for many years, including Max Planck, Niels Bohr, and notably Robert Millikan. Starting in 1905, for 10 years Millikan worked to disprove Einstein's theory. Finally, in 1916, Millikan published a long paper on *Physical Review*, entitled *A Direct Photoelectric Determination of Planck's h* [61]. The conclusion reads as follows:

> 1. Einstein's photoelectric equation has been subject to very searching tests and it appears in every case to predict exactly the observed results.
>
> 2. Planck's h has been photoelectrically determined with a precision of about .5 percent.

In 1923, Millikan received a Nobel Prize "for his work on the elementary charge of electricity and on the photoelectric effect."

An interesting fact in the history of science is that in the same paper Millikan emphatically rejected Einstein's theory of photons. He said that Einstein's photon hypothesis "may well be called reckless first because an electromagnetic disturbance which remains localized in space seems a violation of the very conception of an electromagnetic disturbance, and second because it flies in the face of the thoroughly established facts of interference." Millikan wrote that Einstein's photoelectric equation, although accurately representing the experimental data, "cannot in my judgment be looked upon at

Figure 2.7 Albert Einstein and Robert Millikan. Both Einstein and Millikan won a Nobel Prize for their contributions to the photoelectric effect. Photograph taken in 1930 when Robert Millikan invited Albert Einstein to a conference in California. Original photograph courtesy of Smithsonian Museum, slightly cleaned up by the author.

present as resting upon any sort of a satisfactory theoretical foundation [61]." In 1950, at age 82, in his autobiography [62], Millikan reversed his position and admitted that his experiments

> proved simply and irrefutably, I thought, that the emitted electron that escapes with the energy $h\nu$ gets that energy by the direct transfer of $h\nu$ units of energy from the light to the electron, and hence scarcely permits of any other interpretation than that which Einstein had originally suggested, namely that of the semi-corpuscular or photon theory of light itself.

2.4.3 Wave–Particle Duality

The earlier objections to Einstein's theory of photons was related to an even more profound problem: the wave–particle duality of all particles. At the beginning of the twentieth century, electrons were described by classical mechanics as being similar to billiard balls. Einstein's theory seemed to imply that photons are also like billiard balls and that the photoelectric effect is a collision of the billiard balls with electrons. Such a picture is not only hard to conceive but also in direct conflict with the well-established interference phenomenon of light.

The paradox was resolved after Louis de Broglie extended Einstein's postulate that light can be both wave and particle to *all particles*, including the electron. According to de Broglie, a particle with a momentum $\mathbf{p}$ should also be a plane wave with a wavevector $\mathbf{k}$ such that

$$\mathbf{p} = \hbar\mathbf{k}, \tag{2.114}$$

where $\hbar = h/2\pi$, Planck's constant divided by 2π, is often called Dirac's constant. According to the theory of de Broglie, a better picture of the photoelectric effect is that light radiation as a plane wave interacts with the electron in the electrode, which is also a plane wave, but the energy transferred to the electron must be quantized to satisfy the Einstein equation

$$\epsilon = h\nu \equiv \hbar\omega, \tag{2.115}$$

where $\omega = 2\pi\nu$ is the circular frequency of the light wave. This is an essential concept regarding the understanding of solar cells and solar photochemistry, which we will discuss in the corresponding chapters.

2.5 Einstein's Derivation of Blackbody Formula

Based on the concept of photons and the interaction of photons with matter, Einstein made a very simple derivation of the blackbody radiation formula. The key of his derivation is the introduction of *stimulated emission*, which gave birth to the laser, an acronym for light amplification by stimulated emission of radiation, and provides a better understanding of the interaction between solar radiation and atomic systems.

Einstein studied a simple two-state atomic system; see Fig. 2.8. The radiation field is represented by an energy density $\rho(\nu)$, where ν is the frequency. The atomic system has two states with an energy difference $h\nu$. According to Maxwell–Boltzmann statistics, the ratio of the populations of the two states is

$$\frac{N_2}{N_1} = e^{-h\nu/k_\mathrm{B}T}. \tag{2.116}$$

Einstein assumed three transition coefficients: the absorption coefficient B_{12}, the spontaneous emission coefficient A, and the stimulated emission coefficient B_{21}. The rate equations are

$$\frac{dN_2}{dt} = B_{12}N_1\rho(\nu) - B_{21}N_2\rho(\nu) - AN_2, \tag{2.117}$$

$$\frac{dN_1}{dt} = -B_{12}N_1\rho(\nu) + B_{21}N_2\rho(\nu) + AN_2. \tag{2.118}$$

At equilibrium, both dN_1/dt and dN_2/dt should vanish. Therefore,

$$\frac{N_2}{N_1} = \frac{B_{12}\rho(\nu)}{A + B_{21}\rho(\nu)} = e^{-h\nu/k_\mathrm{B}T}. \tag{2.119}$$

The coefficients should not depend on temperature. At high temperature, the power density should be high, and the right-hand side of Eq. 2.119 should approach unity.

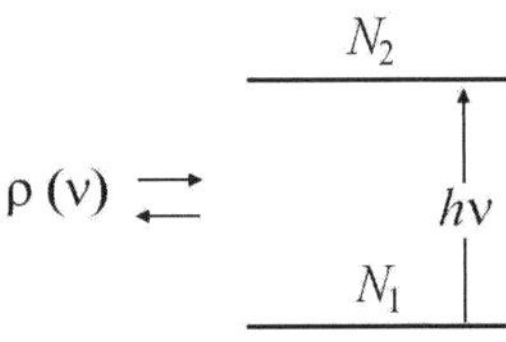

Figure 2.8 Einstein's derivation of blackbody radiation formula. The radiation field $\rho(\nu)$ interacts with a two-level atomic system. Three interaction modes are assumed: absorption, to lift the atomic system from state 1 to state 2; spontaneous emission and stimulated emission, the atomic system decays from state 2 to state 1, giving out energy to the radiation field.

Therefore, one must have

$$B_{12} = B_{21} = B. \tag{2.120}$$

The absorption coefficient B_{12} equals the stimulated emission coefficient B_{21}, which can be represented by a single coefficient B. Under any temperature, the power density distribution of radiation is then

$$\rho(\nu) = \frac{A}{B}\frac{1}{e^{h\nu/k_B T} - 1}. \tag{2.121}$$

For radiations of low photon energy, Eq. 2.121 reduces to

$$\rho(\nu) \to \frac{A}{B}\frac{k_B T}{h\nu}. \tag{2.122}$$

It should be identical to the Rayleigh–Jeans formula. Comparing with Eq. 2.90, we find the ratio of coefficients A and B,

$$\frac{A}{B} = \frac{8\pi h\nu^3}{c^3}. \tag{2.123}$$

Finally, Planck's formula is recovered,

$$\rho(\nu) = \frac{8\pi h\nu^3}{c^3}\frac{1}{e^{h\nu/k_B T} - 1}. \tag{2.124}$$

Problems

2.1. Show that the capacitance C of a parallel-plate capacitor with vacuum as the dielectric is

$$C = \frac{\varepsilon_0 A}{d} \text{ [F]}, \tag{2.125}$$

where A is the area and d is the distance between the electrodes.

2.2. Show that the capacitance C of a parallel-plate capacitor with a medium of relative dielectric constant ε_r is

$$C = \frac{\varepsilon_0 \varepsilon_r A}{d} \text{ [F]}. \tag{2.126}$$

Calculate the capacitance of a capacitor with $A = 1\,\text{m}^2$ and $s = 1\,\text{mm}$ for glass and silicon.

2.3. Show that the inductance L of an inductor made of a long solenoid of N loops with cross-sectional area A and length l is

$$L = \frac{\mu_0 N^2 A}{l} \text{ [H]}. \tag{2.127}$$

2.4. Show that the speed of light v in a medium of relative dielectric constant ε_r is

$$v = \frac{c}{\sqrt{\varepsilon_r}}. \tag{2.128}$$

Calculate the speed of light v in glass and silicon (the relative dielectric constants ε_r for glass and silicon are 2.25 and 11.7, respectively).

2.5. The refractive index of window glass is $n = 1.50$. How much light power is lost when going through a sheet of glass at normal incidence? (*Hint*: there are two glass–air interfaces.)

2.6. The radius of the Sun is $R = 6.96 \times 10^8$ m, and the distance between the Sun and Earth is $D = 1.5 \times 10^{11}$ m. The solar constant is 1366 W/m^2. Estimate the surface temperature of the Sun. (*Hint*: use the Stefan–Boltzmann law.)

2.7. What is the magnitude of the electric field intensity of the sunlight just outside the atmosphere of Earth?

2.8. What is the electric field intensity of the electron in a hydrogen atom at the distance of one Bohr radius from the proton?

2.9. Derive the blackbody radiation spectral density per unit wavelength in unit of micrometers.

2.10. Using the blackbody radiation formula per unit wavelength, derive the Wien displacement law in micrometers.

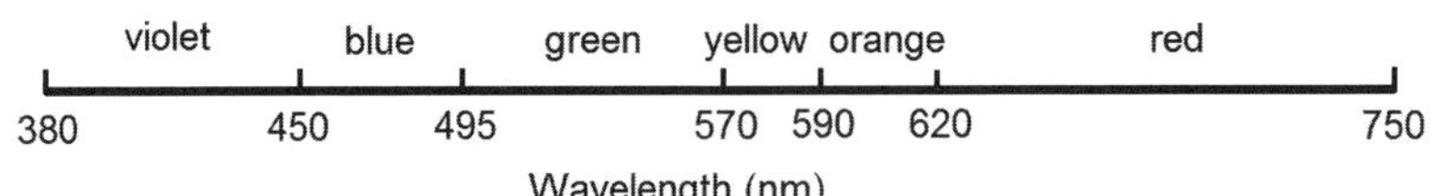

Figure 2.9 Wavelengths of visible lights.

2.11. The wavelengths of visible light with different colors in nanometers are shown in Fig. 2.9. Compute the frequencies and energy values of the photons, in both joules and electron volts.

2.12. What is the solar constant of Venus? Assume that the Sun is a blackbody emitter at 5800 K and the mean Venus-Sun distance is 1.08×10^{11} m.

2.13. To compute the blackbody irradiation for photon energy from ϵ_0 to infinity, an easy-to-use formula can be obtained by introducing $x_0 = \epsilon_0/k_B T$ and expanding the denominator of Eq. 2.99 into

$$\begin{aligned} U(T,\epsilon_0) &= \frac{2\pi(k_B T)^4}{c^2\,h^3}\int_{x_0}^{\infty}\frac{e^{-x}\,x^3\,dx}{1-e^{-x}} \\ &= \frac{2\pi(k_B T)^4}{c^2\,h^3}\sum_{n=1}^{\infty}\int_{x_0}^{\infty} e^{-nx}\,x^3\,dx. \end{aligned} \tag{2.129}$$

Prove that

$$U(T,\epsilon_0) = \frac{2\pi(k_B T)^4}{c^2\,h^3}\sum_{n=1}^{\infty} e^{-nx_0}\left[\frac{x_0^3}{n}+\frac{3x_0^2}{n^2}+\frac{6x_0}{n^3}+\frac{6}{n^4}\right], \tag{2.130}$$

with

$$x_0 = \frac{\epsilon_0}{k_B T}. \tag{2.131}$$

2.14. Assuming that the Sun is a blackbody emitter at 5800 K, what fraction of solar radiation is green (wavelength between 495 and 570 nm)?

2.15. Assuming that the Sun is a blackbody emitter at 5800 K, what fraction of solar radiation has photon energy greater than 1.1 eV?

Chapter 3

Origin of Solar Energy

Since the late-nineteenth century, measurement of solar radiation power density has been diligently pursued. After the advancement of artificial satellites, solar radiation data outside the atmosphere became available. From the data accumulated to date, variation in the solar radiation power density at the average position of Earth outside the atmosphere has not exceeded 0.1% over a century. Although not a physical constant, this quantity, S, is often called the *solar constant*,

$$S = 1366 \pm 3\,\mathrm{W/m^2}. \tag{3.1}$$

The total radiation power of the Sun, $L_\odot$, or the *solar luminosity*, can be evaluated using the solar constant and the average distance between the Sun and Earth, $A_\odot$, the *astronomical unit of length*,

$$A_\odot = 1.5 \times 10^{11}\,\mathrm{m}, \tag{3.2}$$

which gives (see Fig. 3.1)

$$L_\odot = 4\pi A_\odot^2 S = 3.84 \times 10^{26}\,\mathrm{W}. \tag{3.3}$$

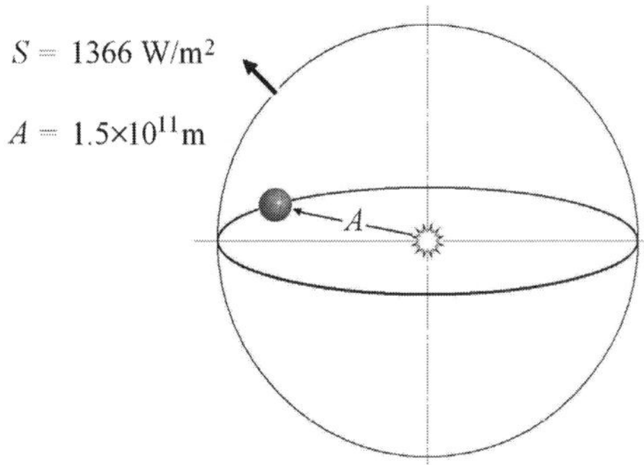

Figure 3.1 Luminosity of the Sun. The average radiation power density of sunlight outside the atmosphere of Earth, the *solar constant*, is $S = 1366\ \mathrm{W/m^2}$. The average distance between the Sun and Earth, the *astronomical constant*, is $A_\odot = 1.5 \times 10^{11}$ m. The luminosity of the Sun, or the total power of solar radiation, is then $L_\odot = 3.84 \times 10^{26}$ W.

The total power of solar radiation is about twenty trillion times the energy consumption of the entire world, 1.6×10^{13} W. For the purpose of utilizing solar poinof solar radiation can one expect?

3.1 Basic Parameters of the Sun

For centuries, the Sun has been the subject of intense study by astronomers. The basic parameters and the basic structure of the Sun are well understood [7, 63, 68, 79]. Here is a brief summary.

3.1.1 Distance

The distance between the Sun and Earth has been measured using triangulation and *radar echoes.* Because the orbit of Earth around the Sun is an ellipse, the distance is not constant. Around January 3, at *perihelion*, the distance is at a minimum, 1.471×10^{11} m. Around July 3, at *aphelion*, the distance is at a maximum, 1.521×10^{11} m. The average distance, $A_\odot = 1.5 \times 10^{11}$ m (see Eq. 3.2) is treated as a basic parameter in astronomy.

Besides the distance in meters, the average time for light to travel between the Sun and Earth, the *light time for the astronomical unit of distance* $\tau_\odot$, has been determined with high accuracy. For all practical applications in solar energy, the approximate value, accurate to 0.2%, is worth memorizing:

$$\tau_\odot = 500\,\text{s}. \tag{3.4}$$

3.1.2 Mass

The mass of the Sun is measured using Kepler's law and the orbital parameters of the planets. Again, for all practical applications, an approximate value, accurate to 1%, is worth memorizing:

$$m_\odot = 2 \times 10^{30}\,\text{kg}. \tag{3.5}$$

It is 333,000 times the mass of Earth. Due to radiation and solar wind, the Sun loses 10^{17} kg of its mass per year, which is negligible. Practically, the mass of the Sun is a constant over its lifetime.

3.1.3 Radius

The radius of the Sun is measured by the angular diameter of the visible disc. Because the distance between the Sun and Earth is not constant, the angular diameter of the Sun is not constant. It ranges from $31.6'$ to $32.7'$. The average value is $32'$, or $0.533°$. Using the average distance $A_\odot$, the radius of the Sun is then

$$r_\odot = \frac{1.5 \times 10^{11} \times 0.533}{2 \times 57.3}\,\text{m} \cong 6.96 \times 10^8\,\text{m}. \tag{3.6}$$

Calculated from the radius, the volume of the Sun is $1.412 \times 10^{27}\text{m}^3$, and its average density is 1.408 g/cm^3.

3.1.4 Emission Power

The emission power of the surface of the Sun, in watts per square meter, can be calculated from Eqs. 3.3 and 3.6:

$$U_\odot = \frac{L_\odot}{4\pi r_\odot^2} = \frac{3.84 \times 10^{26}}{4\pi\,(6.96 \times 10^8)^2} \cong 63.1\,\text{MW/m}^2. \tag{3.7}$$

3.1.5 Surface Temperature

Considering the Sun as a blackbody, the temperature can be calculated from the Stefan–Boltzmann law. Using Eqs. 3.7, 2.99, and 2.101,

$$T_\odot = \left[\frac{U_\odot}{\sigma}\right]^{1/4} = \left[\frac{63.1 \times 10^6}{5.67 \times 10^{-8}}\right]^{1/4} \cong 5800\,\text{K}. \tag{3.8}$$

In the literature, the surface temperature of the Sun varies from one source to another, from 5600 to 6000 K. The difference is insignificant. Throughout this book, we will use 5800 K as the nominal surface temperature of the Sun. Not only is this easy to remember, but it is exactly equal to 0.5 eV, which is convenient for the treatment of solar cells.

Table 3.1: Chemical Composition of the Sun

Element	Z	Molecular Weight	Abundance (% of Number of Atoms)	Abundance (% of Mass)
Hydrogen	1	1.008	91.2	71.0
Helium	2	4.003	8.7	27.1
Oxygen	8	16.000	0.078	0.97
Carbon	6	12.011	0.043	0.40
Nitrogen	7	14.007	0.0088	0.096
Silicon	14	28.086	0.0045	0.099
Magnesium	12	24.312	0.0038	0.076
Neon	10	20.183	0.0035	0.058
Iron	26	55.847	0.0030	0.14

—emphSource: Ref. [79].

3.1.6 Composition

The spectrum of solar radiation does not precisely match the spectrum of blackbody radiation. The fine structure of the solar spectrum provides evidence of its chemical composition (Table 3.1). Actually, the second most abundant element in the Sun, helium, which was discovered from the spectrum of the solar radiation, comes from the Greek word *helios*.

3.2 Kelvin–Helmholtz Time Scale

For centuries, the origin of sunlight has been a fundamental inquiry of all cultures. In the middle of the nineteenth century, the field of thermodynamics matured. The first law of thermodynamics states that energy can only change its form and can neither be created nor be destroyed. The immense solar radiation is apparently draining enormous amounts of energy from the Sun. In the 1850s, Sir William Thomson (Fig. 3.2), one of the founders of thermodynamics, made a thorough study on the origin of the Sun's energy based on the knowledge of physics at that time [81]. He argued that chemical energy could not be the answer, because even if the Sun is made of coal and burns completely, it can only supply a few thousand years of sunlight. The only explanation based on the knowledge of that time was the gravitational contraction of the Sun, also independently proposed by German physicist Hermann von Helmholtz; see Fig. 3.3(a). When two meteorites with masses m_1 and m_2 approach from infinity to a distance r,

Figure 3.2 Sir William Thomson. Irish-born Scottish physicist (1824-1907), aka Lord Kelvin, one of the founders of thermodynamics. He attributed the origin of the Sun's energy to gravitational interaction, and asserted that the Sun could not shine for more than 30 million years. He used his theory to refuse Sir Charles Lyell's geology, and especially Charles Darwin's evolutionary biology; both require that the lifetime of the Sun must be at least one billion years. The discrepancy was resolved by Albert Einstein and Hans Bethe in the 20th century. Artwork by Hubert Herkomer, courtesy of Smithsonian Museum.

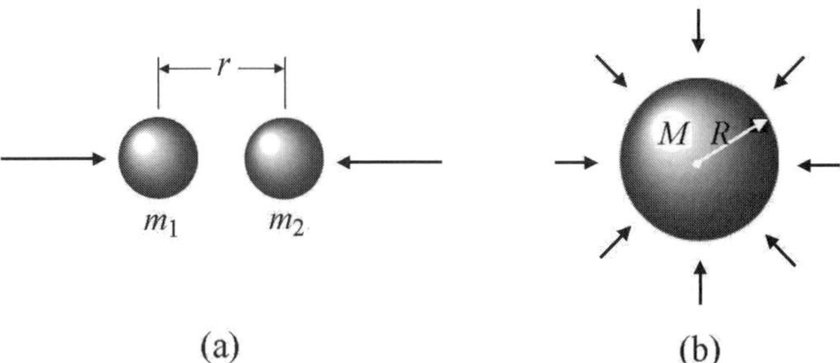

Figure 3.3 The Kelvin-Helmholtz model. A model of the origin of solar energy based on gravitational contraction. (a) When two meteorites approach each other from faraway, the gravitational potential energy is reduced and the kinetic energy is increased. (b) If the Sun is formed by coalition of a large number of meteorites, the gravitational energy can be converted into kinetic energy, and then radiated as heat.

the potential energy is decreased by

$$U = -\frac{Gm_1m_2}{r}, \tag{3.9}$$

where $G = 6.67 \times 10^{-11}\mathrm{N\cdot m^2/kg^2}$ is the gravitational constant. Because of gravitational attraction, the system gains a kinetic energy

$$E_\mathrm{K} = \frac{G\,m_1\,m_2}{r}. \tag{3.10}$$

Assuming that the Sun is formed by the coalescence of large number of meteorites into a sphere of radius $r_\odot$, as shown in Fig. 3.3(b), kinetic energy can be gained by gravitational attraction. The value of the energy gain depends on the distribution of the mass. By analogy to Eq. 3.10, we can make a good order-of-magnitude estimate of the energy gain,

$$E_\mathrm{K} \cong \frac{G\,M_\odot^2}{r_\odot}. \tag{3.11}$$

Obviously it is an overestimate. For example, by calculating the energy gain with a uniform mass distribution, a numerical factor of 0.6 appears. However, Eq. 3.11 gives an upper limit. Based on the Kelvin–Helmholtz assumption, the lifetime of the Sun can be estimated by dividing that energy gain by the rate of energy loss, or the solar luminosity $L_\odot$:

$$\tau_\mathrm{KH} \cong \frac{G\,M_\odot^2}{r_\odot\,L_\odot} = \frac{6.67\times10^{-11}\times\left(2\times10^{30}\right)^2}{6.96\times10^{8}\times3.84\times10^{26}} \approx 10^{15}\,\mathrm{s}. \tag{3.12}$$

The quantity τ_KH, about 30 million years, is called the *Kelvin–Helmholtz time scale.* Based on this calculation, Lord Kelvin asserted that the Sun's life span must not exceed 30 million years [81].

At that time, geology was already well established based on two century's accumulation of discoveries and studies of fossils as well as geological strata, summarized by Sir Charles Lyell in the 1830s in a three-volume monograph *Elements of Geology* [55]. Geological evidence showed that Earth existed for more than one billion years under basically uniform sunlight conditions. Charles Darwin explained the sequence of discoveries in paleontology by evolution through natural selection in his 1859 monumental monograph *The Origin of Species* [20]. According to Lyell and Darwin [55, 20], countless evidence in paleontology has shown that the Sun has shined consistently for at least 1 billion years. The assertion of Lord Kelvin that the Sun's life expectancy is about 20 million years is in direct conflict with the findings in geology and biology.

The discrepancy is due to the limitation in the theory of physics in the Victorian period. On April 27, 1900, Lord Kelvin gave a lecture at the Royal Institution of Great Britain entitled *Nineteenth-Century Clouds over the Dynamical Theory of Heat and Light* [82]. Kelvin mentioned that the "beauty and clearness of theory" was overshadowed by "two clouds": the null result of the Michelson–Morley experiment and the difficulties in explaining the Stefan–Boltzmann law of blackbody radiation based on classical statistical mechanics. Kelvin believed that these two problems were minor and could be resolved within the framework of classical physics. Nevertheless, two years before his death, in 1905, these "two clouds" evolved into a perfect storm in theoretical physics with the emergence of relativity and quantum theory, which completely overturned the Kelvin–Helmholtz theory on the origin of solar energy.

The mystery of the origin of solar energy was resolved after Albert Einstein established the relation between energy and mass [26], and Hans Bethe's 1938 detailed account of nuclear fusion as the origin of stellar energy [9, 8]. The time scale of Bethe matches perfectly with the estimate of the Sun's age based on the discoveries of Lyell and Darwin.

3.3 Energy Source of the Sun

The answer to the source of stellar energy resides in the last of five 1905 papers by Einstein, entitled *Does the Inertia of a Body Depends Upon its Energy Content?* In contemporary notations, Einstein's statement is as follows [26]:

> If a body gives off energy ΔE in the form of radiation, its mass diminishes by $\Delta E/c^2$. Because whether the energy withdrawn from the body becomes radiation or else makes no difference, we might make a more general conclusion that the mass of a body is a measure of its energy content; if the energy changes by ΔE, the mass changes accordingly by $\Delta E/c^2$.

Accordingly, if the initial mass of the body is m_0 and its final mass is m_1, with $\Delta m = m_0 - m_1$, radiation can be emitted with energy

$$\Delta E = \Delta m c^2. \tag{3.13}$$

Using the vast experimental data on nuclear reaction already available at that time, in 1938, Hans Bethe (Fig. 3.4) made a thorough study of all possible nuclear reactions which could generate stellar energy [8, 9]. He concluded that, for stars with mass similar to or less than the Sun, the proton–proton chain dominates. For stars with mass much larger than the Sun, the process catalyzed by carbon, nitrogen, and oxygen (the carbon chain) dominates.

3.3.1 The $p - p$ Chain

The mass of proton is 1.672623×10^{-27} kg, and the mass of a helium nucleus (the alpha particle) is 6.644656×10^{-27} kg. Every time four proton fuse into one alpha particle, there is an excess of mass,

$$\Delta m = (4 \times 1.672623 - 6.644656) \times 10^{-27}\text{kg} = 4.5836 \times 10^{-29}\text{kg}. \tag{3.14}$$

Using Einstein's equation, the excess energy is

$$\Delta E = 4.5836 \times 10^{-29} \times \left(2.99792 \times 10^{8}\right)^{2} = 4.11952 \times 10^{-12}\text{J}, \tag{3.15}$$

or 25.7148 MeV. In other words, every hydrogen atom participating in the reaction generates 6.4287 MeV of energy.

However, as shown below, part of the energy is radiated in the form of neutrinos, for which the Sun and Earth are transparent. Therefore, the radiation energy generated

Figure 3.4 Hans Albrecht Bethe. German-born American physicist (1906–2005) and Nobel laureate for his 1938 theory on nuclear fusion as the origin of stellar energy. A versatile theoretical physicist, Bethe also made important contributions to quantum electrodynamics, nuclear physics, solid-state physics, and astrophysics. Bethe left Germany in 1933 when the Nazis came to power and he lost his job at the University of Tübingen. He moved first to England, then to the United States in 1935, and joined the faculty at Cornell University, a position which he occupied for the rest of his career. During World War II, he was appointed by John Oppenheimer as the Director of Theoretical Division of the Manhattan Project at Los Alamos laboratory. During 1948–1949 he was a visiting professor at Columbia University. Photograph courtesy of Mickael Okoniewski. Taken at Cornell University on December 19, 1996.

in the nuclear reaction is reduced:

$$
\begin{aligned}
{}^{1}\mathrm{H} + {}^{1}\mathrm{H} &\longrightarrow {}^{2}\mathrm{D} + \mathrm{e}^{+} + \nu + 0.164\,\mathrm{MeV}, && (3.16)\\
{}^{2}\mathrm{D} + {}^{1}\mathrm{H} &\longrightarrow {}^{3}\mathrm{He} + \gamma + 5.49\,\mathrm{MeV}, && (3.17)\\
{}^{3}\mathrm{He} + {}^{3}\mathrm{He} &\longrightarrow {}^{4}\mathrm{He} + 2\,{}^{1}\mathrm{H} + 12.85\,\mathrm{MeV}. && (3.18)
\end{aligned}
$$

The overall reaction is

$$4\,{}^{1}\mathrm{H} \longrightarrow {}^{4}\mathrm{He} + 2\mathrm{e}^{+} + 2\nu + 2\gamma + 24.16\,\mathrm{MeV}. \tag{3.19}$$

3.3.2 Carbon Chain

For stars with mass greater than the Sun, the carbon chain, or the CNO chain, dominates the energy generation. Note that the carbon and nitrogen nuclei only act as catalysts. The net result is again combining four protons to form a helium nucleus, or an alpha particle.

$$
\begin{aligned}
{}^{12}\mathrm{C} + {}^{1}\mathrm{H} &\longrightarrow {}^{13}\mathrm{N} + \gamma + 1.95\,\mathrm{MeV}, && (3.20)\\
{}^{13}\mathrm{N} &\longrightarrow {}^{13}\mathrm{C} + \mathrm{e}^{+} + \nu + 1.50\,\mathrm{MeV}, && (3.21)\\
{}^{13}\mathrm{C} + {}^{1}\mathrm{H} &\longrightarrow {}^{14}\mathrm{N} + \gamma + 7.54\,\mathrm{MeV}, && (3.22)\\
{}^{14}\mathrm{N} + {}^{1}\mathrm{H} &\longrightarrow {}^{15}\mathrm{O} + \gamma + 7.35\,\mathrm{MeV}, && (3.23)\\
{}^{15}\mathrm{O} &\longrightarrow {}^{15}\mathrm{N} + \mathrm{e}^{+} + \nu + 1.73\,\mathrm{MeV}, && (3.24)\\
{}^{15}\mathrm{N} + {}^{1}\mathrm{H} &\longrightarrow {}^{12}\mathrm{C} + {}^{4}\mathrm{He} + 4.96\,\mathrm{MeV}. && (3.25)
\end{aligned}
$$

The overall reaction is

$$4\,{}^{1}\mathrm{H} \longrightarrow {}^{4}\mathrm{He} + 2\mathrm{e}^{+} + 2\nu + 3\gamma + 25.03\,\mathrm{MeV}. \tag{3.26}$$

3.3.3 Internal Structure of the Sun

Observing from Earth, the Sun is a sphere of roughly equal temperature, 5800 K. The internal structure of the Sun is not observable. However, it can be inferred using the laws of physics. A simple model of the structure of the Sun is based on an approximation that the Sun is made of spherical layers of uniform materials. The only spatial parameter is the radial distance from the center. A balance between the gravitational force and radiation pressure determines its density and temperature distribution. The source of the energy is a nuclear reaction as presented in Section 3.3. The parameters of the model are determined by comparing the output of the model with observations.

The current standard model is shown in Fig 3.5 [7, 68]. The core of the Sun, with a density on the order of 100 g/cm^3 and a temperature of $10 \times 10^6 - 15 \times 10^6$ K, is the engine of energy creation. Here the nuclei and electrons combine to form a dense and superhot plasma. The radiation energy generated in the core, mostly gamma rays, travels to the surface through the radiation zone. The density of the radiation zone

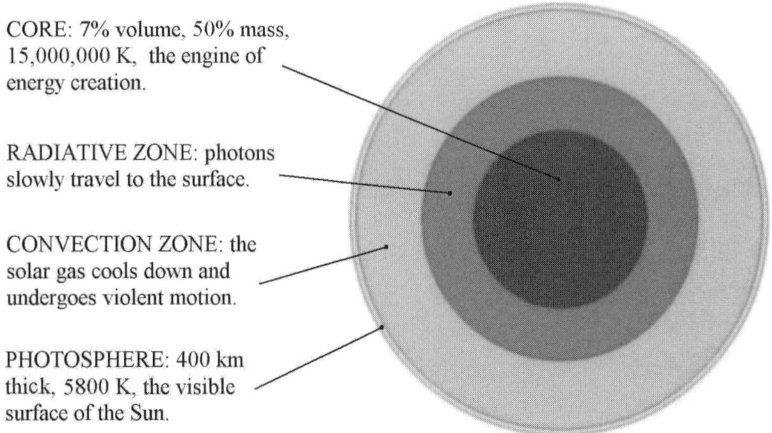

Figure 3.5 Internal structure of the Sun. The core of the Sun, with a density on the order of 100 g/cm^3 and temperature of $10 \times 10^6 - 15 \times 10^6$ K, is the engine of energy creation. The radiation energy generated in the core travels to the surface through the radiative zone. Through the convection zone, the radiation energy is emitted by the photosphere. The boundaries between adjacent spheres are gradual rather than well defined.

decreases from about 5 to about 1 g/cm^3. It takes a photon about 170 thousand years to travel through that region. The convection zone, with a density less than that of water, is a zone of violent motion. The photosphere, a plasma of hydrogen and helium with a density much lower than that of Earth's atmosphere but totally opaque, is the blackbody radiator that emits sunlight.

Problems

3.1. Based on the current hydrogen reserve in the Sun and the energy output, if the efficiency of generating radiation from the $p-p$ chain is 90%, how many years can the Sun keep burning?

3.2. Assuming that the nuclear radius R follows the simple formula $R = r_0 A^{1/3}$, where A is the atomic mass number (number of protons Z plus number of neutrons N) and $r_0 = 1.25 \times 10^{15}$ m, how much kinetic energy is required for a proton (hydrogen nucleus) to touch a carbon nucleus?

3.3. If the kinetic energy of the hydrogen nucleus is $E_k = 1/2k_BT$, to have 0.1% of hydrogen nuclei reach the energy threshold for a CNO reaction, what is the minimum temperature to initiate the nuclear fusion?

Chapter 4

Tracking Sunlight

Because of the rotation and the orbital motion of Earth around the Sun, the apparent position of the Sun in the sky changes over time. To utilize the solar energy efficiently, we must understand the apparent motion of the Sun. The accurate theory of the solar system in astronomy and the data in *The Astronomical Almanac* can be overwhelmingly complicated. In this chapter, we present a simple model which results in formulas easily programmable on a microcomputer and accurate enough for solar energy utilization. Analytic formulas are derived and presented.

4.1 Rotation of Earth: Latitude and Longitude

Figure 4.1 shows the apparent motion of the stars in the night sky. This motion is due to the rotation of Earth on its axis. For solar energy applications, we can consider

Figure 4.1 The night sky. By orienting a camera toward the sky in the night and exposing for some time, the stars seem to rotate around the celestial North Pole. Photo taken by Robert Knapp, Portland, Oregon. See www.modernartphotograph.com. Courtesy of Robert Knapp.

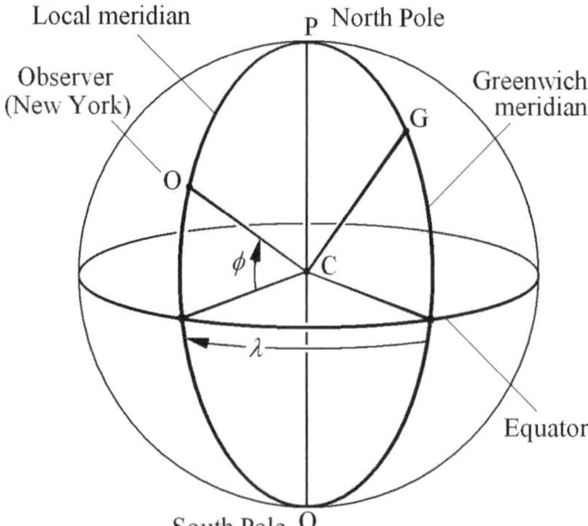

Figure 4.2 Latitude and longitude. The zero point of latitude, the prime meridian, is defined as the meridian passing through the Royal Greenwich Observatory. East of the prime meridian is the Eastern Hemisphere, and in the west is the Western Hemisphere. Similarly, north of the equator is the Northern Hemisphere, and south of the equator is the Southern Hemisphere. The position of the observer is identified by its *latitude* ϕ and *longitude* λ, as marked on the map, and can be determined using GPS. The conversion of sign is eastward is positive and westward is negative. For example, the latitude of New York City is $\phi = 40°47'$ N, or +0.712 rad, and its longitude is $\lambda = 73°58'$ W, or -1.29 rad.

Earth as a perfect sphere rotating with a constant angular velocity on a fixed axis. The axis of rotation of Earth crosses the surface of Earth at two points: the *North Pole* and the *South Pole*. The great circle perpendicular to the axis is the *equator*. A location on Earth can be specified by two coordinates, the *latitude* ϕ and the *longitude* λ, as marked on the map and can be determined using GPS (the Global Positioning System). The longitude specifies a *meridian* (a half great circle passing through the two poles and the location). While the latitude is uniquely defined by the poles and the equator, the longitude requires an origin as the zero point, the *prime meridian*. The prime meridian was chosen by the International Meridian Conference held in October 1884 in Washington, D.C., as the meridian passing through a marked point in the Royal Greenwich Observatory near London. Therefore, the prime meridian is often called the *Greenwich meridian*. Figure 4.2 shows the definition of longitude and latitude.

4.2 Celestial Sphere

From the point of view of an observer on Earth, the Sun, as well as any star, is located on a sphere of a large but undefined radius. The imaginary sphere is the *celestial sphere*. There are two commonly used coordinate systems to describe the position of an astronomical object on the celestial sphere, the *horizon system* and the *equatorial*

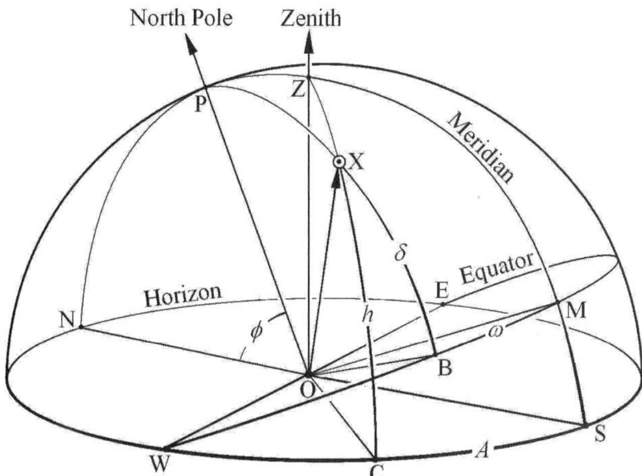

Figure 4.3 Celestial sphere and coordinate transformation. The horizon system defines the position of a celestial body X as directly perceived by the observer. The angular distance of a celestial body X to the horizon is its *height* h, also called *altitude* or *elevation*. The other coordinate is *azimuth* A. The zero point of the azimuth is defined as the south point of the horizon. In the equatorial coordinate system, the North Pole is the reference point. The celestial equator is the basic plane. The distance of a celestial body to the equator is its *declination* δ. The other coordinate is *hour angle* ω. The zero point of the hour angle is defined as the meridian; the half great circle passes through the north celestial pole and the zenith. To convert the coordinates from the horizon system to the equatorial system and *vice versa*, identities in spherical trigonometry are applied on a spherical triangle formed by vertices Z (zenith), P (north celestial pole), and celestial body X.

system.

The extension of the center of Earth and an observer O into the sky is pointing to the *zenith* Z (see Fig.4.2), and the plane perpendicular to that line or the corresponding great circle is the *horizon*. The horizon divides the sphere into two hemispheres. The upper hemisphere is visible to the observer, and the lower hemisphere is hidden under the horizon. The angular distance of a celestial body above the horizon is its *height* h. In the astronomy literature, the term *altitude* or *elevation* is also used. Obviously, the height of the North Pole P equals the geographical latitude of the observer, ϕ.

To completely identify the position of a celestial body, X, we need another reference point. The great circle connecting the zenith with the North Pole is called the *meridian*. It intersects the horizon at point S, the *south point of the horizon*. To identify the position of a celestial body with respect to the south point, we draw a great circle through the zenith and the star, which intersects the horizon at point C. The angle $\widehat{SC}$ is defined as the *azimuth* A, or the horizontal direction of the celestial body. Regarding the utilization of solar energy, we take the definition of azimuth to be westward. Therefore, the azimuth of the Sun always increases over time.

The horizon system defines the position of a celestial body as directly perceived by the observer. However, because Earth is round and rotating on its axis, those

Table 4.1: Notation in Positional Astronomy

Quantity	Notation	Definition
Latitude	ϕ	Geographical coordinate
Longitude	λ	Geographical coordinate
Height	h	Also called altitude or elevation
Azimuth	A	Horizontal direction or bearing
Declination	δ	Angular distance to the equator
Hour angle	ω	In radians, westward
Sunset hour angle	ω_s	In radians, always positive
East-west hour angle	ω_{ew}	In radians, always positive
Right ascension	α	Absolute celestial coordinate
Mean ecliptic longitude	l	On ecliptic plane
True ecliptic longitude	θ	On ecliptic plane
Eccentricity of orbit	e	Currently ≈ 0.0167
Obliquity of ecliptic	ε	Currently $\approx 23.44°$

coordinates depend on the location of the observer and vary over time. In the *equatorial coordinate system*, on the other hand, the position of the Sun is relatively independent of the location of the observer. The coordinates of the Sun in the horizon system can be obtained using a coordinate transformation from its coordinates in the horizontal system.

In the equatorial system, the coordinate equivalent to the latitude of Earth is the *declination*, δ, and the coordinate equivalent to the longitude of Earth for a fixed observer is the *hour angle*, ω; see Fig. 4.3.

As shown, the declination is the angular distance of the celestial body to the celestial equator. It is positive for the stars to the north of the celestial equator and negative for the stars to the south of the celestial equator.

As we mentioned previously, the great circle connecting the zenith and the celestial pole is the *meridian*, which is the reference point equivalent to the Greenwich meridian in geography, which intersects the equator at point M. The great circle connecting the celestial pole and the star is called the *hour circle*, which intersects the equator at point B. The angle $\widehat{MB}$ is the *hour angle* ω, equivalent to longitude on Earth. The convention is, if the celestial body is to the west of the meridian, the hour angle is positive. This is natural and convenient for the position of the Sun, as its hour angle defines the *solar time*.

Another coordinate frequently used in astronomy in place of the hour angle is the *right ascension* α, which takes the *vernal equinox* as the reference point. We will introduce it in a later section.

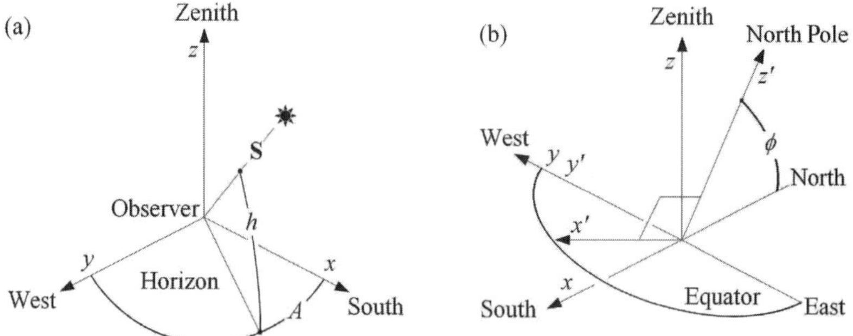

Figure 4.4 Coordinate transformation in Cartesian coordinates. (a) Cartesian coordinates for the horizon system use the conventions that south is x, west is y, and the zenith is z. The position of a celestial body is determined by two angles, height h and azimuth angle A. (b) Cartesian coordinates for the equatorial system use the convention that z'-axis points to the North Pole, while the east-west axis y'-axis is identical to that in the horizon system. The x'-axis is perpendicular to both. The position of a celestial body is determined by declination δ and hour angle ω, see Fig. 4.3.

4.2.1 Coordinate Transformation: Cartesian Coordinates

The standard method of coordinate transformation in positional astronomy is using *spherical trigonometry*, a brief summary of which is presented in-pdimensional Cartesian coordinates and spherical polar coordinates. Treating a two-dimensional problem with three-dimensional methods is overkill. However, because most physicists and engineers are familiar with such an approach, it may be easier to understand.

The Cartesian coordinates for the horizon system is shown in Fig. 4.4(a). Because all celestial bodies move from east to west, the convention is, south is x, west is y, and the zenith is z. A unit vector pointing to a celestial body is determined by two angles. The polar angle is the height h. The azimuth angle A takes south as the zero point. As shown in Fig. 4.4(a), using the three unit vectors **i**, **j**, and **k** pointing to x, y, and z, respectively, a unit vector **S** pointing to the Sun is given as

$$\mathbf{S} = \mathbf{i}\cos h\cos A + \mathbf{j}\cos h\sin A + \mathbf{k}\sin h. \tag{4.1}$$

The Cartesian coordinates for the equatorial system and its relation to the horizon system are shown in Fig. 4.4(b). The y'-axis is identical to that in the horizon system. The z'-axis points to the North Pole, and the x'-axis is perpendicular to both. The unit vectors in the equatorial system pointing to the x', y', and z'-axes are $\mathbf{i}'$, $\mathbf{j}'$, and $\mathbf{k}'$, respectively. Using declination δ and hour angle ω defined in Fig. 4.3, a unit vector **S** pointing to the Sun in terms of $\mathbf{i}'$, $\mathbf{j}'$, and $\mathbf{k}'$ is given as

$$\mathbf{S} = \mathbf{i}'\cos\delta\cos\omega + \mathbf{j}'\cos\delta\sin\omega + \mathbf{k}'\sin\delta. \tag{4.2}$$

From Fig. 4.4(b), the transformations between the two sets of unit vectors, **i**, **j**, **k** and $\mathbf{i}'$, $\mathbf{j}'$, $\mathbf{k}'$ are

$$\begin{aligned}\mathbf{i}' &= \mathbf{i}\,\sin\phi + \mathbf{k}\,\cos\phi,\\ \mathbf{j}' &= \mathbf{j},\\ \mathbf{k}' &= -\mathbf{i}\,\cos\phi + \mathbf{k}\,\sin\phi,\end{aligned} \tag{4.3}$$

and

$$\begin{aligned}\mathbf{i} &= \mathbf{i}'\,\sin\phi - \mathbf{k}'\,\cos\phi,\\ \mathbf{j} &= \mathbf{j}',\\ \mathbf{k} &= \mathbf{i}'\,\cos\phi + \mathbf{k}'\,\sin\phi,\end{aligned} \tag{4.4}$$

where ϕ is the (geographical) latitude of the observer; see Figs 4.2 and 4.4. Using Eqs. 4.1 and 4.2, we obtain

$$\begin{aligned}\mathbf{S} = \mathbf{i}\,&(\cos\delta\,\cos\omega\,\sin\phi - \sin\delta\,\cos\phi)\\ &+\mathbf{j}\,\cos\delta\,\sin\omega\\ &+\mathbf{k}\,(\cos\delta\,\cos\omega\,\cos\phi + \sin\delta\,\sin\phi),\end{aligned} \tag{4.5}$$

and

$$\begin{aligned}\mathbf{S} = \mathbf{i}'\,&(\cos h\,\cos A\,\sin\phi + \sin h\cos\phi)\\ &+\mathbf{j}'\,\cos h\,\sin A\\ &+\mathbf{k}'\,(-\cos h\,\cos A\,\cos\phi + \sin h\,\sin\phi).\end{aligned} \tag{4.6}$$

Comparing Eqs. 4.5 and 4.6 with Eqs. 4.1 and 4.2, we obtain the transformation formulas for the two sets of angles:

$$\cos h\cos A = \cos\delta\,\cos\omega\,\sin\phi - \sin\delta\,\cos\phi, \tag{4.7}$$

$$\cos h\sin A = \sin\omega\,\cos\delta, \tag{4.8}$$

$$\sin h = \cos\delta\,\cos\omega\,\cos\phi + \sin\delta\,\sin\phi; \tag{4.9}$$

and

$$\cos\delta\,\cos\omega = \cos h\,\cos A\,\sin\phi + \sin h\,\cos\phi, \tag{4.10}$$

$$\cos\delta\,\sin\omega = \cos h\,\sin A, \tag{4.11}$$

$$\sin\delta = -\cos h\,\cos A\,\cos\phi + \sin h\,\sin\phi. \tag{4.12}$$

4.2.2 Coordinate Transformation: Spherical Trigonometry

The coordinate transformation formulas can be easily obtained using formulas in spherical trigonometry; see Fig. 4.3. We should focus our attention on the spherical triangle PZX, with three arcs $p = \widehat{ZX}$, $z = \widehat{XP}$, and $x = \widehat{PZ}$. As seen from Fig. 4.3, the relations between the elements of the spherical triangle and the quantities of interest are

$$\begin{aligned} P &= \omega, \\ Z &= 180^\circ - A, \\ p &= 90^\circ - h, \\ z &= 90^\circ - \delta, \\ x &= 90^\circ - \phi. \end{aligned} \tag{4.13}$$

First, consider the case of given declination δ and hour angle ω in the equatorial system to find height h and azimuth A in the horizon coordinate system. The latitude of the observer's location ϕ is obviously a necessary parameter. Using the cosine formula

$$\cos p = \cos x \,\cos z + \sin x \,\sin z \,\cos P, \tag{4.14}$$

with Eqs. 4.13, we obtain

$$\sin h = \sin\delta \,\sin\phi + \cos\delta \,\cos\phi \,\cos\omega. \tag{4.15}$$

Further, using the sine formula

$$\frac{\sin Z}{\sin z} = \frac{\sin P}{\sin p}, \tag{4.16}$$

we find

$$\cos h \sin A = \sin\omega \,\cos\delta. \tag{4.17}$$

Finally, applying Formula C in Appendix B,

$$\sin x \cos Z = \cos z \,\sin p - \sin z \,\cos p \,\cos X, \tag{4.18}$$

and using Eqs. 4.13, we find

$$\cos h \cos A = \sin\phi \,\cos\delta \,\cos\omega - \cos\phi \,\sin\delta. \tag{4.19}$$

Equations 4.15, 4.18 and 4.19 are identical to Eqs. 4.7, 4.8, and 4.9.

Next, we consider the case of given height h and azimuth A in the horizon coordinate system to find declination δ and hour angle ω in the equatorial system. Again, the latitude of the observer's location ϕ is a necessary parameter.

Using the cosine formula

$$\cos z = \cos p \,\cos x + \sin p \,\sin x \,\cos Z, \tag{4.20}$$

we have

$$\sin\delta = -\cos h \,\cos A \,\cos\phi + \sin h \,\sin\phi. \tag{4.21}$$

By rearranging Eq. 4.18, we find

$$\cos\delta\,\sin\omega = \cos h\,\sin A. \tag{4.22}$$

Similar to the derivation of Eq. 4.19, using formula C,

$$\sin z \cos P = \cos p\,\sin x - \sin p\,\cos x\,\cos Z, \tag{4.23}$$

we obtain

$$\cos\delta\,\cos\omega = \cos\,h\,\cos A\,\sin\phi + \sin\,h\,\cos\phi. \tag{4.24}$$

Those equations are identical to Eqs. 4.10–4.12.

4.3 Treatment in Solar Time

Since the prehistory era, human activities have been revolving around the apparent motion of the Sun across the sky. The *solar time* $t_\odot$, which is based on the hour angle of the Sun, is an intuitive measure of time and was used for thousands of years in all cultures of the world. As we will discuss in Section 4.4, because the apparent motion of the Sun is nonuniform and depends on location, it is not an accurate measure of time. The difference between solar time and standard time as used in everyday life, even with proper alignment, can be more than ± 15 min.

To make an estimate of solar radiation, sometimes high accuracy is not required. Thus, the simple and intuitive solar time is widely used in the solar energy literature. For example, to compute the integrated values of total insolation over a day or a year, the time shift is irrelevant. The concepts become simple. For example, at solar noon, when the Sun is passing the meridian, solar time is zero. Sunrise time, which is always negative, equals sunset time in magnitude. In this case, the hour angle of the Sun is a measure of time. In other words, if $t_\odot$ is the solar time on an 24-h scale, the hour angle of the Sun ω_s in radians is

$$\omega_s = \pi\frac{t_\odot - 12}{12}, \tag{4.25}$$

and

$$t_\odot = 12 + 12\frac{\omega_s}{\pi}. \tag{4.26}$$

4.3.1 Obliquity and Declination of the Sun

The orbital plane of Earth around the Sun, the *ecliptic*, is at an angle called the *obliquity* ϵ from the equator. From the point of view of an observer on Earth, the Sun is moving in the *ecliptic plane*; see Fig. 4.5. On a time scale of centuries, the obliquity angle varies over time. Currently, $\epsilon = 23.44°$. This is what causes the seasons.

Related to the motion of the Sun over a calendar year, there are four cardinal points. At the *vernal equinox*, the trajectory of the Sun intersects the celestial equator,

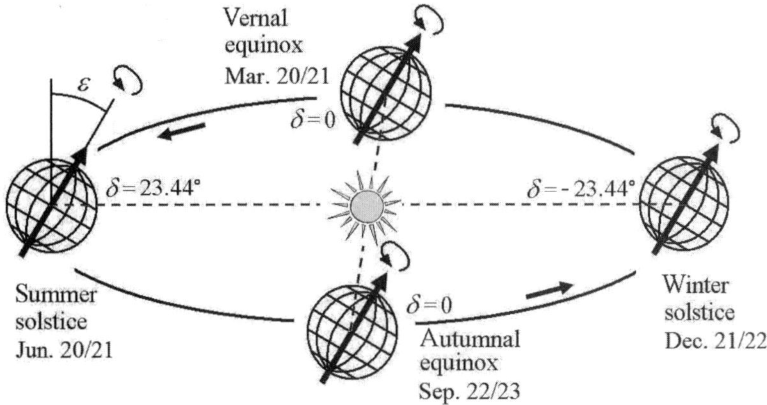

Figure 4.5 Obliquity and the seasons. The rotational axis and the orbital plane of Earth has a tilt angle ϵ, the *obliquity*, which is the origin of the seasons. It causes a periodic difference between solar time and civil time.

heading north. At the *summer solstice*, the trajectory of the Sun reaches its northern most point, which is about 23.44° above the celestial equator. At the *autumnal equinox*, the trajectory of the Sun intersects the celestial equator, heading south. At the *winter solstice*, the trajectory of the Sun reaches its lowest point, which is about 23.44° below the celestial equator. The dates and times of these four cardinal points vary year by year.

In line with the concept of solar time, the motion of the Sun along its orbital can be described by the *mean longitude* l. At the vernal equinox, $l = 0$. At the summer solstice, $l = \pi/2$, or 90°. At the autumnal equinox, $l = \pi$, or 180°. At the winter solstice, $l = 3\pi/2$, or 270°.

An accurate formula for the declination of the Sun is presented in Section 4.4.7. Here we present a simple approximation by assuming that the declination varies sinusoidally with the mean longitude l and consequently is linear according to the number of the day in a year. The error could be as large as 1.60°, but is acceptable in many applications:

$$\delta \approx \varepsilon \sin l = \varepsilon \sin\left(\frac{2\pi\,(N-80)}{365.2422}\right), \tag{4.27}$$

where ε is the obliquity of the ecliptic, currently $\varepsilon = 23.44°$; and N is the number of the day counting from January 1, which can be computed using the formula

$$N = \mathrm{INT}\left(\frac{275 \times M}{9}\right) - K \times \mathrm{INT}\left(\frac{M+9}{12}\right) + D - 30, \tag{4.28}$$

where M is the month number, D is the day of the month, and $K = 1$ for a leap year, $K = 2$ for a common year. A leap year is defined as divisible by 4, but not by 100, except if divisible by 400. INT means taking the integer part of the number. This

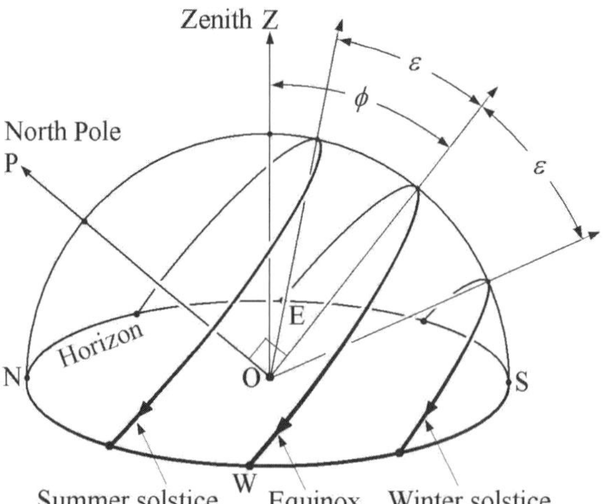

Figure 4.6 Apparent motion of the Sun. Earth is rotating on its axis OP eastward. The apparent motion of the Sun is moving westward. Because of obliquity, on different days of a year, the declination of the Sun varies. At the winter solstice, the declination of the Sun reaches its minimum, $-\varepsilon$. At the summer solstice, the declination of the Sun reaches its maximum, $+\varepsilon$. At the vernal equinox or autumnal equinox, the declination of the Sun is zero, and the Sun is moving on the celestial equator.

formula can be verified directly. The number 80 in Eq. 4.27 is the number of the day of the vernal equinox, March 20 or 21. The actual date varies from year to year. It also differs between leap year and common year. Using Eq. 4.28, it can be shown that this date varies from 79 to 81. The most common number of days of vernal equinox is 80.

The apparent motion of the Sun from an observer on Earth is shown in Fig. 4.6. Earth is rotating on its axis OP eastward because of its spin. Therefore, apparently, the Sun is moving westward. Due to obliquity, on different days of a year, the declination of the Sun varies. At the winter solstice, the declination of the Sun reaches its minimum, $-\varepsilon$. At the summer solstice, the declination of the Sun reaches its maximum, $+\varepsilon$. At the vernal equinox or autumnal equinox, the declination of the Sun is zero, and the Sun is moving on the celestial equator.

4.3.2 Sunrise and Sunset Time

Using the treatments in the previous section, we will give an example using the time of sunrise (or sunset) in solar time. The condition of sunrise is the time when the height of the Sun h is zero. According to Eq. 4.15, the condition is

$$\sin\delta\,\sin\phi + \cos\delta\,\cos\phi\,\cos\omega = 0, \tag{4.29}$$

or

$$\cos\omega_s = -\tan\delta\,\tan\phi. \tag{4.30}$$

For each value of the cosine, there are multiple values of the angle ω. For example, at the equator, $\phi = 0$, the times of sunrise and sunset are always at 6:00 am and 6:00 pm solar time. The duration of the day is always 12 hours. At the North Pole or South Pole, $\tan\phi = \infty$, there is never a sunrise or a sunset. In the temperate zones and the Torrid Zone, the sunrise and sunset times in terms of the 24-h solar time of the Nth day of the year is determined by

$$t_s = 12 \mp \frac{12}{\pi} \arccos\left[-\tan\delta \, \tan\phi\right]. \tag{4.31}$$

In the frigid zones, where

$$\phi > 90^\circ - \varepsilon, \tag{4.32}$$

there is a period of time in a year when the Sun never rises or never sets.

Another time of the day which is useful for the computation of solar radiation is the solar time when the Sun crosses the East–West great circle. The condition is where the azimuth $A = \pm\pi/2$, or $\cos A = 0$. From Eq. 4.7, one finds

$$\cos\omega_{\text{ew}} = \tan\delta \, \cot\phi, \tag{4.33}$$

or, in terms of solar time in hours,

$$t_{\text{ew}} = 12 \pm \frac{12}{\pi} \arccos\left[\tan\delta \, \cot\phi\right]. \tag{4.34}$$

4.3.3 Direct Solar Radiation on an Arbitrary Surface

For a surface of arbitrary orientation, with polar angle β and azimuth angle γ, the unit vector of its norm $\mathbf{N}$ is given se

$$\mathbf{N} = \mathbf{i}\,\sin\beta\,\cos\gamma + \mathbf{j}\,\sin\beta\,\sin\gamma + \mathbf{k}\,\cos\beta. \tag{4.35}$$

The convention of the sign of the angle is the same as the hour angle: The origin of the azimuth is south and westward it is positive. Combining Eq. 4.35 with Eqs. 4.7, 4.8, and 4.9, the cosine between the norm of the surface and the solar radiation is

$$\begin{aligned} \cos\theta = \mathbf{N}\cdot\mathbf{S} &= \sin\beta\,\cos\gamma\,(\cos\delta\,\cos\omega\,\sin\phi - \sin\delta\,\cos\phi) \\ &+ \sin\beta\,\sin\gamma\,\cos\delta\,\sin\omega \\ &+ \cos\beta\,(\cos\delta\,\cos\omega\,\cos\phi + \sin\delta\,\sin\phi), \end{aligned} \tag{4.36}$$

or, rearranging

$$\begin{aligned} \cos\theta = &\sin\delta\,(\sin\phi\,\cos\beta - \cos\phi\,\sin\beta\,\cos\gamma) \\ &+ \cos\delta\,(\cos\phi\,\cos\beta\,\cos\omega + \sin\phi\,\sin\beta\,\cos\gamma\,\cos\omega \\ &+ \sin\beta\,\sin\gamma\,\sin\omega). \end{aligned} \tag{4.37}$$

For a surface facing south with $\gamma = 0$, Eq. 4.37 simplifies to

$$\cos\theta = \sin(\phi - \beta)\,\sin\delta + \cos(\phi - \beta)\cos\delta\,\cos\omega. \tag{4.38}$$

Consider special cases as follows: For a horizontal surface, $\beta = 0$,

$$\cos\theta = \sin\phi\,\sin\delta + \cos\phi\cos\delta\,\cos\omega. \tag{4.39}$$

At the North Pole, where $\phi = \pi/2$,

$$\cos\theta = \sin\delta. \tag{4.40}$$

And at the equator, where $\phi = 0$,

$$\cos\theta = \cos\delta\,\cos\omega. \tag{4.41}$$

For a vertical surface facing south, $\beta = \pi/2$ and $\gamma = 0$,

$$\cos\theta = -\cos\phi\,\sin\delta + \sin\phi\cos\delta\,\cos\omega. \tag{4.42}$$

At the North Pole, where $\phi = \pi/2$,

$$\cos\theta = \cos\delta\,\cos\omega. \tag{4.43}$$

And at the equator, where $\phi = 0$,

$$\cos\theta = \sin\delta. \tag{4.44}$$

Of particular importance is a surface with *latitude tilt*, or $\beta = \phi$. Equation 4.38 is greatly simplified:

$$\cos\theta = \cos\delta\,\cos\omega. \tag{4.45}$$

The surface can get high radiation energy over the entire year, because $\cos\delta$ is always greater than 0.93.

4.3.4 Direct Daily Solar Radiation Energy

An important application in terms of solar time is the computation of direct solar radiation energy H_{D} on a surface on a clear day. The effect of clouds and scattered sunlight will be treated in Chapter 5. In a clear day, on a surface perpendicular to the sunlight, the power is $1\,\mathrm{kW/m^2}$, and the total radiation energy in an hour is $1\,\mathrm{kWh/m^2}$. When the sunlight is tilted with an angle θ, the radiation energy is reduced to $\cos\theta \times 1\,\mathrm{kWh/m^2}$. Therefore, the daily direct solar radiation energy in units of kilowatt-hours per square meter is the integration of $\cos\theta$ over 24 h.

Consider first a vertical surface facing south in the northern temperate zone, namely, $\beta = \pi/2$ and $\gamma = 0$. From Eq. 4.36, one obtains

$$\cos\theta = \cos\delta\,\cos\omega\,\sin\phi - \sin\delta\,\cos\phi. \tag{4.46}$$

During days between the vernal equinox and the autumnal equinox, sunlight can shine on the south surface only when the Sun locates in the southern half of the sky. The direct daily solar radiation energy H_D in kilowatt-hours per square meter is

$$\begin{aligned} H_\mathrm{D} &= \frac{12}{\pi}\cos\delta\,\sin\phi\int_{-\omega_\mathrm{ew}}^{\omega_\mathrm{ew}}\cos\omega\,d\omega - \frac{24}{\pi}\,\omega_\mathrm{ew}\,\sin\delta\,\cos\phi \\ &= \frac{24}{\pi}\left(\cos\delta\,\sin\phi\,\sin\omega_\mathrm{ew} - \omega_\mathrm{ew}\,\sin\delta\,\cos\phi\right). \end{aligned} \tag{4.47}$$

During the days between the autumnal equinox and the vernal equinox of the next year, the available sunlight is limited from sunrise to sunset. The daily solar radiation energy in kilowatt-hours per square meter is

$$\begin{aligned} H_\mathrm{D} &= \frac{12}{\pi}\cos\delta\,\sin\phi\int_{-\omega_s}^{\omega_s}\cos\omega\,d\omega - \frac{24}{\pi}\,\omega_s\,\sin\delta\,\cos\phi \\ &= \frac{24}{\pi}\left(\cos\delta\,\sin\phi\,\sin\omega_s - \omega_s\,\sin\delta\,\cos\phi\right). \end{aligned} \tag{4.48}$$

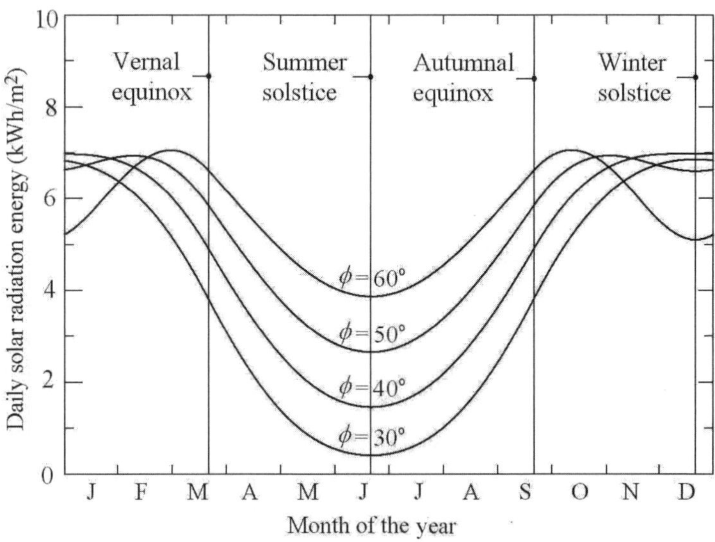

Figure 4.7 Daily solar radiation energy on a vertical surface facing south. In the winter, the surface enjoys almost full sunlight, except for places with very high latitude. In the summer, the solar radiation is much weaker. From the point of view of passive solar buildings, south-facing windows are highly preferred. For solar photovoltaic applications, south-facing panels are much more efficient in the winter.

The direct daily solar radiation energy on a surface facing south for four latitudes in the northern temperate zone of the entire year is shown in Fig. 4.7. As shown, in the winter time, the surface enjoys almost full sunlight, except for places with very high latitude, where the radiation energy is reduced because of the late sunrise and early sunset. In the summer, the solar radiation is much weaker because of tilting. From the point of view of passive solar buildings, South-facing windows are highly preferred. For solar photovoltaics applications, south-facing panels are much more efficient in the winter. However, the overall efficiency is not high.

Next, consider a west-facing surface, with $\beta = \pi/2$ and $\gamma = \pi/2$. From Eq. 4.36, one obtains

$$\cos\theta = \cos\delta\, \sin\omega. \tag{4.49}$$

Sunlight starts at noon with $\omega = 0$ and disappears at sunset. Therefore, the daily solar radiation energy in kilowatt-hours per square meter is

$$\begin{aligned} H_{\mathrm{D}} &= \frac{12}{\pi}\cos\delta \int_0^{\omega_s} \sin\omega\, d\omega \\ &= \frac{12}{\pi}(\cos\delta + \sin\delta\, \tan\phi). \end{aligned} \tag{4.50}$$

where Eq. 4.30 is used. As shown in Fig. 4.8, in the summer, the solar radiation is much stronger than in the winter. From the point of view of passive solar buildings,

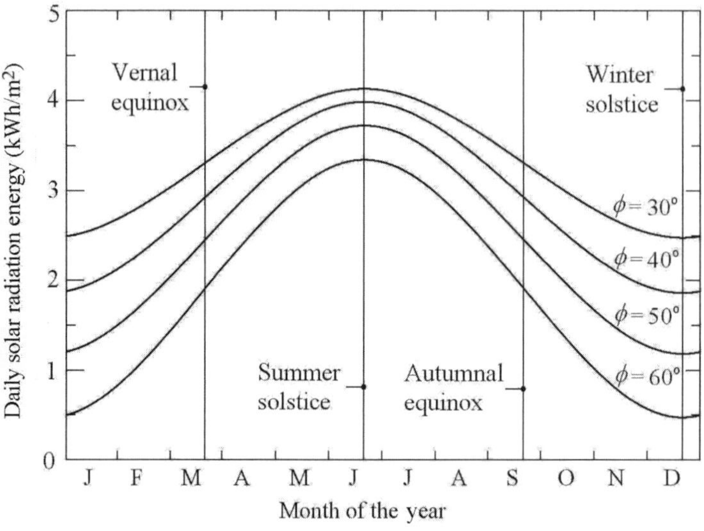

Figure 4.8 Daily solar radiation energy on a vertical surface facing west. In the summer, the solar radiation is much stronger than in the winter. From the point of view of passive solar buildings, west-facing windows must be avoided whenever possible.

west-facing windows must be avoided whenever possible. For solar photovoltaics applications, west-facing panels only work in the summer, but the overall efficiency is only one-half of the optimum orientation; see below.

For rooftop applications in large cities, to avoid structural damages from wind, horizontal placement of solar panels surface is often used. Since $\beta = 0$, using Eq. 4.36,

$$\cos\theta = \cos\delta\,\cos\omega\,\cos\phi + \sin\delta\,\sin\phi. \tag{4.51}$$

By integrating the cosine over time from sunrise to sunset, the daily radiation energy is

$$\begin{aligned} H_{\mathrm{D}} &= \frac{12}{\pi}\cos\delta\,\cos\phi\int_{-\omega_s}^{\omega_s}\cos\omega\,d\omega + \frac{24}{\pi}\,\omega_s\,\sin\delta\,\sin\phi \\ &= \frac{24}{\pi}\left(\cos\delta\,\cos\phi\,\sin\omega_s + \omega_s\,\sin\delta\,\sin\phi\right). \end{aligned} \tag{4.52}$$

The variation of the radiation energy over a year is shown in Fig. 4.9. As shown, in the summer, especially in locations with low latitude, the radiation energy is strong. However, in winter, especially in locations of higher latitude, the daily radiation energy is weak.

A much better choice for solar panel placement is *latitude tilt.* From Eq. 4.45, between the vernal equinox and the autumnal equinox, the daily radiation energy is

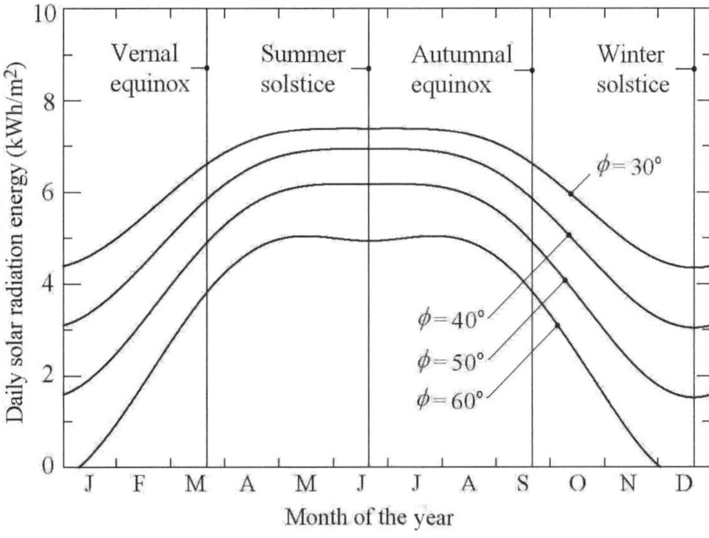

Figure 4.9 Daily solar radiation energy on a horizontal surface. In the summer, especially in locations with low latitude, the radiation energy is strong. However, in the winter, especially in locations of higher latitude, the daily radiation energy is weak.

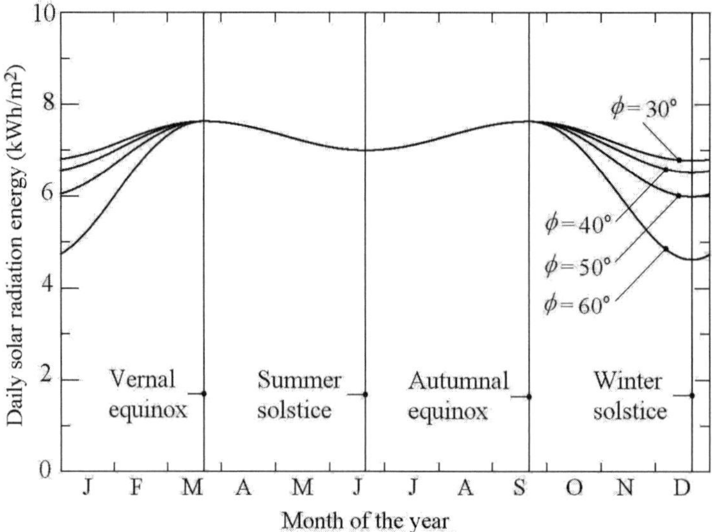

Figure 4.10 Daily solar radiation energy on a latitude-tilt surface. Solar panels placed on a latitude-tilt surface enjoy almost the maximum solar radiation over the entire year.

only limited by the angle of incidence,

$$\begin{aligned} H_{\mathrm{D}} &= \frac{12}{\pi} \cos\delta \int_{-\pi/2}^{\pi/2} \cos\omega \, d\omega \\ &= \frac{24}{\pi} \cos\delta . \end{aligned} \tag{4.53}$$

As shown in Fig. 4.10, in this period of the year, the daily radiation energy is independent of latitude. However, between the autumnal equinox and the vernal equinox of the next year, sunrise is later than 6:00 am and sunset is earlier than 6:00 pm. The daily radiation energy is

$$\begin{aligned} H_{\mathrm{D}} &= \frac{12}{\pi} \cos\delta \int_{-\omega_s}^{\omega_s} \cos\omega \, d\omega \\ &= \frac{24}{\pi} \cos\delta \sin\omega_s . \end{aligned} \tag{4.54}$$

Also shown in Fig 4.10, the radiation energy is reduced and depends on the latitude of the location, but not by much. Over the entire year, maximum solar radiation energy is obtained.

If the surface is allowed to follow the apparent motion of the Sun, the daily radiation energy can be further enhanced. Consider a solar panel mounted on an axis parallel to

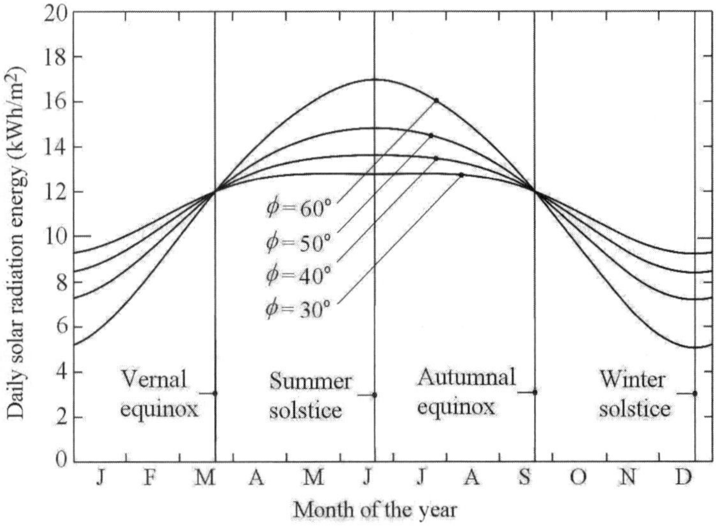

Figure 4.11 Daily solar radiation energy on a surface with tracking. Daily radiation received by a solar panel mounted on an axis parallel to the axis of Earth and the rotates uniformly one turn per day, similar to that on Plate 7. The average daily solar radiation over the entire year is independent of latitude, see Table 4.2.

the axis of Earth and that rotates uniformly one turn per day. The daily solar radiation energy is

$$H_{\rm D} = \frac{24}{\pi} \cos\delta\,\omega_s. \tag{4.55}$$

As shown in Fig. 4.11, the daily solar radiation on a surface with single-axis tracking is substantially higher than for fixed surfaces. The advantage is more apparent in the average direct daily solar radiation $\overline{H}_{\rm D}$ over a year, as shown in Table 4.2. The solar radiation on a surface with tracking is more than 50% higher than all fixed surfaces. Nevertheless, because of the effect of shadows, it does not save horizontal area.

Table 4.2: Average Daily Solar Radiation on Various Surfaces

$\overline{H}_{\rm D}$ in kWh/day at Latitude	30°	40°	50°	60°
Vertical, facing south	3.72	4.57	5.25	5.60
Vertical, facing west or east	3.31	2.93	2.46	1.91
Horizontal	6.25	5.43	4.39	3.11
Latitude tilt facing south	7.27	7.21	7.08	6.77
Optimum single-axis tracking	11.50	11.50	11.50	11.50

4.3.5 The 24 Solar Terms

As shown in previous sections, the motion of the Sun in a calendar year is determined by the Sun's *ecliptic longitude* l. The vernal equinox at $l = 0°$ or $360°$, the summer solstice at $l = 90°$, the autumnal equinox at $l = 180°$, and the winter solstice at $l = 270°$ are the four *cardinal points*. To study the position of the Sun over a year, more points are needed. In traditional Western calendars, including the Julian calendar and the Gregorian calendar, the definitions of the 12 months are not based on natural phenomena. The calendars are not accurately synchronized with the motion of the Sun. Therefore, it is inaccurate and inconvenient to use calendar dates to specify the motion of the Sun.

In East Asia, however, for over two thousand years, a purely solar-based calendar system has been used: the 24 *solar terms*. It is defined solely on the basis of the orbital motion of Earth around the Sun. Similar to dividing a mean solar day into 24 h,

Table 4.3: The 24 Solar Terms

No.	Name	In Pinyin	l (deg)	Approx. date
0	Winter solstice	Dōngzhì	270°	December 22
1	Minor cold	Xiǎohán	285°	January 6
2	Major cold	Dàhán	300°	January 20
3	Spring commences	Lìchūn	315°	February 4
4	Rain water	Yǔshǔi	330°	February 19
5	Insect awakes	Jīngzhé	345°	March 6
6	Vernal equinox	Chūnfēn	0°	March 21
7	Pure brightness	Qīngmíng	15°	April 5
8	Grain rain	Gǔyǔ	30°	April 20
9	Summer commences	Lìxià	45°	May 6
10	Grain forms	Xiǎomǎn	60°	May 21
11	Grain in ear	Mángzhòng	75°	June 6
12	Summer solstice	Xiàzhì	90°	June 21
13	Minor heat	Xiǎoshǔ	105°	July 7
14	Major heat	Dàshǔ	120°	July 23
15	Autumn commences	Lìqiū	135°	August 8
16	Heat recedes	Chǔshǔ	150°	August 23
17	White dew	Báilǔ	165°	September 8
18	Autumnal equinox	Qiūfēn	180°	September 23
19	Cold dew	Hánlǔ	195°	October 8
20	Frost descents	Shuāngjiàng	210°	October 23
21	Winter commences	Lìdōng	225°	November 7
22	Minor snow	Xiǎoxuě	240°	November 22
23	Major snow	Dàxuě	255°	December 7

the solar term system divides a solar year (the time between two consecutive vernal equinoxes) into 24 equal parts; see Table 4.3. Each year, the exact date and *time of day* of each of the 24 solar terms is published in the Almanac. The date and time for the four cardinal points in years 2011 through 2020 is shown in Table 4.4. Because the rough coordination of the date in the Gregorian calendar with vernal equinox, the dates differ for only one or two days from year to year. Since the *ecliptic longitude* l of the Sun at each solar term is well defined, by using the formula of the Sun's declination (Eq. 4.27), the declination of the Sun at the 24 solar terms can be calculated,

$$\delta \approx \varepsilon \sin l = \varepsilon \sin \left(\frac{(S-6)\,\pi}{12} \right), \tag{4.56}$$

where S is the order of the solar term, shown in Table 4.3. The number 6 is taken because in Table 4.3, vernal equinox is number 6.

Traditionally, in the Asian calendar, the first solar term is defined as *spring commences*. It is equivalent to defining 3 am early morning as the starting time of a day. This tradition was established probably because spring commences means the starting of agricultural activities of a year. The logical starting solar term is the *winter solstice*, which is equivalent to defining midnight as the starting time of a day.

4.4 Treatment in Standard Time

For several millennia, all over the world, humans have been using the motion of the Sun for time keeping, the *solar time*. However, because the motion of the Sun is not uniform, solar time shows significant deviation from the time defined by uniform motion, for example, the rotation of Earth, the pendulum, or an atomic clock, represented by the motion of the fictitious mean sun. The difference could be as much as 16 min or more. In order to make a sunlight tracking system based on standard time, the difference must be taken into account to reasonable accuracy. In this section, we will present a treatment which is simple to understand and program and yet accurate enough for solar energy utilization.

4.4.1 Sidereal Time and Solar Time

To very high accuracy, the angular velocity of Earth's rotation is a constant. Therefore, the time interval between two consecutive passages of a given fixed star over an observer's meridian is a constant, which is an accurate measure of time, the *sidereal day*. The word sidereal was derived from the Latin *sideus*, which means "star". However, because Earth also has an orbital motion around the Sun, the duration of the solar day is different from the sidereal day; see Fig. 4.12. For example, at midnight, an observer sees a distant fixed star at its meridian. At the time the same star passes the meridian again, Earth has rotated about the Sun by an angle

$$\Delta\phi = \frac{360^\circ}{365.2422} = 59'08''. \tag{4.57}$$

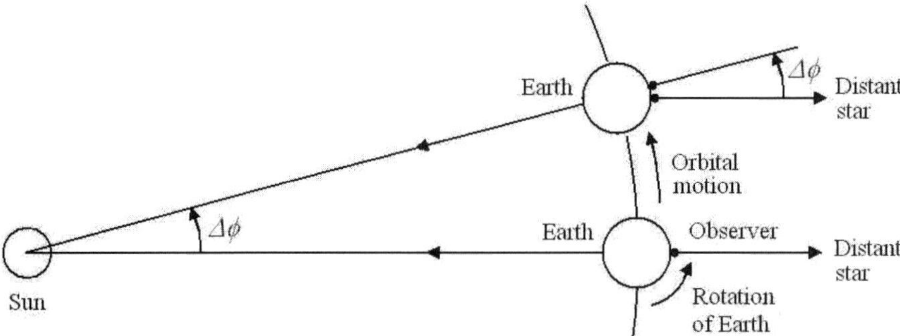

Figure 4.12 Sidereal time and solar time. Because Earth has a rotation on its axis and an orbital motion around the Sun, the duration of the solar day is different from the sidereal day. A solar day is 0.273% longer than the sidereal day.

Therefore, a solar day is 0.273% longer than the sidereal day. Because solar time is taken as time as we know, and remember that a 360° angle of rotation corresponds to 24 h, it equals

$$24^h \text{mean solar time} = \frac{24^h \times 366.2422}{365.2422} = 24^h 3^m 56^s \text{ sidereal time.} \tag{4.58}$$

the length of one sidereal day is

$$24^h \text{ sidereal time} = \frac{24^h \times 365.2422}{366.2422} = 23^h 56^m 04^s \text{ mean solar time.} \tag{4.59}$$

In the above equations, we introduced the term *mean solar time*. Here we are using a simplified model for the orbital motion of Earth around the Sun, namely, a perfect circle on the sidereal equator with a uniform angular speed. From the point of view of an observer on Earth, such a fictitious Sun of uniform motion along the celestial equator is called the *mean sun*.

From Fig. 4.12, the actual value of time depends on the location of the observer. Specifically, it depends on the longitude of the observer. To make a universal time, a standard longitude must be selected. Similar to the case of the origin of longitude, the standard longitude for the universal time is selected as the *prime longitude* located at Greenwich. The time starting at midnight at Greenwich is called Greenwich mean time (GMT), or more often, *universal time* (UT).

The world is divided into *time zones*. Each zone has a definition of time which differs mostly an integer number of hours from the GMT or UT. For example, Eastern standard time (EST) is defined as UT−5h. In the summer, Eastern daylight saving time (EDT) is defined as UT−4h.

4.4.2 Right Ascension of the Sun

As discussed in section 4.2, in the equatorial coordinate system, the position of a star can be characterized by the declination and the hour angle. The declination of a star does not vary over time. However, because of the rotation of Earth, the hour angle of a star varies over time. Uusing the *right ascension* α instead of the hour angle, the coordinates of each fixed star are, to a high degree of accuracy, fixed.

Similar to the case of longitude on Earth, a fixed point on the celestial sphere should be chosen as the reference point. The point universally chosen is the *vernal equinox* Υ, the point where the Sun passes the celestial equator while heading north; see Fig. 4.13. The angle between the intersection of the meridian passing through the star S with the equator B and the vernal equinox Υ is called the right ascension of the star. The sign convention is opposite to that of the hour angle: It is measured eastward. For the treatment of the motion of the Sun, this convention is natural, because it increases with the number of the day in the year;

4.4.3 Time Difference Originated from Obliquity

As mentioned previously, there are two origins of the difference between the motion of the true sun and the mean sun. The first is the obliquity ε. Time is measured by the angle of the mean sun on the equator, but the true sun is moving on the ecliptic plane. Assuming that the Sun is moving uniformly on the ecliptic circle, its *mean longitude* l is a linear function of time. The mean longitude of the Sun increases by 2π during a calendar year; see Fig. 4.13. Taking the number of the day N as the unit of time and the vernal equinox Υ as the origin, which is March 20 or 21, approximately the 80st day of the year, the mean longitude in radians is

$$l = \frac{2\pi\,(N - 80)}{365.2422}. \tag{4.60}$$

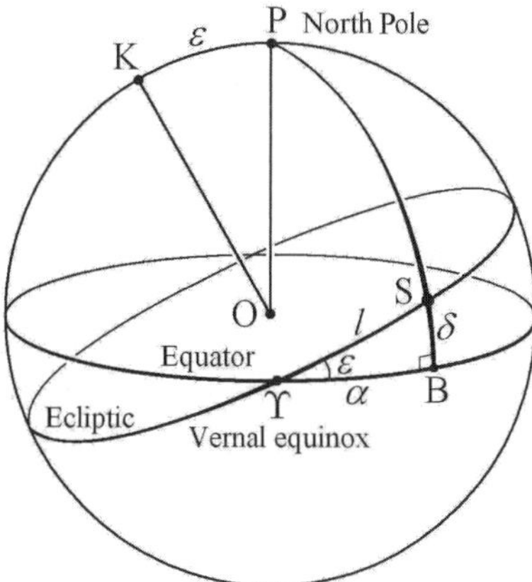

Figure 4.13 Obliquity and equation of time. Because of obliquity, even if Earth is orbiting the Sun with a uniform angular speed, the projection of the speed of the apparent motion of Sun on the equator is not uniform. It gives rise to a term of the difference between mean solar time and true solar time with a periodicity of half a year.

The number of the day N can be computed using Eq. 4.28. The accurate date and time for vernal equinox is shown in Table 4.4.

Because of obliquity, the projection of the longitude l on the celestial equator, the right ascension α, is not linear over time. Using the formula for a rectangular spherical triangle, we find the relation

$$\tan\alpha = \cos\varepsilon \, \tan l. \tag{4.61}$$

Currently, $\varepsilon \approx 23.44°$, and $\cos\varepsilon \approx 0.917$, very close to 1. Using a trigonometry identity

$$\cos\varepsilon = \frac{1 - \tan^2\frac{\varepsilon}{2}}{1 + \tan^2\frac{\varepsilon}{2}}, \tag{4.62}$$

Eq. 4.61 can be written as

$$\frac{\tan l - \tan\alpha}{\tan l + \tan\alpha} = \tan^2\frac{\varepsilon}{2}. \tag{4.63}$$

Using the obvious relation

$$\frac{\tan l - \tan\alpha}{\tan l + \tan\alpha} = \frac{\sin l \cos\alpha - \cos l \sin\alpha}{\sin l \cos\alpha + \cos l \sin\alpha} = \frac{\sin(l - \alpha)}{\sin(l + \alpha)}, \tag{4.64}$$

we find

$$\frac{\sin(l - \alpha)}{\sin(l + \alpha)} = \tan^2\frac{\varepsilon}{2}. \tag{4.65}$$

Because $l - \alpha$ is a small quantity, and hence $\sin(l - \alpha) \approx l - \alpha$ and $\sin(l + \alpha) \approx \sin(2l)$, we obtain a formula up to the first order of ε^2,

$$l - \alpha \approx \tan^2\frac{\varepsilon}{2}\,\sin(2l) \approx 0.043\,\sin(2l). \tag{4.66}$$

4.4.4 Aphelion and Perihelion

The second difference between the fictitious mean sun and the true sun is that the orbit of Earth is elliptical rather than circular. The distance between Earth and the Sun is farthest at *aphelion* and closest at *perihelion*. Aphelion is derived from the Greek words *apo* ("away from") and *helios* ("sun"), while perihelion includes the Greek word *peri* ("near"). The date and the hour in the day of each aphelion and perihelion for 2011 to 2020 are shown in Table 4.4. (The accuracy of the time for aphelion and perihelion is only up to the hour, whileas the accuracy of equinoxes and solstices is up to the minute). The distance of the Earth and the Sun is about 3% further at aphelion than it is at perihelion. Because radiation intensity is inversely proportional to the square of distance, the solar radiation power in early January is 6% stronger than that in early July.

Table 4.4: Cardinal Points in Years 2011 through 2020

Year	Perihelion	Aphelion	Equinoxes	Solstices
2011	Jan 3, 19h	Jul 4, 15h	Mar 20, 23:21	Jun 21, 17:16
			Sep 23, 09:05	Dec 22, 05:30
2012	Jan 5, 01h	Jul 5, 04h	Mar 20, 05:14	Jun 20, 23:09
			Sep 22, 14:49	Dec 21, 11:12
2013	Jan 2, 05h	Jul 5, 15h	Mar 20, 11:02	Jun 21, 17:16
			Sep 22, 20:44	Dec 22, 05:30
2014	Jan 4, 12h	Jul 4, 00h	Mar 20, 16:57	Jun 21, 16:38
			Sep 23, 02:29	Dec 22, 04:48
2015	Jan 4, 07h	Jul 4, 15h	Mar 20, 22:45	Jun 21, 17:16
			Sep 23, 08:21	Dec 22, 05:30
2016	Jan 2, 23h	Jul 4, 16h	Mar 20, 04:30	Jun 20, 22:34
			Sep 22, 14:21	Dec 21, 10:44
2017	Jan 4, 14h	Jul 3, 20h	Mar 20, 10:29	Jun 21, 04:24
			Sep 22, 20:02	Dec 21, 16:28
2018	Jan 3, 06h	Jul 6, 17h	Mar 20, 16:15	Jun 21, 10:07
			Sep 23, 01:54	Dec 21, 22:23
2019	Jan 3, 05h	Jul 4, 22h	Mar 20, 21:58	Jun 21, 15:54
			Sep 23, 07:50	Dec 22, 04:19
2020	Jan 5, 08h	Jul 4, 12h	Mar 20, 03:50	Jun 20, 21:44
			Sep 22, 13:31	Dec 21, 10:02

4.4.5 Time Difference Originated from Eccentricity

The eccentricity of the orbit of Earth around the Sun gives rise to a second term in the equation of time. According to Kepler's first law, the orbit of Earth around the Sun is an ellipse, and the position of the Sun is at a focus of the ellipse. From the point of view of Earth, the Sun is orbiting Earth along an ellipse,

$$r = \frac{p}{1 + e\cos(\theta - \theta_0)}, \tag{4.67}$$

where r is the instantaneous distance between the Sun and Earth; e is the eccentricity of the ellipse, currently $e = 0.0167$; θ is the true longitude of the Sun along the ecliptic; θ_0 is the true longitude of the perihelion; and p is a constant.

According to Kepler's second law, the radius vector of the ellipse sweeps out equal

Figure 4.14 Eccentricity of Earth's orbit: Kepler's laws. According to Kepler's laws, Earth is moving around the Sun on an elliptical orbit. It gives rise to a term of the difference between mean solar time and true solar time with periodicity of a year.

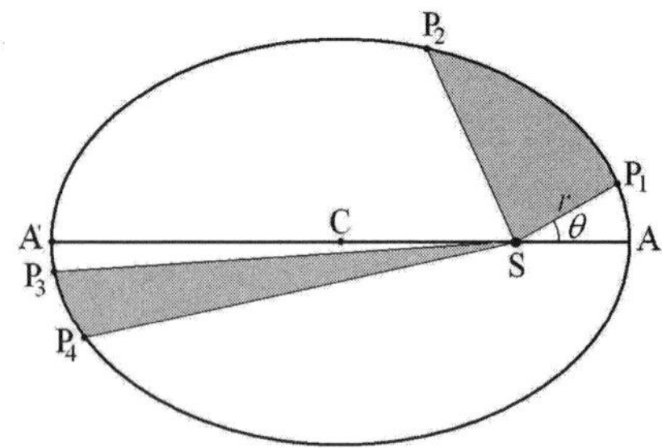

areas in equal times:

$$\frac{1}{2}\int_0^t r^2\, d\theta = \frac{1}{2}\int_0^t \frac{p^2}{[1+e\cos(\theta-\theta_0)]^2}\, d\theta \propto t. \tag{4.68}$$

Because the eccentricity e is small, the integrand of Eq. 4.68 can be expanded into a power series,

$$t \propto \int_0^t [1-2e\cos(\theta-\theta_0)]\, d\theta = \theta - 2e\sin(\theta-\theta_0). \tag{4.69}$$

For an entire year, the longitude increases by 2π, and the sine function returns to its original value.

Now, we relate the true longitude of the Sun, θ, with the mean longitude l discussed in the last section, which is proportional to time. For a calendar year, both quantities increase by 2π. Therefore,

$$l = \theta - 2e\sin(\theta-\theta_0). \tag{4.70}$$

If the eccentricity e is negligibly small, we have

$$\theta = l. \tag{4.71}$$

For a first-order approximation, we substitute θ in the second term of Eq. 4.70 with l, which yields

$$\theta = l + 2e\sin(l-l_0), \tag{4.72}$$

where l_0 is the longitude of perihelion. The second term in Eq. 4.72 arises from the eccentricity of the orbit.

4.4.6 Equation of Time

By combining Eq. 4.72 with Eq. 4.66, we obtain a complete expression of the equation of time (ET) up to the first order of obliquity and the first order of eccentricity, $e = 0.0167$,

$$\begin{aligned}\text{ET} = l - \alpha &= \tan^2\frac{\varepsilon}{2}\sin(2l) - 2e\sin(l-l_0)\\ &\approx 0.043\sin(2l) - 0.0334\sin(l-l_0).\end{aligned} \tag{4.73}$$

The unit of angles, l and α, is in radians. A complete circle is 2π. To convert the unit to time as we know, we notice that the mean Sun revolves around Earth every 24 hours. The time it takes for the mean Sun to catch up the true Sun, or *vice versa*, following Eq. 4.73, is

$$\begin{aligned}\mathrm{ET(min)} &= \frac{24\times 60}{2\,\pi}\,[0.043\sin(2l) - 0.0334\sin(l-l_0)] \\ &= 9.85\sin(2l) - 7.65\sin(l-l_0).\end{aligned} \tag{4.74}$$

In Eq. 4.74, as usual, the time difference is expressed in minutes. Using the approximate formula for the mean longitude l, Eq. 4.27, an explicit expression of the equation of time can be obtained,

$$\mathrm{ET} = \left[9.85\,\sin\left(\frac{4\pi\,(N-80)}{365.2422}\right) - 7.65\,\sin\left(\frac{2\pi\,(N-3)}{365.2422}\right)\right]\,\mathrm{(min)}. \tag{4.75}$$

Here the date of perihelion is assumed to be January 3. The number of the day in a year N can be computed using Eq. 4.28. Equation 4.75 is sufficiently accurate to deal with problems in sunlight tracking. A chart is shown in Fig. 4.15. Remember that hour angle ω is to the negative of right ascension α, the equation of time ET should be added to the hour angle of the Sun, which means that if ET is positive, the real Sun is faster than the mean Sun.

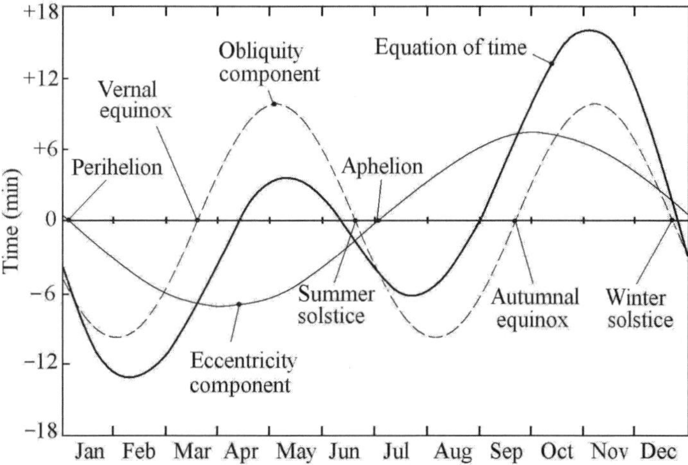

Figure 4.15 Equation of time. Thick solid curve, the difference between mean solar time and true solar time, the so-called equation of time has two terms. The first term, the thin solid curve, has a periodicity of a year, originated from the eccentricity of the orbit, which starts at aphelion. The second term, the dashed curve, originated from the obliquity of the ecliptic, has a periodicity of half a year, and starts at the vernal equinox.

Equation 4.75 can be used to convert standard time to solar time. Standard time is defined by an offset Δ from UT, almost always an integer number of hours. For example, Eastern Standard Time (EST) is defined by UT–5h, or $\Delta = -5$; Eastern Daylight Saving Time (EDT) is defined by UT–4h, or $\Delta = -4$. The solar time $t_\odot$ is

$$t_\odot = \text{UT} + \frac{1}{15}\lambda - \Delta + \text{ET}, \tag{4.76}$$

where λ is the longitude of the observer, each hour corresponds to 15° degree; and ET is the equation of time given by Eq. 4.73.

By setting $t_\odot = 0$, the standard time for solar noon, T_0, can be determined from Eq. 4.76,

$$T_0 = \Delta - \frac{1}{15}\lambda - \text{ET}. \tag{4.77}$$

As an example, New York City, where $\lambda = -73°58'$, corresponds to minus 4 h and 56 min. For EST, $\Delta = -5$. At EST noon, the mean solar time is 12:04 in the afternoon. In mid-November, Eq. 4.75 gives ET = 16 min, which means the real Sun is 16 min faster than the mean Sun. Therefore, at 12:00 EST, solar time is 12:20 in the afternoon. The Sun is 5° west of the meridian.

4.4.7 Declination of the Sun

In Section 4.3.1, we presented an approximate equation for the declination of the Sun (Eq. 4.27). Here, we derive a more accurate equation for the declination δ of the Sun. The reference point of the Sun's declination is the March equinox, where by definition the declination is zero and the ecliptic longitude is also zero. If the Sun's longitude runs linear over time, the sine formula gives

$$\sin\delta = \sin\varepsilon \sin l. \tag{4.78}$$

However, the ecliptic motion of Earth gives rise to an additional term,

$$\sin\delta = \sin\varepsilon \sin\left\{l + 2e\left[\sin(l - l_0) - \sin l_0\right]\right\}. \tag{4.79}$$

Consequently,

$$\delta = \arcsin\left(\sin\varepsilon \sin\left\{l + 2e\left[\sin(l - l_0) - \sin l_0\right]\right\}\right). \tag{4.80}$$

4.4.8 Analemma

As a combined result of Equation of Time and the variation of declination, the apparent position of the Sun at a given time of the day varies over the date in a year, which forms a well-defined trajectory on the sky, the *analemma*. The trajectory can be recorded by fixing a camera towards the Southern sky (in the Northern hemisphere), taking one picture at the same time of the day everyday, then superimpose the pictures of the sunny days. A "8"-like pattern is revealed. Plate 2 is compiled by Greek astronomer Anthony Ayiomamitis from 47 clear-day photos taken in 2003 near the Temple of Apollo, Corinth, Greece.

Problems

4.1. For a place in the frigid zone with latitude ϕ that satisfies $\phi > \pi/2 - \varepsilon$, determine the starting day and ending day in a year that the Sun never sets.

4.2. For a place in the frigid zone with latitude ϕ that satisfies $\phi > \pi/2 - \varepsilon$, determine the starting day in a year and ending day in the next year that the Sun never rises.

4.3. For a south-facing window of area A, calculate the total solar radiation for any day in the year.

Hint: For days after the vernal equinox and before the autumnal equinox, the time of radiation is between the two points where the Sun crosses the east–west great circle. Otherwise the Sun is at the west side of the building.

4.4. On gage C3 of the 2009 *Astronomical Almanac*, the leading terms of the equation of time in units of seconds are (using our notations)

$$\mathrm{ET} = -108.5\sin l + 596.0\sin 2l - 428.2\cos l, \tag{4.81}$$

where

$$l = 279^\circ.791 + 0.985647N, \tag{4.82}$$

where N is the number of days counted from January 1, and l is the mean longitude of the Sun.

Questions:

1. What is the meaning of the number 0.985647?
2. What is the meaning of the phase angle 279°.791?

4.5. Write the sine and cosine terms in Eq. 4.81 in the form of sin $(l - l_0)$, where l_0 is a constant phase angle. Explain the meaning of the constant l_0.

4.6. Determine the sunset time (in solar time) and the length of daytime of New York City on New Years Day, Memorial Day, Labor Day, and Thanksgiving Day.

4.7. What is the ratio of solar power on a vertical surface facing south, a horizontal surface, and a latitude-tilt surface, at solar noon of an equinox (vernal equinox or autumnal equinox) for New York City?

4.8. In New York City, where the latitude is 40°47', on Memorial Day (May 25) and Thanksgiving Day (Nov 27) of 2009, at solar noon, determine the declination of the Sun, the height of the Sun, and the power density of direct sunlight in watts per square meter on a horizontal surface at that time.

4.9. In New York City, where the latitude is 40°47' and longitude is -73°58', on Memorial Day (May 25) and Thanksgiving Day (Nov 27) of 2009, determine the civil time (EST, or EDT if necessary) of solar noon (the time the Sun passes the local meridian).

4.10. For a surface of arbitrary orientation β and γ, find out the starting time and ending time of solar radiation on the surface.

4.11. For a surface facing south, $\gamma = 0$ but $\beta \neq 0$, find out the starting time and ending time of solar radiation on the surface.

4.12. For a surface facing south, $\gamma = 0$ but $\beta \neq 0$, find out the direct daily radiation.

Chapter 5

Interaction of Sunlight with Earth

In this chapter, we study the interaction of sunlight with Earth. Due to the effect of the atmosphere, about one-half of sunlight is reflected, scattered, or absorbed before reaching the ground. The available sunlight varies with location and time. Furthermore, the sunlight absorbed by the ground penetrates into Earth, is stored as heat, and becomes *shallow geothermal energy*, a significant component of renewable energy.

5.1 Interaction of Radiation with Matter

In this section, the general physical phenomena of the interaction of radiation with matter are described.

5.1.1 Absorptivity, Reflectivity, and Transmittivity

When a ray of radiation falls on a piece of matter, in general, part of the radiation is *reflected*, another part is *absorbed*, and yet another part is *transmitted*. To describe and characterize the interaction of radiation with matter, the following three dimensionless coefficients are introduced:

$A(\lambda)$, *absorptivity* or *absorptance*: fraction of incident radiation of wavelength λ absorbed

$R(\lambda)$, *reflectivity* or *reflectance*: fraction of incident radiation of wavelength λ reflected

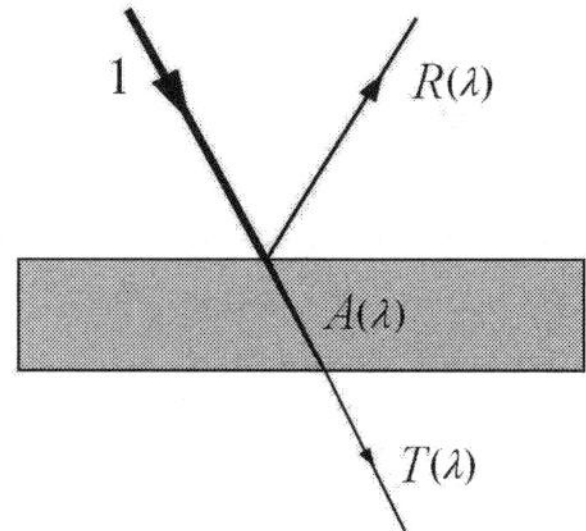

Figure 5.1 Absorptivity, reflectivity, and transmittivity. Part of the radiation falling on a piece of matter is *absorbed*, another part is *reflected*, and the rest is *transmitted*. Three dimensionless coefficients are introduced: *absorptivity* $A(\lambda)$, *reflectivity* $R(\lambda)$, and *transmittivity* $T(\lambda)$. Conservation of energy requires that $A(\lambda) + R(\lambda) + T(\lambda) = 1$.

$T(\lambda)$, *transmittivity* or *transmittance*: fraction of incident radiation of wavelength λ transmitted

If the nature of the radiation does not change or its wavelength stays unchanged, conservation of energy requires that incident radiation be absorbed, reflected, or transmitted; see Fig. 5.1. Therefore

$$A(\lambda) + R(\lambda) + T(\lambda) = 1. \tag{5.1}$$

For opaque surfaces, the transmittivity is zero. Conservation of energy requires that incident radiation be either absorbed or reflected,

$$A(\lambda) + R(\lambda) = 1. \tag{5.2}$$

5.1.2 Emissivity and Kirchhoff's Law

When matter is heated, it will radiate according to Planck's law (Fig. 5.2). The actual power density of radiation also depends on the nature of the surface. However, it never exceeds that of a blackbody. The actual radiation from a surface as a fraction of blackbody radiation at a given wavelength is called its *emissivity* $E(\lambda)$. It is less than 1, except it equals exactly 1 for a blackbody. Based on classical thermodynamics, Kirchhoff showed that *the emissivity of a surface at a given wavelength must equal its absorptivity*. At thermal equilibrium, the radiation energy emitted must equal the radiation energy absorbed. Otherwise heat can transfer from a cold reservoir to a hot reservoir, which violates the second law of thermodynamics (see chapter 6). Therefore, the absorptivity at a given wavelength must equal the emissivity at the same wavelength,

$$E(\lambda) = A(\lambda). \tag{5.3}$$

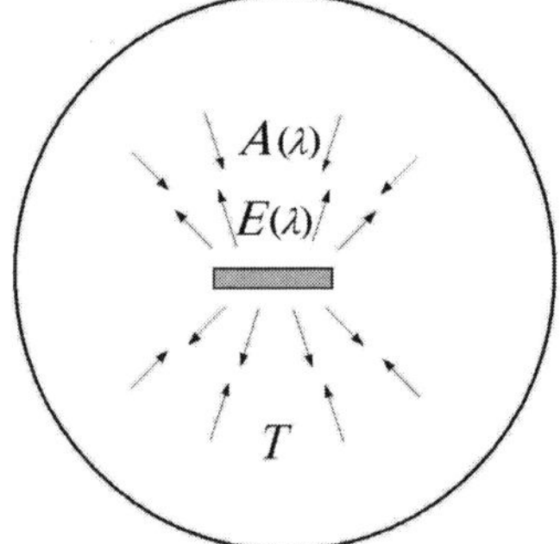

Figure 5.2 Emissivity and absorptivity. When heated, matter radiates electromagnetic waves. The maximum radiation power spectrum follows Planck's law. For the general case, the radiation from a surface as a fraction of blackbody radiation at a given wavelength is called its *emissivity*, $E(\lambda)$. At thermal equilibrium, the radiation energy emitted must equal the radiation energy absorbed. Consequently, $E(\lambda) = A(\lambda)$.

5.1.3 Bouguer–Lambert–Beer's Law

An empirical relationship between the absorption of light and the property of the absorbing medium was discovered by Pierre Bouguer before 1729, then was formulated by Johann Heinrich Lambert in 1760 in his monograph "Photometria." It states that the light intensity depends exponentially on the thickness of the optical path z,

$$I_\lambda(z) = I_\lambda(0)\, e^{-A(\lambda)\, z}, \tag{5.4}$$

where $A(\lambda)$ is the absorption coefficient of the medium at wavelength λ, which has a dimension of inverse length and z is the optical path.

The empirical relation was further developed by August Beer in 1852 to correlate the absorption coefficient with the *concentration of absorbing particles*, known as the Beer's law or Bouguer–Lambert–Beer's law,

$$I_\lambda(z) = I_\lambda(0)\, e^{-N\sigma(\lambda)\, z}, \tag{5.5}$$

where N is the number of absorbing particles per unit volume and $\sigma(\lambda)$ is the *absorption cross section* of the absorbent at wavelength λ. An intuitive proof is shown in Fig. 5.3. For a thin slice in the absorption path of thickness dz and cross sectional area S, the fractional area dS occupied by the absorbing particles is

$$dS = N\sigma(\lambda)\, S\, dz. \tag{5.6}$$

Obviously, a proportion dS/S of radiation energy is blocked by the absorbing particles,

$$\frac{dI_\lambda(z)}{I_\lambda(z)} = -\frac{dS}{S} = -N\sigma(\lambda)\, dz. \tag{5.7}$$

Integrating over z and using the initial condition at $z = 0$, one obtains

$$I_\lambda(z) = I_\lambda(0)\, e^{-N\sigma(\lambda)\, z}. \tag{5.8}$$

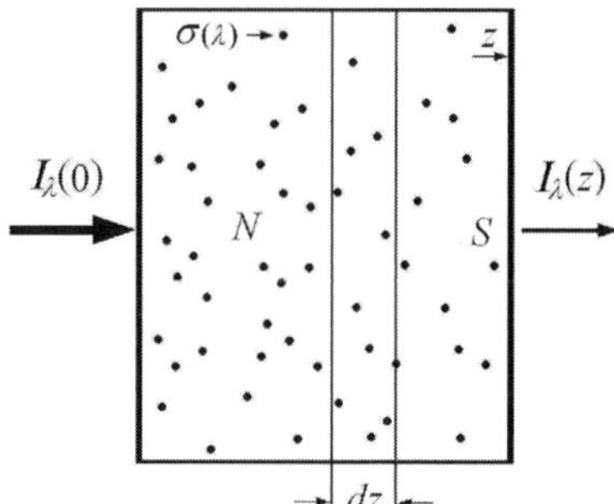

Figure 5.3 Bouguer–Lambert–Beer's law. Variation of light intensity with the concentration of the absorbing particles and the length of the optical path plays an important role in the study of the effect of the atmosphere on sunlight.

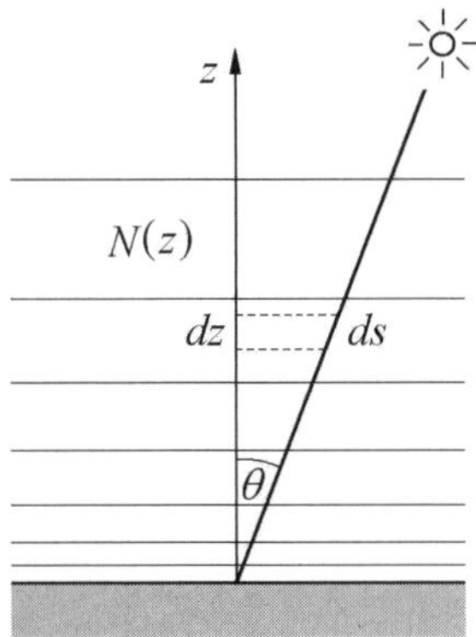

Figure 5.4 Attenuation of sunlight at azimuth θ. Variation of light intensity with the concentration of the absorbing particles and the length of the optical path plays an important role in the study of the effect of the atmosphere on sunlight.

Comparing with Eq. 5.4, one finds $A(\lambda) = N\sigma(\lambda)$. If the concentration of particles is not uniform over the optical path, which can be described by a concentration distribution $N(z)$, the above relation can be extended to

$$I_\lambda(z) = I_\lambda(0)\,\exp\left(-\int_0^z \sigma(\lambda)\,N(z)\,dz\right). \tag{5.9}$$

Now we look into a situation relevant to the attenuation of sunlight by the atmosphere; see Fig. 5.4. In the atmosphere, to a good approximation, the distribution of molecules and particles is a function of height z. If the Sun is at the zenith, according to Eq. 5.9, the absorbance over the entire atmosphere is

$$\alpha(\lambda, 0) = \frac{I_\lambda(0)}{I_\lambda(\infty)} = \exp\left(-\int_0^\infty \sigma(\lambda)\,N(z)\,dz.\right). \tag{5.10}$$

If the azimuth of the Sun is θ — notice that $dz = ds\cos\theta$ (see Fig. 5.4) — the absorbance becomes

$$\begin{aligned}\alpha(\lambda, \theta) &= \exp\left(-\int_0^\infty \sigma(\lambda)\,N(z)\,ds\right)\\ &= \frac{1}{\cos\theta}\exp\left(-\int_0^\infty \sigma(\lambda)\,N(z)\,dz\right)\\ &= \frac{\alpha(\lambda, 0)}{\cos\theta}.\end{aligned} \tag{5.11}$$

5.2 Interaction of Sunlight with Atmosphere

The interaction of sunlight with the atmosphere has been studied extensively by climate scientists. Here is a summary. Approximately, 30% of solar radiation is reflected or

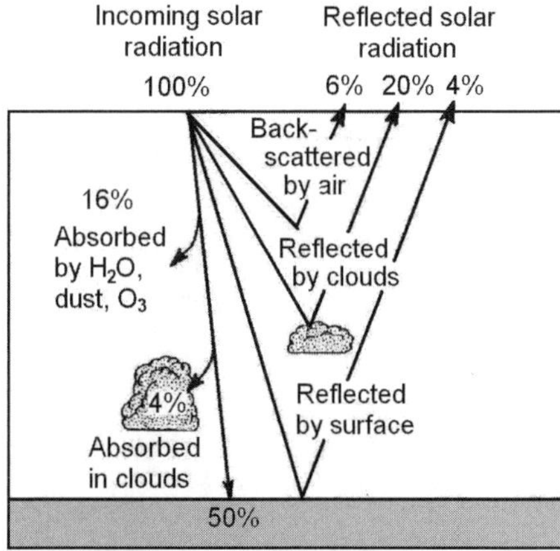

Figure 5.5 Interaction of sunlight with atmosphere. Approximately 30% of solar radiation is reflected or scattered back to space immediately, 20% is absorbed by the atmosphere and cloud, and 50% is absorbed by Earth. (See Ref. [66], p. 94).

scattered back to space, see Fig. 5.5. Six percent is scattered by air; 20% is reflected by clouds; 4% is reflected by the surface of Earth; and 20% is absorbed by the atmosphere: 16% is absorbed by water vapor, dust, and O_3. Another 4% is absorbed by the cloud. The solar radiation thus absorbed heats up the atmosphere. Fifty percent is absorbed by the solid surface of Earth. The total radiation energy received by the atmosphere and the solid Earth is about 70%.

Earth should be in thermal equilibrium with the surroundings. Indeed, the 70% of solar energy entering the atmosphere and Earth is transferred back to space as heat radiation.

5.2.1 AM1.5 Reference Solar Spectral Irradiance

According to several decades of careful measurements, the power density of solar radiation outside the atmosphere is 1366 W/m^2. On the surface of Earth, due to scattering and absorption, even under a perfectly clear sky, when the Sun is right at the zenith, solar radiation is reduced by about 22%. Because, on average, the Sun should have an azimuth angle with the horizon, the reduction should be on average more than 22%.

To standardize the measurement of solar energy applications, in 1982, the American Society for Testing and Materials (ASTM) started to promulgate *Standard Tables of Reference Solar Spectral Irradiance at Air Mass 1.5.* The standard was revised in 2003 as ASTM G173-03. A separate standard for zero air mass was promulgated in 2006 to become ASTM E490-06. An extension to different tilt angles was promulgated in 2008 as ASTM G197-08. In both ASTM G173-03 and G197-08, atmospheric and climatic conditions are identical. These standards represent the solar radiation under reasonable cloudless atmospheric conditions favorable for the computerized simulation,

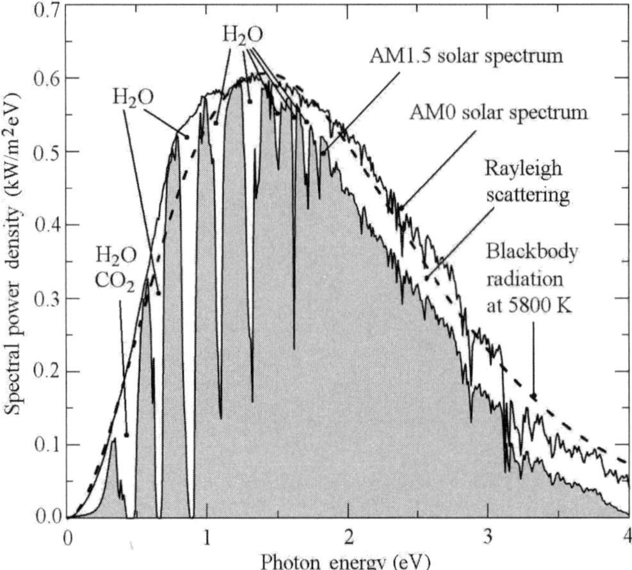

Figure 5.6 AM0 and AM1.5 solar radiation spectra. The AM0 spectrum is the solar radiation outside the atmosphere. It is approximately a blackbody radiation at 5800 K. The integrated power density is 1.366 kW/m^2. The AM1.5 spectrum shows a number of atmospheric effects. On the blue side, there is a broad-band reduction of power density due to Rayleigh scattering from molecules and dust particles. On the infrared side, water vapor contributes the most absorption, followed by carbon dioxide.

comparative rating, or experimental testing of fenestration systems. The U.S. standard was adapted by the International Organization for Standardization (ISO) to become ISO 9845-1, 1992.

The meaning of *air mass* (AM) is as follows. When the Sun is at the zenith under reasonable cloudless atmospheric conditions, the absorption due to the atmosphere is defined as air mass 1. In most cases, the zenith angle of the Sun is not zero. ASTM chose as the standard the condition when the absorption is 1.5 times the normal air mass, abbreviated as AM1.5. According to Eq. 5.11, the standard zenith angle is

$$\theta = \arccos\frac{1}{1.5} = 48.19°. \tag{5.12}$$

Also see Fig. 5.4. The integrated power density of the AM1.5 solar radiation is 1 kW/m^2. The quantity 1 kW/m^2 is defined as a unit of radiation, called *one sun.* We shall use this unit throughout the book.

The spectral irradiance or the radiation power spectrum for AM0 and AM1.5 solar radiation is shown in Fig. 5.6. The AM0 spectrum is the solar radiation outside the atmosphere. It is approximately a blackbody radiation at 5800 K. The integrated power density is 1.366 kW/m^2. The AM1.5 spectrum shows a number of atmospheric effects.

On the blue side, there is a broad-band reduction of power density due to Rayleigh scattering from molecules and dust particles. The probability of Rayleigh scattering is proportional to the inverse fourth power of the wavelength of the radiation, and thus the short-wavelength radiation is reduced heavily. On the infrared side, water vapor contributes the most absorption followed by carbon dioxide. Plate 1 is a colored verion of Fig. 5.6. A table of the data is shown in Appendix E.

5.2.2 Annual Insolation Map

In Chapter 4, we discussed the variation of direct solar radiation as a function of time (day in a year and time in a day) and location (latitude and longitude). Because of the interaction of solar radiation with the atmosphere, the actual solar radiation received at the surface is always less, and the percentage of reduction depends on the location. A frequently used representation is the *annual insolation map*; see Plates 3–5. There are two conventions:

1. Annual radiation energy in kilowatt-hours per square meter. The standard solar radiation is defined as one sun, or 1 kW/m^2. Therefore, the insolation is often expressed in hours per year. The number ranges from more than 2000 h/year (Sahara desert, part of Outback of Australia, part of South Africa) to less than 600 h/year (Greenland, northern parts of Siberia, Finland, and Canada).

2. Average diurnal radiation energy in kilowatt-hours per square meter over a year. Similarly, it is often expressed in h/day. The number ranges from more than 6 h/day to less than 2 h/day.

There is an obvious relation between those two conventions:

$$\text{Annual insolation} = 365.2422 \times \text{average diurnal insolation}. \tag{5.13}$$

Plate 3 is an average diurnal insolation map of the world, Plate 4 is an average diurnal insolation map of the United States, and Plate 5 is an annual insolation map of Europe.

5.2.3 Clearness Index

In most cases the insolation shown on the maps is smaller than the solar radiation derived in Chapter 4. The ratio is the *clearness index*,

$$\overline{K}_{\mathrm{T}} = \frac{\overline{H}}{\overline{H}_0}, \tag{5.14}$$

where $\overline{H}_0$ is the average diurnal insolation under cloudless conditions from Chapter 4; and $\overline{H}$ is the observed average diurnal insolation, typically over a year or a month.

From a practical point of view, monthly averaged diurnal insolation is the most useful data for the application of solar energy and is also widely measured at many solar observation stations in the world. The clearness index $\overline{K}_{\mathrm{T}}$, in percent, is also collected for many cities in the world. Table 5.1 shows the clearness index at several U.S. cities, from the sunniest city on Earth, Yuma, Arizona, to one of the wettest cities, Hilo, Hawaii. The last column is the annual average clearness index.

Table 5.1: Monthly Clearness Index $\overline{K}_T$ [%] of Selected U.S. Cities

City	Jan	Feb	Mar	Apr	May	Jun	Jul	Aug	Sep	Oct	Nov	Dec	Ave
Yuma, AZ	83	87	91	94	97	98	92	91	93	93	90	83	91
Las Vegas, NV	74	77	78	81	85	91	84	86	92	84	83	75	82
Los Angeles, CA	70	69	70	67	68	69	80	81	80	76	79	72	73
Denver, CO	67	67	65	63	61	69	68	68	71	71	67	65	67
New York, NY	49	56	57	59	62	65	66	64	64	61	53	50	59
Seattle, WA	27	34	42	48	53	48	62	56	53	36	28	24	45
Hilo, HI	48	42	41	34	31	41	44	38	42	41	34	36	39

5.2.4 Beam and Diffuse Solar Radiation

Most solar radiation data are collected with a *pyranometer* to measure the global solar irradiance from a hemisphere. A typical pyranometer is shown in Fig. 5.7. The centerpiece of the instrument is a dishlike blackbody absorber (1), covered by a protective glass dome (2). The radiation received by the absorber generates a voltage which is proportional to the heat and is connected to a voltmeter through the cable (3). No battery and electronics are needed. By design, the instrument receives direct (beam) sunlight and the diffuse sunlight from the entire sky.

The solar radiation on a surface is always a mixture of direct (beam) sunlight and diffuse sunlight. In any practical application, direct (beam) sunlight and diffuse sunlight behave differently. For example, for concentrated solar applications, only direct sunlight is used. On the other hand, for flat solar thermal or photovoltaic receivers, diffuse sunlight plays a significant role. It is important to know the proportion of these two.

In general, the cloudier the climate, the higher the proportion of diffuse radiation. There should also be a quantitative correlation between the clearness index and the ratio between beam sunlight and diffuse sunlight. The problem was studied in detail by Liu and Jordan [52], and a large body of literature has been accumulated. The central issue is to estimate the ratio of diffuse radiation averaged over a period of time

Figure 5.7 Pyranometer. The center piece of the instrument is a dishlike blackbody absorber (1), covered by a protective glass dome (2). It generates a voltage proportional to the radiation received by the absorber over the entire hemisphere, then outputs it through the cable (3).

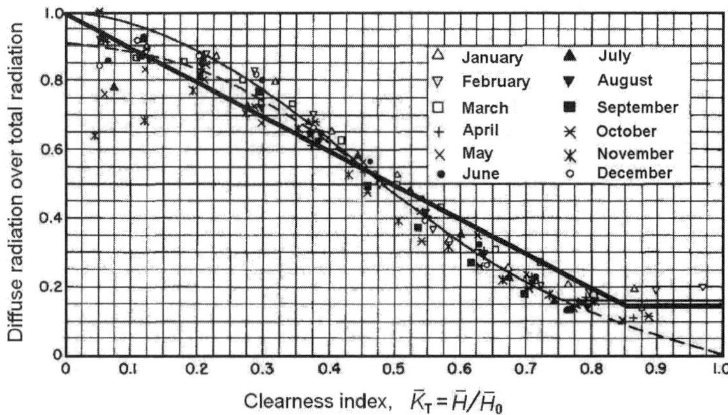

Figure 5.8 Correlation between diffuse radiation and clearness index. Data points are from Liu and Jordan [52]. The thick straight lines represent the simplified model, Eqs. 5.15 – 5.17. The curves represent third-order polynomials by Liu and Jordan [52]. Later data and analysis also support a simple linear relation, see Ref. [54].

$\overline{H}_d$ over the total averaged radiation $\overline{H}$. Here we present a conceptually simple model for a rule-of-thumb estimate. It has three components.

First, even for a perfectly clear day, diffuse radiation from the sky is appreciable. Experiments show that the average proportion of diffuse radiation is about 15% of total radiation. Therefore, if the clearness is greater than 85%, the ratio is 15%:

$$\text{if } \overline{K}_\text{T} > 0.85, \text{ then } \frac{\overline{H}_d}{\overline{H}} = 0.15, \tag{5.15}$$

where $\overline{K}_\text{T}$ is defined in Eq. 5.14.

Second, for a heavily cloudy sky, all the radiation is diffuse.

$$\text{if } \overline{K}_\text{T} \approx 0, \text{ then } \frac{\overline{H}_d}{\overline{H}} = 1. \tag{5.16}$$

Third, for the cases in between, the simplest mathematical representation is a linear interpolation,

$$\text{If } 0.85 > \overline{K}_\text{T} > 0, \text{ then } \frac{\overline{H}_d}{\overline{H}} = 1 - \overline{K}_\text{T}. \tag{5.17}$$

The conceptually and mathematically very simple model fits the data published by Liu and Jordan [52] reasonably well and is in agreement with several later publications. See Fig 5.8. The thick straight lines represent the simplified model, Eqs 5.15 – 5.17. The curves represent more complicated formulas by Liu and Jordan.

5.3 Penetration of Solar Energy into Earth

It is an experimental fact that in most solids the heat flux q_z along direction z is proportional to the temperature gradient in the solid along that direction,

$$q_z = -k\left(\frac{\partial T}{\partial z}\right), \tag{5.18}$$

where k is the thermal conductivity, a constant depending on the material nature of the solid; and the negative sign is a result of that fact that the heat flows in the direction of decreasing temperature. In the International System of Units (SI), the unit for heat flux is watts per square meter. The unit for thermal conductivity is watts per meter Kelvin. A table of thermal conductivities of commonly used materials in construction is given as Table 5.2.

Consider a one-dimensional system in which the temperature and heat flux are a function of position z and time t. Practically, the system could be a bar with uniform cross section A. The temperature and the heat flux are uniform within each cross section. If the density of the material is ρ and its heat capacitance is c_p, the change in the heat content of a slab with thickness Δz with temperature is

$$\Delta Q = \rho c_p A\,\Delta T\,\Delta z. \tag{5.19}$$

Combining Eqs 5.18 and 5.19, we find the differential equation for the temperature distribution,

$$\frac{\partial T}{\partial t} = \frac{k}{\rho c_p}\frac{\partial^2 T}{\partial z^2}. \tag{5.20}$$

Let

$$\alpha = \frac{k}{\rho c_p}. \tag{5.21}$$

Table 5.2: Thermal Property of Earth

Material	Density ρ (10^3kg/m^3)	Heat capacity c_p (10^3J/kg·K)	Thermal conductivity k (W/m·K)	Coefficient α (10^{-6}m^2/s)
Limestone	2.18	0.91	1.5	0.626
Granite	3.0	0.79	3.5	1.47
Earth (wet)	1.7	2.1	2.5	0.70
Earth (dry)	1.26	0.795	0.25	0.25

Source : American Institute of Physics Handbook,
American Institute of Physics, New York, 3rd Ed., 1972; and Ref. [31].

Equation 5.20 then becomes

$$\frac{\partial T}{\partial t} = \alpha \frac{\partial^2 T}{\partial z^2}. \tag{5.22}$$

We seek a solution of Eq. 5.22 for a semi-infinite space $z \geq 0$ with boundary conditions

$$T = T_0 + \Delta T \cos \omega t, \qquad z = 0, \tag{5.23}$$

$$T = T_0, \qquad z = \infty, \tag{5.24}$$

where ΔT is the amplitude of the temperature variation at $z = 0$ and ω is its circular frequency. In the case we are considering, that is, the annual variation of temperature, the circular frequency is

$$\omega = \frac{2\pi}{(365.25 \times 86400)} \approx 2 \times 10^{-7}\mathrm{s}^{-1}. \tag{5.25}$$

Introducing a dimensionless temperature,

$$\Theta = \frac{T - T_0}{\Delta T}, \tag{5.26}$$

Equation 5.22 becomes

$$\frac{\partial \Theta}{\partial t} = \alpha \frac{\partial^2 \Theta}{\partial z^2} \tag{5.27}$$

with boundary conditions

$$\Theta = \cos \omega t, \qquad z = 0, \tag{5.28}$$

$$\Theta = 0, \qquad z = \infty. \tag{5.29}$$

Equation 5.27 can be resolved much easier in complex numbers using the Euler relation $e^{ix} = \cos x + i \sin x$, or $\cos x = \mathrm{Re}[e^{ix}]$. The boundary conditions now become

$$\Theta = \mathrm{Re}\left[e^{-i\omega t}\right], \qquad z = 0, \tag{5.30}$$

$$\Theta = 0, \qquad z = \infty, \tag{5.31}$$

and Eq. 5.27 can be resolved using the *Ansatz*,

$$\Theta = \mathrm{Re}\left[e^{\lambda z - i\omega t}\right]. \tag{5.32}$$

Obviously, the suggested solution (Eq. 5.32), satisfies the boundary condition at $z = 0$ (Eq. 5.30). The constant λ can be determined by the differential equation Eq. 5.27. In fact, it gives

$$\alpha \lambda^2 = -i\omega. \tag{5.33}$$

There are two solutions to Eq. 5.33,

$$\lambda = \pm \left[\sqrt{\frac{\omega}{2\alpha}} - i\sqrt{\frac{\omega}{2\alpha}}\right]. \tag{5.34}$$

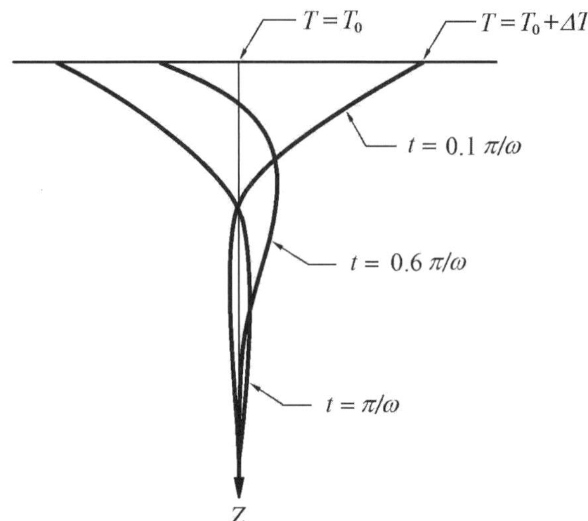

Figure 5.9 Penetration of solar energy into Earth. On the ground, the average temperature is T_0, and the amplitude of annual temperature variation is ΔT. The heat penetrates into the ground with a time delay. At a certain depth, the profile of annual temperature variation is *reversed*.

Because of the boundary condition at $z = \infty$, only the negative sign is admissible. Finally,

$$\Theta = \exp\left\{-\sqrt{\frac{\omega}{2\alpha}}z\right\} \cos\left(\sqrt{\frac{\omega}{2\alpha}}z - \omega t\right). \tag{5.35}$$

Or, using Eq. 5.26, the solution is

$$T = T_0 + \Delta T \exp\left\{-\sqrt{\frac{\omega}{2\alpha}}z\right\} \cos\left(\sqrt{\frac{\omega}{2\alpha}}z - \omega t\right). \tag{5.36}$$

For a numerical example, we use $T_0 = 15°\text{C}$, $\Delta T = 12°\text{C}$, and limestone. From Eq. 5.25 and Table 5.2, one finds $\sqrt{\omega/2\alpha} = \sqrt{2 \times 10^{-7}/2 \times 0.626 \times 10^{-6}} \approx 0.33\,\text{m}^{-1}$. Taking the number of the month as a parameter,

$$T = 15° + 12°\, e^{-0.33\,z} \cos\left(0.33\,z - \frac{2\pi \times (\text{month} - 7)}{12}\right). \tag{5.37}$$

At a distance z where $\sqrt{\omega/2\alpha} = \pi$, the value of the cosine function becomes its negative. For example, according to Eq. 5.37, when $0.33z = \pi$, or $z = 9.5$ m, the tempeture in the summer is the lowest and the temperature in the winter is the highest. The profile of the annual temperature variation is *reversed*; see Fig. 5.9. Therefore, the stored solar energy can be effectively utilized. We will discuss the details of its implementation in chapter 6.

Problems

5.1. Modeling the Sun as a blackbody radiator of $T_\odot = 5800$ K, calculate Earth's surface temperature $T_\oplus$, assuming that the temperature is uniform over its entire surface which is also a blackbody radiator.

5.2. As in the previous problem, assume the Sun as a blackbody radiator of $T_\odot = 5800$ K. If the absorptivity of Earth is 0.65 over the entire spectral range and its entire surface, what is Earth's surface temperature $T_\oplus$?

5.3. By direct substitution, prove that

$$u(z,t) = \frac{A}{\sqrt{t}} \exp\left\{\frac{-z^2}{4\alpha t}\right\} \tag{5.38}$$

is a solution of the one-dimensional heat conduction equation

$$\frac{\partial u}{\partial t} = \alpha \frac{\partial^2 u}{\partial z^2}. \tag{5.39}$$

5.4. Because the heat conduction equation is linear, if $u(z,t)$ is a solution, then the infinite integral of $u(z,t)$ is also a solution. Prove that

$$u(z,t) = A \operatorname{erf} \frac{z}{2\sqrt{\alpha t}} \tag{5.40}$$

is a solution of Eq. 5.39, where the error function $\operatorname{erf}(x)$ is defined as

$$\operatorname{erf}(x) = \frac{2}{\sqrt{(\pi)}} \int_0^x e^{-\xi^2} d\xi. \tag{5.41}$$

5.5. If the surface temperature of Earth suddenly changes from 0°C to T_0, how long does it take for the interior of Earth to be in equilibrium with its surface, which means it reaches 99% of the surface temperature? Make estimates for depths of 10 m, 100 m, and 1000 m, where the ground is made of granite.

5.6. Using the theory of solar energy penetration into Earth, determine the effect of the average daily variation of earth temperature. Assume that the amplitude of daily temperature is $\Delta T = 5$°C and the ground is made of limestone:

1. At what depth the phase of temperature profile is reversed (i.e., cooler at about 3pm and warmer at about 3am)?
2. What is the ratio of the amplitude of temperature variation at the depth of temperature profile reversal versus the temperature amplitude at the surface?

5.7. The average temperature of Oklahoma City is 28°C in July and 4°C in January. The temperature profile is approximately a sinusoidal curve during the entire year. If the ground is made of granite, what is the monthly temperature variation (list the values for each month) 10 m beneath the ground?

Chapter 6

Thermodynamics of Solar Energy

Thermodynamics is a branch of physics devoted to the study of energy and its transformation. It was established in the eighteenth and nineteenth centuries in attempts to better understand the underlying principles of heat engine, which converts heat energy to mechanical energy, such as the steam engine and the internal combustion engine. In the middle of nineteenth century, thermodynamics evolved into a logically consistent system starting with a few axioms, or laws, from which the entire theory could be deduced. In the twentieth century, the theory was extended to refrigeration and heat pumps as well as other forms of energy, such as electric, magnetic, elastic, chemical, electrochemical, and nuclear. At the core is the first law of thermodynamics regarding the conservation of energy and the second law of thermodynamics regarding the conversion of thermal energy to other forms of energy. In many textbooks of thermodynamics, there is also a zeroth law and a third law. The zeroth law is a self-evident definition of temperature. The third law has limited applications, most of which are unrelated to the study of solar energy. The theory of thermodynamics is macroscopic in nature, dealing directly with measurable physical quantities. The corresponding microscopic theory is statistical physics, where the laws of thermodynamics are derived from an atomic point of view.

In this chapter, we will present the basic concepts of thermodynamics that are essential for the understanding of solar energy. Complete presentations of thermodynamics, including the details of the third law of thermodynamics, can be found in standard textbooks.

6.1 Definitions

In this section, we introduce several basic definitions. The object of investigation by thermodynamics is called a system. A typical system is a uniform body of substance with a well-defined boundary, such as a piece of solid, a volume of liquid, a package of gas, and a surface layer. The physical objects outside the system boundary are called the surroundings. A state of a system represents the totality of its macroscopic properties. Two types of physical quantities are present in thermodynamics: the extensive quantity and the intensive quantity. An extensive quantity is proportional to the volume or mass of the system and is additive, such as mass, volume, energy, and entropy.

Intensive quantities are independent of the volume or mass of the system and are not additive, such as temperature, density, pressure, and the specific values of the extensive quantities such as energy density. A thermodynamic system can be connected to the surroundings or isolated. There is no heat or work exchange between an isolated system and its surroundings. The state of a system can go through a process where at least one of the physical quantities is changing over time. Two processes are of particular interest: the isothermal process, where the temperature of the system does not change, and the adiabatic process, where there is no heat change between the system and its surroundings.

An infinitesimal quantity of *work*, dW, is defined as the product of the force F acting on the boundary of the system and the length dL it moves in the direction of the force. Because the force F is a product of pressure P and area A, it is convenient to write work as

$$dW = F\,dL = PA\,dL = P\,dV, \tag{6.1}$$

where dV is the infinitesimal change of the volume of the system. The occurrence of the negative sign is because, when the pressure is positive, the force is pointing to the inside of the system. When the system expands under a positive pressure, it does work to the surroundings. The total work during a process from state 1 to state 2 is then

$$W = \int_1^2 F\,dL = \int_1^2 P\,dV. \tag{6.2}$$

Heat is defined as the energy transferred to the system across the boundary without moving the surface. If the temperature of the surroundings is higher than that of the system, the heat is transferred to the system, which is denoted as positive. Similarly, the total heat transferred to the system during a process from state 1 to state 2 is

$$Q = \int_1^2 dQ. \tag{6.3}$$

Temperature is defined by the zeroth law of thermodynamics, which asserts as follows:

> *When two systems are in thermodynamic equilibrium with a third system, the two systems are in thermodynamic equilibrium with each other, and all three systems have the same temperature.*

The zeroth law tells us how to compare temperature, but it does not define the scale of temperature. In thermodynamics, there are two independent definitions of temperature. The first one is defined by Lord Kelvin based on the Carnot cycle, which bears his name; (see Section 2.3.2). The second definition is based on the properties of ideal gas; (see Section 2.5.1). The two definitions are equivalent within a constant multiplier.

6.2 First Law of Thermodynamics

The first law of thermodynamics asserts that energy can be converted from one form to another but can never be created or annihilated. A succinct presentation is as follows:

> *It is impossible to build a perpetual motion that generates energy from nothing.*

The first law of thermodynamics is by no means trivial, regarding the fact that each year, numerous patent applications on perpetual motion are still received by the patent offices of countries all over the world, even in the era of high technology. Searching on Google for "perpetual motion", you would be surprised by the exotic new designs of perpetual motion proposed by overambitious inventors.

In his autobiography, Max Planck described how his physics teacher taught him about the concept of energy: A construction worker lifted a brick and put it at the top of the building. His work increased the energy of the brick. But the increase of energy was in the form of potential energy and not explicit. One day, the brick fell down from the top. As it almost reached the ground, the brick moved fast, which had an explicit kinetic energy. Finally, the brick hit the ground and converted the energy to heat. In the first step, the construction worker did the work as

$$W = fh = mgh, \tag{6.4}$$

where m is the mass of the brick and $g = 9.81m/s^2$ is the gravitational acceleration. The gravitational force $f = mg$. And h is the height, or the distance along the direction of force the brick moves. The increase of potential energy ΔE equals the work performed on the system,

$$\Delta E = W = mgh. \tag{6.5}$$

Just before the brick hit the ground, its velocity v is given as

$$v = \sqrt{2gh}, \tag{6.6}$$

which satisfies the law of conservation of energy,

$$\Delta E = \frac{1}{2}mv^2 = mgh. \tag{6.7}$$

The unit of energy is the product of the unit of force, the newton, and the unit of length, the meter. The unit of energy, newton-meter, or joule, is named after English physicist James Prescott Joule, who did the first experiment to demonstrate the equivalence of mechanical work and heat in 1844. A schematic of Joule's experiment is shown in Fig. 6.1.

In Joule's experiment, at the beginning, the weight is positioned at a predetermined distance from its equilibrium position. The initial temperature of the water, T_0, is measured. Then, by setting the weight to move and waiting to the end of motion,

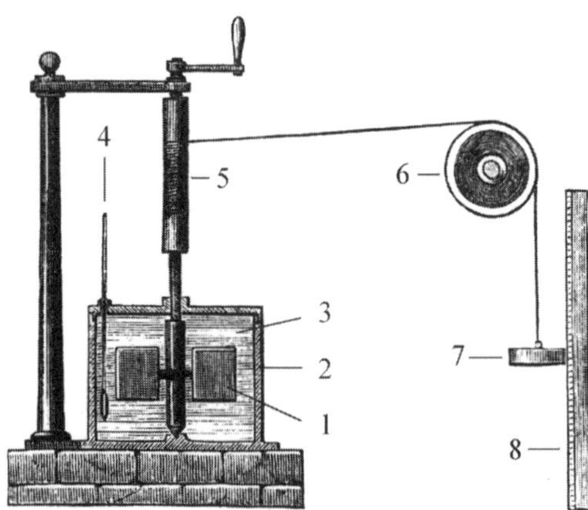

Figure 6.1 Joule's experiment. A paddle-wheel 1 is placed inside an insulated barrel 2 filled with water 3. The temperature is measured by a thermometer 4. The wheel is driven through a spindle 5 and a pulley 6 by a falling weight 7. The height of the weight is measured by the ruler 8. After setting the paddle-wheel to move, the mechanical energy is transformed into heat, which is measured by the thermometer.

the final temperature of the water barrel, T_1, is measured. The heat generated by mechanical disturbance is

$$Q = (T_1 - T_0)M. \tag{6.8}$$

If M is the mass of water in grams, then the heat Q is in calories. The mechanical equivalence of one calorie of heat found by Joule is 4.159 J/cal, very close to the result of modern measurements. As a result of the equivalence of mechanical work and heat, the increment of the energy of a system is the sum of mechanical work and heat,

$$\Delta E = W + Q. \tag{6.9}$$

Energy could transfer as heat as well. By pushing two systems with heat capacity C_1 and C_2 at temperatures T_1 and T_2 into contact, as shown in Fig. 6.5, heat transfers from the hotter system to the cooler system. Assuming that the thermal capacities of the two systems are constant, that is, independent of temperature within the temperature range of interest, eventually, the temperature becomes a single value T_0,

$$T_0 = \frac{1}{C_1 + C_2}(C_1T_1 + C_2T_2). \tag{6.10}$$

The heat transferred from one system to another is

$$Q = \pm\frac{C_1C_2}{C_1 + C_2}(T_1 + T_2). \tag{6.11}$$

In Joule's experiment, mechanical work transforms into heat. However, there no simple way to transform heat back to mechanical work. In the case of heat transfer, heat can spontaneously transfer from a system at a higher temperature to a system at a lower temperature. But heat can never spontaneously transfer from a system at a lower temperature to a system at a higher temperature without expending mechanical work. Such observations lead to the second law of thermodynamics.

6.3 Second Law of Thermodynamics

There are many ways to state the second law of thermodynamics. It can be shown that all those incarnations are equivalent. A succinct formulation, similar to that of Kelvin and Planck, is as follows:

> *It is impossible to build a machine that converts heat to mechanical work from a single source of heat.*

Because the heat in the ocean is unlimited, if *perpetual motion of the second type* could be built, mankind would never have to worry having energy. Another formulation of the second law of thermodynamics, due to Clausius, is as follows:

> *It is impossible to transfer heat from a reservoir at a lower temperature to a reservoir at a higher temperature without spending mechanical work.*

In fact, if a machine to transfer heat from a cold reservoir to a hot reservoir without expending external mechanical energy could be built, everybody on Earth would be able to enjoy free heating and free air conditioning.

6.3.1 Carnot Cycle

The spirit of the second law of thermodynamics can be best understood using the Carnot cycle, proposed by Sadi Carnot in 1824 in the quest for the ultimate efficiency of heat engines [1]. A schematic is shown in Fig. 6.2. The engine consists of two heat reservoirs and a cylinder with a piston filled with a volume of working gas as the thermodynamic system.

The Carnot cycle is an idealization of a heat engine that generates mechanical work by transferring heat from a hot reservoir at temperature T_H to a cold reservoir at temperature T_L. A complete cycle consists of four processes. First, the system is in contact with the hot reservoir, undergoing an isothermal expansion process. The system, always at temperature T_H, gains heat Q_H from the hot reservoir. Second, the system is isolated from the reservoir and is undergoing an adiabatic expansion process. With no heat transfer, the temperature of the system is reduced to T_L. Third, the system is in contact with the cold reservoir, undergoing an isothermal compression process. The system, always at temperature T_L, releases heat Q_L to the cold reservoir. Fourth, the system is isolated from the reservoir and is undergoing an adiabatic compression process. With no heat transfer, the temperature of the system is raised to T_H. A net work W is performed to the surroundings. The first law of thermodynamics requires that

$$Q_H = Q_L + W. \tag{6.12}$$

The efficiency η of a heat engine is defined as the ratio of mechanical work W over the heat energy from the hot reservoir, Q_H:

$$\eta \equiv \frac{W}{Q_H} = 1 - \frac{Q_L}{Q_H}. \tag{6.13}$$

An essential assumption of the Carnot cycle is that the processes are *reversible*. The Carnot cycle in Fig. 6.2 can be operated as a refrigerator or a heat pump; see Fig. 6.3. The thermodynamic system — a body of gas confined in a cylinder with a piston — transfers heat from a cold reservoir to a hot reservoir with a cost of mechanical work. The four processes are as follows: First, the system is in contact with the cold reservoir, undergoing an *isothermal expansion* process. The system, always at temperature T_L, gains heat Q_L from the cold reservoir. Second, the system is isolated from the reservoir and is undergoing an *adiabatic compression* process. With no heat transfer, the temperature of the system is raised to T_H. Third, the system is in contact with the hot reservoir, undergoing an *isothermal compression* process. The system, always at temperature T_H, releases heat Q_H to the hot reservoir. Fourth, the system is isolated from the reservoir and is undergoing an *adiabatic expansion* process. With no heat transfer, the temperature of the system is reduced to T_L.

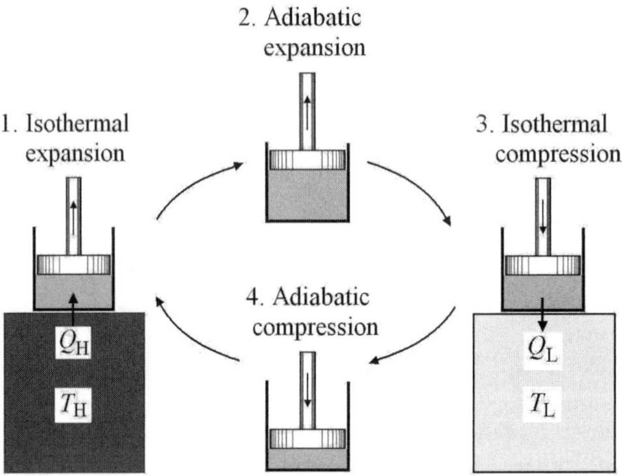

Figure 6.2 Carnot cycle. The thermodynamic system is a quantity of gas confined in a cylinder with a piston. It consists of four reversible processes: (1) an isothermal expansion process at temperature T_H, acquiring a heat Q_H from the hot reservoir; (2) an adiabatic expansion process to reduce the temperature to T_L; (3) an isothermal compression process at temperature T_H, releasing a heat Q_H to the hot reservoir; (4) an adiabatic compression process to raise the temperature to T_H. A net mechanical work W is done to the surroundings.

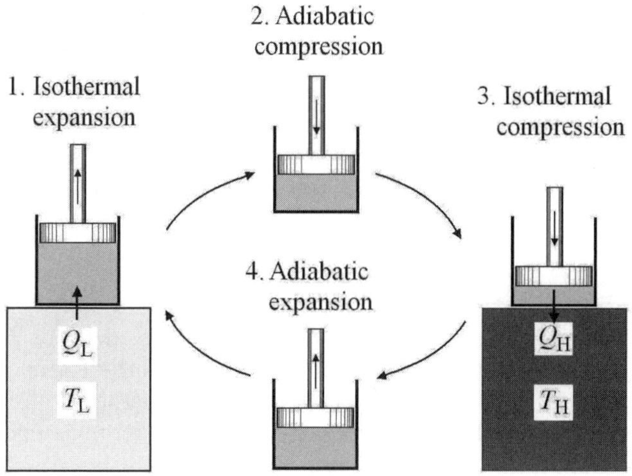

2. Adiabatic compression
1. Isothermal expansion
3. Isothermal compression
4. Adiabatic expansion
Q_L
T_L
Q_H
T_H

Figure 6.3 Reverse Carnot cycle. An idealized representation of a refrigerator or a heat pump, it consists of four processes: (1) an isothermal expansion process at temperature T_L, acquiring a heat Q_L from the cold reservoir; (2) an adiabatic compression process to raise the temperature to T_H;. (3) an isothermal compression process at temperature T_H, releasing a heat Q_H to the hot reservoir; (4) an adiabatic expansion process to reduce the temperature to T_L. A net mechanical work W is required to make the transfer.

Because the Carnot cycle is reversible, the same Carnot machine can function as a heat engine or a heat pump (or equivalently, refrigerator). That fact has a far-reaching consequence: The efficiency of all Carnot cycles depends only on the temperatures of the two heat reservoirs,

$$\eta = 1 - \frac{Q_L}{Q_H} = f(T_H, T_L). \tag{6.14}$$

The ingenious proof, given by Sadi Carnot in 1824, is based on a logical argument as follows: If two Carnot machines have different efficiencies, one can use the Carnot machine with a higher efficiency as the heat engine and the one with a lower efficiency as the heat pump. The combined machine would contradict the second law of thermodynamics. There are two alternative proofs with regard to the two formulations of the second law of thermodynamics.

The first proof is as follows: With the same amount of heat from the hot reservoir, that heat engine can generate more work than is required to pump heat from the cold reservoir to recover the input heat back to the hot reservoir. Therefore, the combined machine can convert heat into mechanical work from a single heat reservoir, which is a *perpetual motion of the second type*. It contradicts the Kelvin–Planck formulation of the second law of thermodynamics.

An alternative proof is as follows: By using all the mechanical work generated with the heat Q_H from the hot reservoir using the more efficient machine, a heat greater than Q_H can be generated to put back into the hot reservoir. The combined machine is capable of transferring heat from a cold reservoir to a hot reservoir without requiring any external mechanical work. It contradicts the Clausius formulation of the second law of thermodynamics. The conclusions of the analysis based on the Carnot cycle is simple but far-reaching: The efficiency of a reversible Carnot cycle, regardless of the nature of the process and the nature of the substance, is uniquely determined by the two temperatures — the temperature of the hot reservoir and the temperature of the cold reservoir. The universality of the Carnot cycle motivated William Thompson (Lord Kelvin) to define the scale of thermodynamic temperature.

6.3.2 Thermodynamic Temperature

In Section 2.1, the condition of equality of temperature was defined. But the *scale* of temperature is yet to be defined. Based on the theory of the Carnot cycle, William Thomson (Lord Kelvin) defined the thermodynamic temperature, also known as the absolute temperature or the Kelvin temperature scale [2], abbreviated K.

For reversible Carnot cycles, the efficiency is defined as

$$\eta \equiv \frac{W}{Q_{\mathrm{H}}} = 1 - \frac{Q_{\mathrm{L}}}{Q_{\mathrm{H}}}. \tag{6.15}$$

Now look at the ratio of Q_{L} and Q_{H}. If the temperature of the hot reservoir equals the temperature of the cold reservoir, then no mechanical work can be generated. Therefore, $Q_{\mathrm{L}} = Q_{\mathrm{H}}$. In other words,

$$\text{If} \quad \frac{Q_{\mathrm{L}}}{Q_{\mathrm{H}}} = 1, \quad \text{then} \quad \frac{T_{\mathrm{L}}}{T_{\mathrm{H}}} = 1. \tag{6.16}$$

Obviously, the temperature scale should also satisfy the following:

$$\text{If} \quad \frac{Q_{\mathrm{L}}}{Q_{\mathrm{H}}} < 1, \quad \text{then} \quad \frac{T_{\mathrm{L}}}{T_{\mathrm{H}}} < 1. \tag{6.17}$$

The simplest temperature scale satisfying these criteria, proposed by Lord Kelvin in 1848, is [2]

$$\frac{T_{\mathrm{L}}}{T_{\mathrm{H}}} = \frac{Q_{\mathrm{L}}}{Q_{\mathrm{H}}}. \tag{6.18}$$

Lord Kelvin showed that this absolute temperature scale is equivalent to the temperature scale based on ideal gas properties [2]. The Kelvin temperature is also identical to the temperature in statistical physics created by Maxwell and Boltzmann; see Appendix D.

In terms of the Kelvin temperature scale, the efficiency of a reversible Carnot cycle (Eq. 6.15), is

$$\eta_{\mathrm{c}} = 1 - \frac{T_{\mathrm{L}}}{T_{\mathrm{H}}}, \tag{6.19}$$

which is the maximum efficiency any heat engine can achieve, often referred to as the Carnot efficiency. As shown, the lower the temperature T_{L}, the greater the efficiency. The efficiency can be artificially close to 1 if T_{L} is sufficiently low. However, the efficiency of any heat engine cannot equal to 1; otherwise the second law of thermodynamics is violated. Although 0 K could never be reached, we can formally define absolute zero temperature by

$$\text{If} \quad Q_L = 0, \quad \text{or} \quad W = Q_H, \quad \text{then} \quad T_L = 0. \tag{6.20}$$

The Celsius temperature scale is defined as a shifted Kelvin scale, with the triple point of water (the state in which the solid, liquid, and vapor phases exist together in equilibrium) defined as 0.01°C. On the Celsius scale, the boiling point of water is found experimentally to be 100.00°C. This definition also fixes the constant factor in the Kelvin scale. The relation between the Kelvin scale and the Celsius scale is

$$\text{K} = {}^\circ\text{C} + 273.15. \tag{6.21}$$

6.3.3 Entropy

Equation 6.18 can be rewritten as

$$\frac{Q_H}{T_H} = \frac{Q_L}{T_L}, \tag{6.22}$$

which has a significant consequence. Actually, Eq. 2.22 can be generalized to any number of steps in a cycle, where in each step an infinitesimal heat δQ is transferred. For a reversible cycle, the cyclic integral is zero,

$$\oint \frac{\delta Q}{T} = 0. \tag{6.23}$$

Therefore, a single-valued *function of the state* can be defined,

$$S_2 - S_1 \equiv \int_1^2 \frac{\delta Q}{T}. \tag{6.24}$$

This function, which plays a central role is thermodynamics, is called *entropy*. However, heat Q is not a function of the state, because the cyclic integral of δQ is, in general, not zero. Each state of the system can have multiple values of the same quantity. However, an infinitesimal amount of heat can be expressed in terms of the state function entropy as

$$\delta Q = T\, dS. \tag{6.25}$$

6.4 Thermodynamic Functions

As discussed in Section 2.1, for the fixed state of a system, the state functions are fixed. Examples of state functions include volume V , mass m, pressure P, temperature T, etc. Heat and the mechanical work, on the other hand, are not state functions. We denote the differentials of heat and mechanical work as δQ and δW.

According to the first law of thermodynamics, the total energy of a system is also a function of the state. In fact, a mathematical representation of the first law of thermodynamics is that the cyclic integral of the sum of mechanical work and heat is zero,

$$\oint (\delta Q + \delta W) = 0. \tag{6.26}$$

The difference of energy between state 1 and state 2 is defined as

$$U_2 - U_1 = \int_1^2 (\delta Q + \delta W). \tag{6.27}$$

Because mechanical work is related to pressure P and volume V as $\delta W = -PdV$, combining eq. 6.27 with Eq. 6.25, the differential of energy is

$$dU = T\,dS - P\,dV. \tag{6.28}$$

6.4.1 Free Energy

The total energy of a system U can be defined as the capability of doing work to the surroundings or the capability of transferring heat to the surroundings. In the analysis of actual problems, sometimes it is necessary to emphasize the portion of energy related to the capability of doing mechanical work. The definition of *free energy* satisfies this purpose:

$$F = U - TS. \tag{6.29}$$

Intuitively speaking, the definition of free energy seems to eliminate the heat component of the total energy and retain the mechanical part. Similar to Eq. 6.25, the differential of free energy is

$$dF = -S\,dT - P\,dV. \tag{6.30}$$

Therefore, more precisely speaking, free energy is a measure of the capability of a state of the system to do mechanical work at a constant temperature.

6.4.2 Enthalpy

Sometimes it is useful to know the portion of the energy of a system related to the capability of delivering heat. The definition of *enthalpy* satisfies that purpose,

$$H = U + PV. \tag{6.31}$$

Intuitively speaking, the definition of enthalpy seems to eliminate the mechanical work component of the total energy and retain the heat content. Similar to Eq. 6.25, the differential of enthalpy is

$$dH = T\,dS + V\,dP. \tag{6.32}$$

More precisely speaking, enthalpy is a measure of the capability of a state of the system to deliver heat at a constant pressure.

When a liquid evaporates under constant pressure, the volume of the gas increases. The heat required for evaporation is the sum of the difference of internal energy U and the work needed to expand the volume, PV. The heat transferred during the phase change is the sum of U and PV, that is, the *enthalpy*,

$$\Delta H = \Delta (U + PV) . \tag{6.33}$$

6.4.3 Gibbs Free Energy

The fourth state function, introduced by Willard Gibbs, is

$$G = U + PV - TS. \tag{6.34}$$

Similarly, the differential of the Gibbs free energy is

$$dG = -S\, dT + V\, dP. \tag{6.35}$$

From Eq. 6.35, it is clear that if the temperature and pressure of a system are kept constant and the quantity of the substance is constant, then the Gibbs free energy is a constant. This property is important in the analysis of chemical reactions, because most chemical reactions occurs under conditions of constant temperature and constant pressure.

First, consider a single-component system in which the quantity of the substance is a variable which by convention is expressed as the number of moles of the substance, N. When the temperature and the pressure of the system are kept constant, the Gibbs free energy can be expressed as

$$dG = -S\, dT + V\, dP + \mu\, dN. \tag{6.36}$$

The quantity μ is the Gibbs free energy per mole of the substance.

6.4.4 Chemical Potential

For a system with more than one component, the expression of the Gibbs free energy (Eq. 6.36), can be generalized to

$$dG = -S\, dT + V\, dP + \sum_i \mu_i\, dN_i, \tag{6.37}$$

where i is the index for the ith component. If there is a chemical reaction in the system, then the composition of the system, represented by the set of molar values N_i, will change. Under constant temperature and constant pressure, the equilibrium condition of the system is

$$dG = \sum_i \mu_i\, dN_i = 0. \tag{6.38}$$

The quantity μ_i is known as the *chemical potential* of the ith component of the system.

6.5 Ideal Gas

Experimentally, it is found that within a large range of temperature and pressure many commonly encountered gases satisfy a universal relation

$$PV = NRT, \tag{6.39}$$

where P is the pressure in pascals, V is the volume in cubic meters, N is the number of moles of the gas, T is absolute temperature in kelvins; and R is a universal gas constant,

$$R = 8.3144 \, \frac{\mathrm{J}}{\mathrm{mol} \cdot \mathrm{K}}. \tag{6.40}$$

In solar energy storage systems, for the gases commonly used, such as nitrogen, oxygen, argon, and methane, at temperatures of interest the ideal gas relation is accurate up to a pressure of 10 MPa, or 100 standard atmospheres pressure.

In the following, we will derive all the thermodynamic functions for the ideal gases. First, we will show that the energy U of an ideal gas depends only on temperature but not on volume. In fact, assuming that $U = U(T, V)$, using Eq. 6.25, the internal energy varies with volume as

$$\left(\frac{\partial U}{\partial V}\right)_T = T\left(\frac{\partial S}{\partial V}\right)_T - P. \tag{6.41}$$

Using the equation of state 6.39, Eq. 6.41 becomes

$$\left(\frac{\partial U}{\partial V}\right)_T = T\left(\frac{\partial P}{\partial T}\right)_V - P = T\frac{P}{T} - P = 0. \tag{6.42}$$

Defining the constant-volume specific heat per mole as

$$C_v \equiv \frac{1}{N}\left(\frac{\partial U}{\partial V}\right)_T, \tag{6.43}$$

the internal energy as a function of temperature is

$$U_2 - U_1 = \int_1^2 NC_v dT. \tag{6.44}$$

If in the temperature interval of interest C_v is a constant,

$$U_2 - U_1 = NC_v(T_2 - T_1). \tag{6.45}$$

To obtain an explicit expression of entropy, we use Eqs. 6.25, 6.39, and 6.43,

$$dS = \frac{dU}{T} + \frac{P\,dV}{T} = NC_v\frac{dT}{T} + NR\frac{dV}{V}. \tag{6.46}$$

Again, assuming that the specific heat C_v is a constant,

$$S_2 - S_1 = N \int_1^2 \left[\frac{C_v\, dT}{T} + R \frac{dV}{V} \right] = N \left[C_v \log \frac{T_2}{T_1} + R \log \frac{V_2}{V_1} \right]. \tag{6.47}$$

Another important quantity is the *constant-pressure specific heat* C_p. For ideal gases, there is a simple relation between C_v and C_p. To heat N moles of gas while keeping the pressure constant, work $P\,\Delta V$ is done to the surroundings. According to the first law of thermodynamics, an additional energy $P\Delta V$ must be supplied. Because for an ideal gas $P\,\Delta V = NR\,\Delta T$, the total energy required to raise the temperature is

$$\Delta Q = NC_v\,\Delta T + NR\,\Delta T = N(C_v + R)\,\Delta T. \tag{6.48}$$

Therefore,

$$C_p = C_v + R. \tag{6.49}$$

The ratio of constant-pressure specific heat and constant-volume specific heat is an important parameter in the study of the adiabatic process, often denoted as γ:

$$\gamma \equiv \frac{C_p}{C_v} = 1 + \frac{R}{C_v}. \tag{6.50}$$

During an adiabatic process, the heat transfer is zero. Therefore,

$$NC_v\,dT = -P\,dV. \tag{6.51}$$

On the other hand, from the equation of state for the ideal gas,

$$NR\,dT = V\,dP + P\,dV. \tag{6.52}$$

Combining Eqs 6.51 and 6.52, we obtain

$$\left(1 + \frac{R}{C_v}\right) P\,dV + V\,dP = 0. \tag{6.53}$$

The solution of the above differential equation is

$$PV^\gamma = \text{const}, \tag{6.54}$$

which is the state equation of an adiabatic process.

In the following, we show that the efficiency of a reversible Carnot cycle of ideal gas is $\eta = 1 - (T_\text{L}/T_\text{H})$, and thus the thermodynamic temperature scale is identical to the temperature scale based on the ideal gas; see Fig. 6.4. In the first process, an isothermal expansion process at temperature T_H from state 1 to state 2, the gas medium receives heat from the hot reservoir,

$$Q_\text{H} = \int_1^2 PdV = NRT_\text{H} \int_1^2 \frac{dV}{V} NRT_\text{H} \log \frac{V_2}{V_1}, \tag{6.55}$$

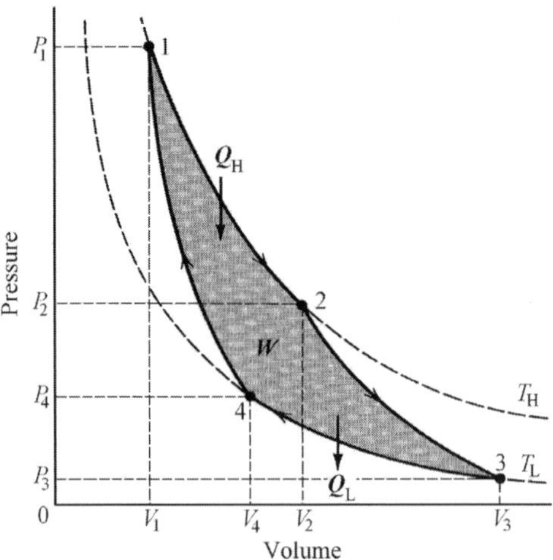

Figure 6.4 Carnot cycle with ideal gas as the system. Curve 1–2 is an isothermal expansion process at T_{H}. Curve 2–3 is an adiabatic expansion process. Curve 3–4 is an isothermal contraction process at T_{L}. Curve 4–1 is an adiabatic contraction process. The gas does work W to the surroundings.

where V_1 and V_2 are the volume of states 1 and 2, respectively. In the third process, an adiabatic contraction process from state 3 to state 4, the gas medium releases heat to the cold reservoir,

$$Q_{\mathrm{L}} = \int_3^4 P\,dV = NRT_{\mathrm{L}} \int_3^4 \frac{dV}{V} NRT_{\mathrm{L}} \log \frac{V_3}{V_4}, \tag{6.56}$$

where V_3 and V_4 are the volumes of state 3 and 4, respectively.

Because processes 1–2 and 3–4 are isothermal, according to the equation of state,

$$\begin{aligned} P_1V_1 &= P_2V_2, \\ P_3V_3 &= P_4V_4. \end{aligned} \tag{6.57}$$

On the other hand, processes 2–3 and 4–1 are adiabatic. According to Eq. 6.54,

$$\begin{aligned} P_2V_2^{\gamma} &= P_3V_3^{\gamma}, \\ P_4V_4^{\gamma} &= P_1V_1^{\gamma}. \end{aligned} \tag{6.58}$$

Combining Eqs 6.57 and 6.58, we have

$$\frac{V_1}{V_2} = \frac{V_3}{V_4}. \tag{6.59}$$

Applying this relation to Eqs. 6.55 and 6.56, we conclude that the thermodynamic temperature scale and the ideal gas temperature scale are identical,

$$\frac{T_{\mathrm{L}}}{T_{\mathrm{H}}} = \frac{Q_{\mathrm{L}}}{Q_{\mathrm{H}}}. \tag{6.60}$$

6.6 Ground Source Heat Pump and Air Conditioning

In Section 5.3, we discussed solar energy stored in the ground: At a depth of about 10 meters, the temperature of Earth is cooler than the average in the summer and warmer than the average in the winter. Tapping into that stored solar energy could greatly reduce the use of energy for space heating and cooling as well as water heating. Although direct use of the stored energy for space cooling in the summer has been practiced for centuries in some places on the world, it allows very little control. The modern method of tapping into that stored solar energy is to use vapor compression heat pump and refrigeration systems [5, 53, 75]. This enables full control to the desired temperature. The energy saving can be as great as 50%.

Table 6.1 shows the accumulated wattage and number of ground source heat-pump installations in several countries up to 2004. Although the United States has the largest installation wattage and number of systems, Sweden has the largest percentage of heat pump installations. This is simply due to economics. In Sweden, the weather is cold, and there are no fossil fuel resources. Nevertheless, most of the electricity in Sweden is from hydropower, which is relatively inexpensive. The combination of cheap electricity and the high efficiency of geothermal systems makes perfect economic sense.

6.6.1 Theory

A schematic of a ground source heat pump is shown in Fig. 6.5. Several meters underground, the temperature is basically constant over the entire year, for example, at $T_0 = 15°C$. In winter, the ambient temperature can be as cold as 0°C. To maintain a comfortable temperature, heat should be supplied into the house. In summer, the ambient temperature can be as hot as 30°C. Heat should be extracted from the house. In Section 6.3.1, we showed that, by supplying mechanical work, heat can be transferred from a cold reservoir to a hot reservoir. Therefore, using a type of machine called *heat*

Table 6.1: Ground Source Heat Pumps in Selected Countries.

Country	Capacity (MWt)	Energy Use (GWh/year)	Number of Units Installed
Austria	275	370	23,000
Canada	435	600	36,000
Germany	640	930	46,400
Sweden	2,300	9,200	230,000
Switzerland	525	780	30,000
Uunited States	6,300	6,300	600,000

Source: Geothermal (Ground-Source) Heat Pumps: A World Overview, GHC Bulletin, September 2004 [53].

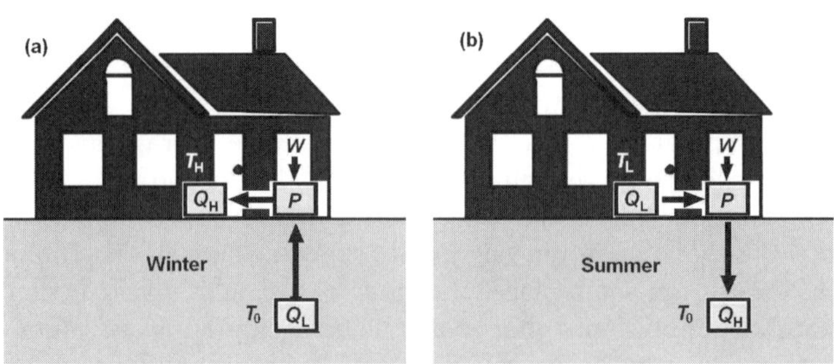

Figure 6.5 Ground source heat pump. The temperature in the ground about 10-m deep, T_0, is basically constant over the year. (a) In the winter, by applying a mechanical work W, a heat pump P transfers heat Q_L from the ground heat reservoir at temperature T_0 to a radiator in the house at temperature T_H. The total heat discharged into the radiator is the sum of the heat extracted from the ground and the work, $Q_\mathrm{H} = Q_\mathrm{L} + W$. (b) In the summer, the heat pump P reverses its function: it transfers heat from the rooms at temperature T_L into the ground heat reservoir at temperature T_0.

pump, it is possible to extract heat from within the ground at 15°C to a floor heating system in the house at $T_\mathrm{H} = 50$°C. In the ideal case, if the machine is reversible, according to Eq. 6.19, the efficiency is

$$\eta_\mathrm{c} = 1 - \frac{T_0}{T_\mathrm{H}} = 1 - \frac{273+15}{273+50} \approx 0.108. \tag{6.61}$$

According to the definition of efficiency (Eq. 6.15), the mechanical work required to transfer heat from the ground at T_0 to the radiator at T_H to become Q_H is

$$W = \eta Q_\mathrm{H} = 0.108 \times Q_\mathrm{H}. \tag{6.62}$$

In other words, mechanical energy equal to a fraction of the heat is needed to transfer heat from the ground to the radiator.

In the summer, assuming that heat Q is extracted from an air-conditioning system at $T_\mathrm{L} = -10$°C into the ground at $T_0 = 15$°C using a reversible Carnot engine. The machine is in fact a refrigerator. The Carnot efficiency is

$$\eta_\mathrm{c} = 1 - \frac{T_\mathrm{L}}{T_0} = 1 - \frac{273-10}{273+15} \approx 0.087. \tag{6.63}$$

Because the heat extracted from the air conditioning system is Q_L in Eq. 6.15, the mechanical work required is

$$W = \frac{\eta}{1-\eta} Q_\mathrm{L} = 0.095 \times Q_\mathrm{L}. \tag{6.64}$$

Theoretically, if the refrigerator as a Carnot engine is reversable, the work needed to cool the space is less than one tenth of the amount of heat taken. Practically, no machine is perfect. The actual work needed is more than that from the theoretical limit.

Gallery

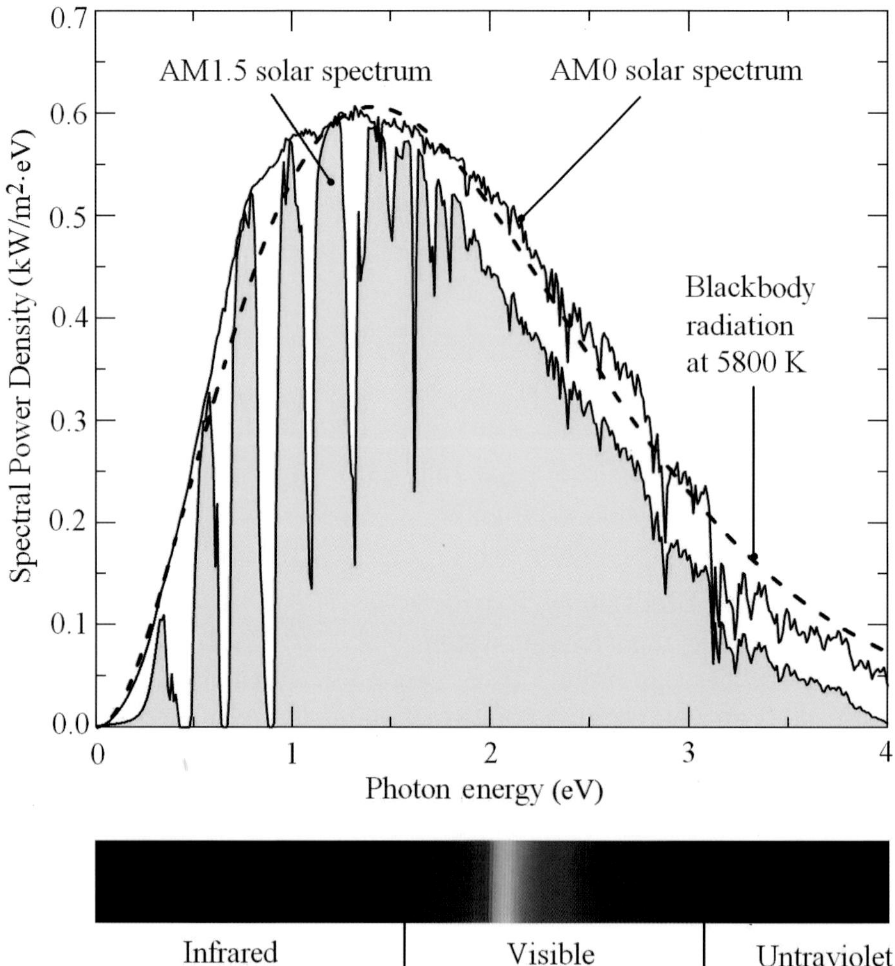

Plate 1. Spectral power density of solar radiation. Top curve: Solar spectrum just outside the atmosphere. The total power density is 1.366 kW/m^2. Curve filled with yellow: the standardized solar spectrum on the surface of Earth, for performance evaluation of solar cells. AM1.5 means air mass at an angle about 37° from the horizontal line on a clear day. The total power density is 1.0 kW/m^2. See Chapter 5 for details. The dashed curve is the solar radiation spectrum at the position of Earth by modeling the Sun as a blackbody radiator at 5800 K. As shown, the solar spectrum just outside the atmosphere, the AM0 spectrum, matches well with the blackbody radiation spectrum at 5800 K, diluted by the distance from the Sun to Earth. The relation with human vision of color spectrum is shown in the bar beneath. As shown, about one half of the solar radiation power is in the visible range. See Chapters 2, 3, and 5. (Source of solar spectrum data: American Society for Testing and Materials.)

Plate 2. Apparent motion of the Sun. By fixing a camera toward the southern sky (in the Northern Hemisphere), taking one picture at the same time of the day for a year, then superimposing the pictures of the sunny days, an "8"-like pattern is recovered, which is called the *analemma*. It is a combined result of equation of time and the variation of declination, see Section 4.4. Plate 2 is compiled by Greek astronomer Anthony Ayiomamitis from 47 clear-day photos taken in 2003 near the Temple of Apollo, Corinth, Greece. Apollo is the God of the Sun in Greek legends. Apollo is also the God of light, truth, music, poetry, and the arts. Courtesy of Anthony Ayiomamitis.

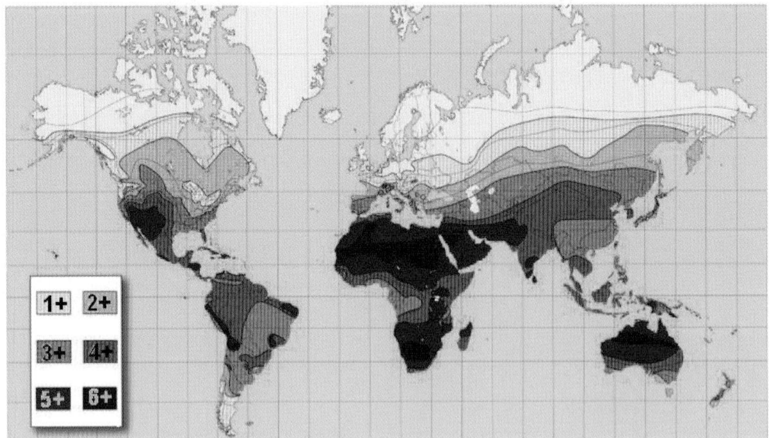

Plate 3. Insolation map of the world. Solar radiation per day on a surface of 1 m^2 in kilojoules averaged over a year. As shown, large areas in Northern Africa have the highest insolation. *Source*: www.bpsolar.com.

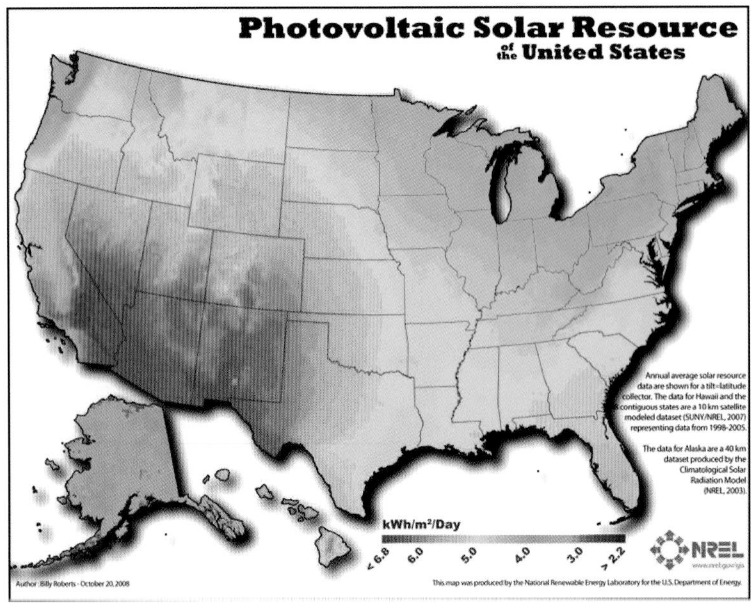

Plate 4. Insolation map of the United States. Solar radiation per day on a surface of 1 m^2 in kilojoules averaged over a year. As shown, large areas in southwest states have the highest insolation. Areas around the Great Lakes and the State of Washington have weak insolation due to cloudiness. *Source*: NREL.

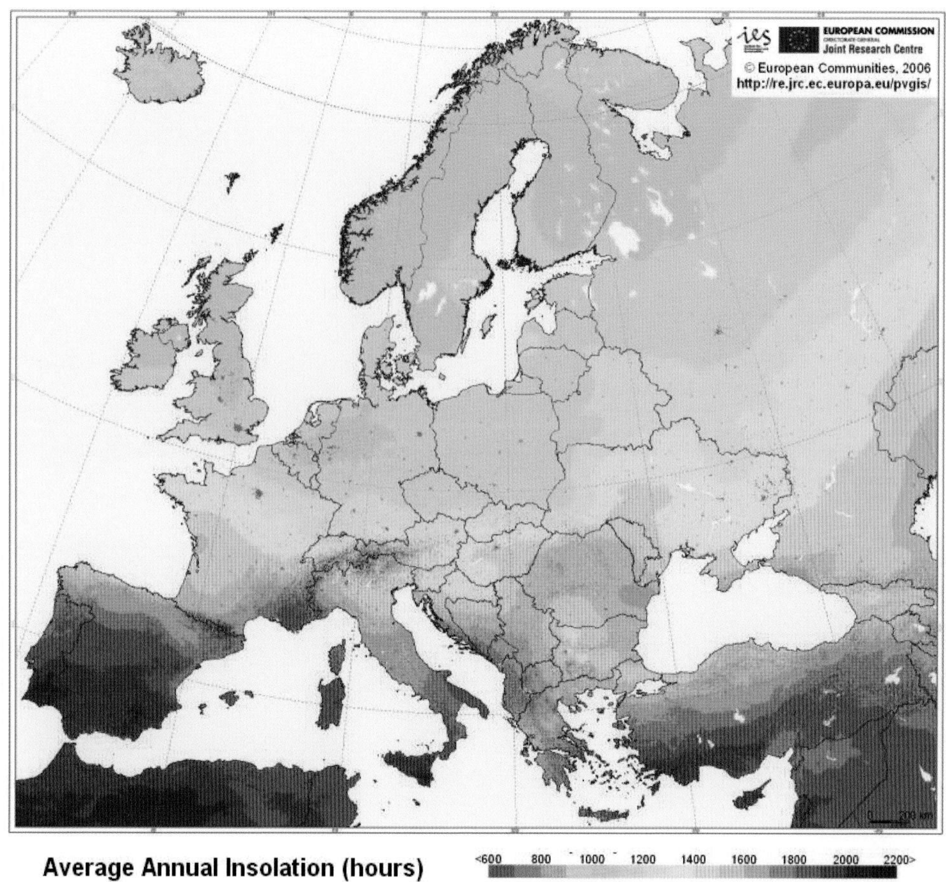

Plate 5. Insolation map of Europe. Total solar radiation on a surface of 1 m^2 over a year in kilojoules. Note that the definition of insolation in this graph is different from the previous ones. Clearly, the total insolation in a calendar year equals 365.25 times the average daily insolation. As shown, the southern parts of Portugal, Spain, and Italy have the highest insolation in Europe. The insolation in Germany is much weaker than those areas. The northern parts of Norway and Finland have rather weak insolation. *Source*: European Commission Joint Research Centre, 2006.

Plate 6. Crystalline silicon solar cells with single-axis tracking. The Nellis Solar Power Plant, located within Nellis Air Force Base in Clark County, Nevada, with a peak power generation capacity of approximately 14 MW. A simple uniaxial sunlight tracking system can increase the power output by 60% from the optimum fixed-orientation system, see Section 4.3.4. Photo courtesy of U.S. Airforce.

Plate 7. Utility-scale CIGS thin-film solar field. Thin film solar cells use much thinner semiconductor materials than crystalline solar cells, thus the cost per watt could be much lower. See Section 9.6. Shown here is a 750 KW CIGS solar panel system installed in Tucson, AZ, manufactured by Nanosolar. Photo courtesy of Nanostar Inc.

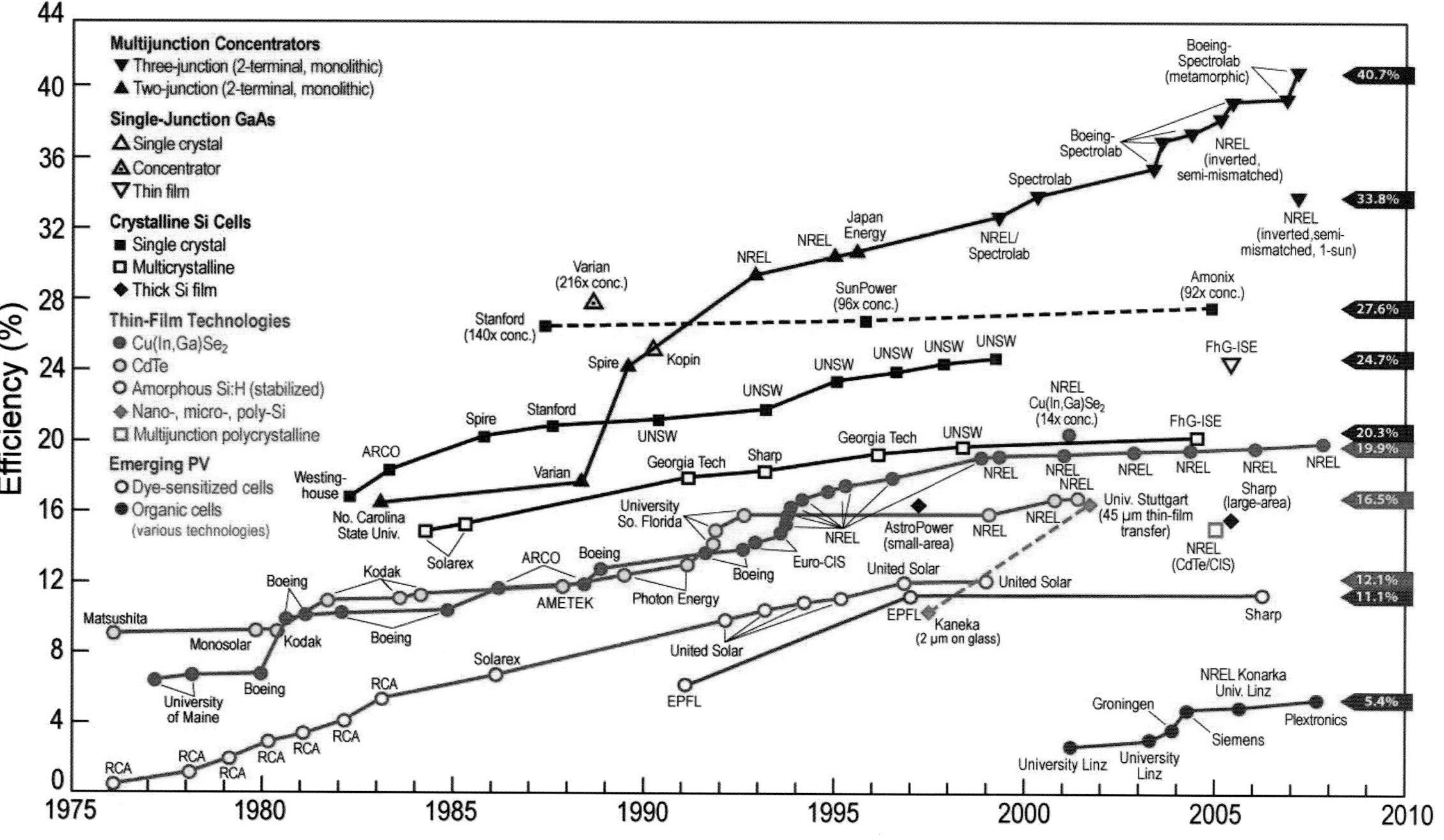

Plate 8. Efficiency of solar cells. *Source*: U.S. National Renewable Energy Laboratory, November 7, 2007.

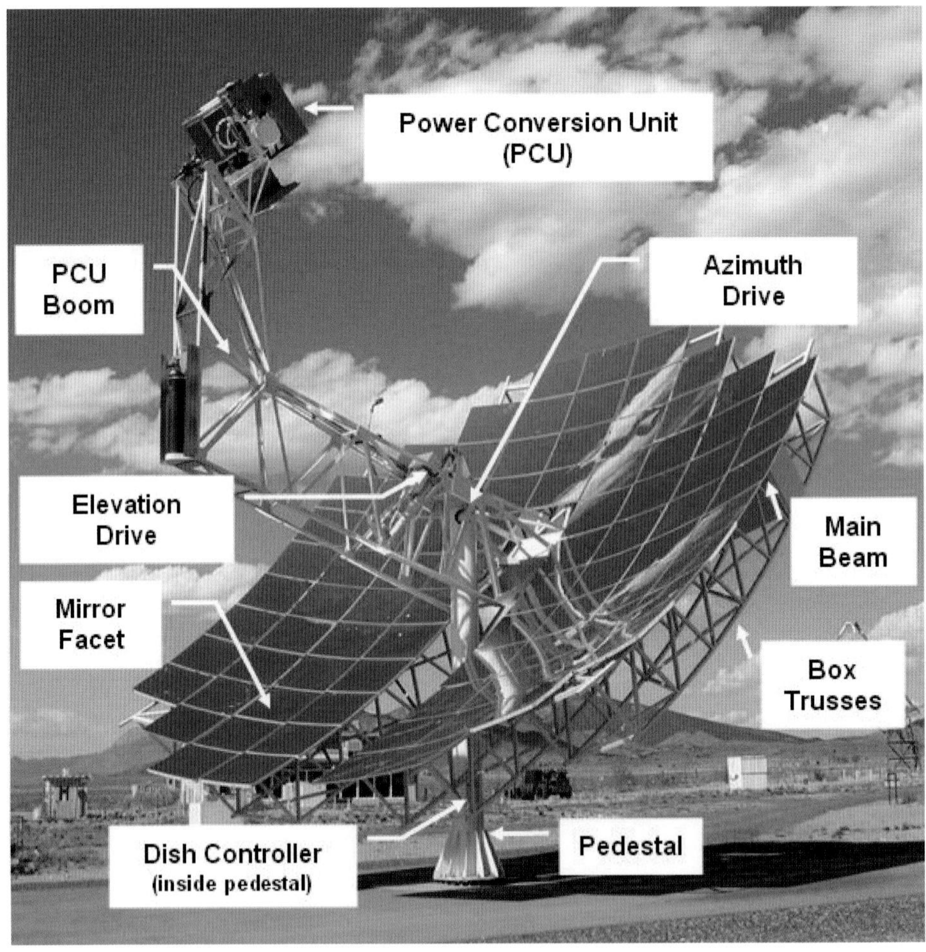

Plate 9. Solar dish Stirling energy system. The Stirling solar energy system developed by Sandia National Laboratories and Stirling Energy Systems (SES) in February 2008 has set a record of the highest solar-to-grid conversion efficiency of 31.25%. The solar dish generates electricity by focusing the radiation from the Sun to a receiver, which transmits the heat energy to a Stirling engine filled with hydrogen. The Stirling engine works best in a cold but sunny weather. See Section 11.4. Photo courtesy of Sandia National Laboratories.

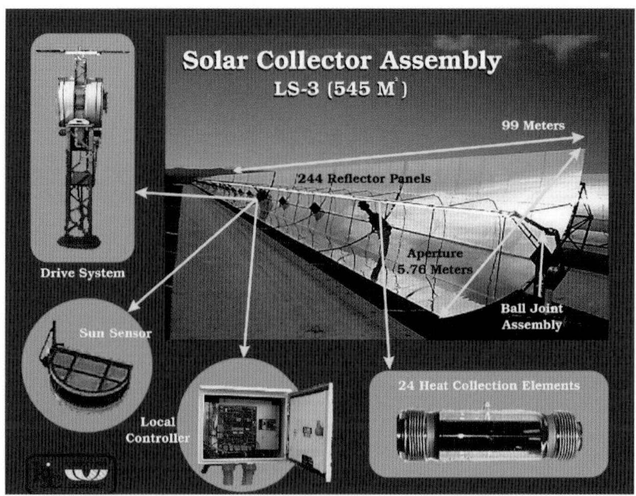

Plate 10. Solar collector assembly of a trough system. The axis of the parabolic troughs are aligned South-North. A tracking mechanism turns the troughs every day from East to West. The solar radiation is focused on the vacuum-tube heat collectors. The collectors heat synthetic oil to 400°C, then generates superheated steam to drive steam turbines. The highest efficiency on record was 20%. See Section 11.4.

Plate 11. Parabolic trough concentrators at Kramer Junction, CA. The system, called the Solar Electricity Generating System (SEGS) installed at Kramer Junction in Mojave desert, California, starting from 1984. The SEGS system is the largest solar electricity generator set on the world. The total capacity is 354 MW. See Section 11.4.

Plate 12. Various types of vacuum tube solar heat collectors. Currently, vacuum tube solar heat collectors are the most used solar heat collectors. They use selective coating to achieve maximum absorption of solar radiation, with minimum loss of heat by radiation. The vacuum between the outer tube and the heated element provides almost perfect thermal insulation. Depending on the ways of transferring heat, there are several types: The direct-flow vacuum tube heat collector allows water to flow by convection into the tubes. The heat siphon collector uses an ingenious phase-change mechanism to transfer heat with very high efficiency and prevents heat from flowing back to the tube (the thermal diode effect). See Section 11.2. Photo taken by the author. Courtesy of Beijing Solar Energy Research Institute.

Plate 13. Fully automatic facility for manufacturing of evacuated tubes. Top: Automatic evacuating machine. Bottom: Automatic conveying line. This facility produces 20 million evacuated tubes each year. Courtesy of Himin Solar Energy Group, Dezhou, China.

Plate 14. Day-and-Night solar water heaters manufactured and installed in 1920s in Maimi. Solar water heaters (see Sections 1.5.1, 11.2, and 11.3) can be very durable. In the 1920s, about 40,000 Day-and-Night solar water heaters were installed in Florida. After 80 years of operation in violent weather conditions, thousands of them are still working properly today. The secret is its simplicity and complete absence of moving parts. Photos taken by the author, August 2010, Miami.

Plate 15. A solar building. As the venue of the 2010 International Solar City Congress, Himin Solar Energy Group designed and built this 75,000-m^2 (800,000-$ft.^2$) building in Dézhōu, Shāndōng province, China, with more than 60% of energy supplied by solar devices. Courtesy of Himin Solar Energy Group, Dezhou, China.

Plate 16. An experimental solar house. As Steven Chu repeatedly advocated, using better building design to take advantage of solar energy, with a small incremental investment, tremendous energy savings can be achieved. Shown here is an experimental single-family house designed by the author. It is a medium-size central-hall colonial house very common in the United States. Nevertheless, the layout, roof design, and window placement are optimized for solar energy utilization. On a sunny winter day, the sunlight through the large south-facing windows warm all the often-occupied rooms such that the thermostat for gas heating is always turned off. In the Summer, the solar panels on the roof drive two central air-conditioning systems to keep the entire house cool. A solar-powered attic fan (on top of the roof, not shown here) further reduces the cooling load. The walls are well insulated and all windows are double pane, air tight, and filled with argon. See Section 13.3. Photo taken by the author.

Plate 17: Deepwater Horizon Explosion. As land-based oil fields in the United States are being essentially depleted, petroleum exploration is moving offshore, to deeper and deeper seas. The cost of drilling is increasing dramatically, and the environment cost is becoming even graver. One example of the engineering and environmental costs is the recent Gulf of Mexico oil spill. On April 20, 2010, the Deepwater Horizon drilling rig, situated about 40 miles (60 km) southeast of the Louisiana coast in the Macondo Prospect oil field, exploded. It killed 11 workers and injured 17 others. It caused the Deepwater Horizon to burn and sink. An estimated 4.9 million barrels (780,000 m^3) of crude oil was released into the water of the Gulf of Mexico. It is now considered the largest environmental disaster in U.S. history. The oil spill severely damaged the environment of the Gulf of Mexico and significantly affected the fishing and tourism businesses of the Gulf states. It cost BP more than $32 billion for the clean-up. To ensure the safety of deep-sea petroleum exploration, the engineering cost will become very high. Photo courtesy of U.S. Coast Guard.

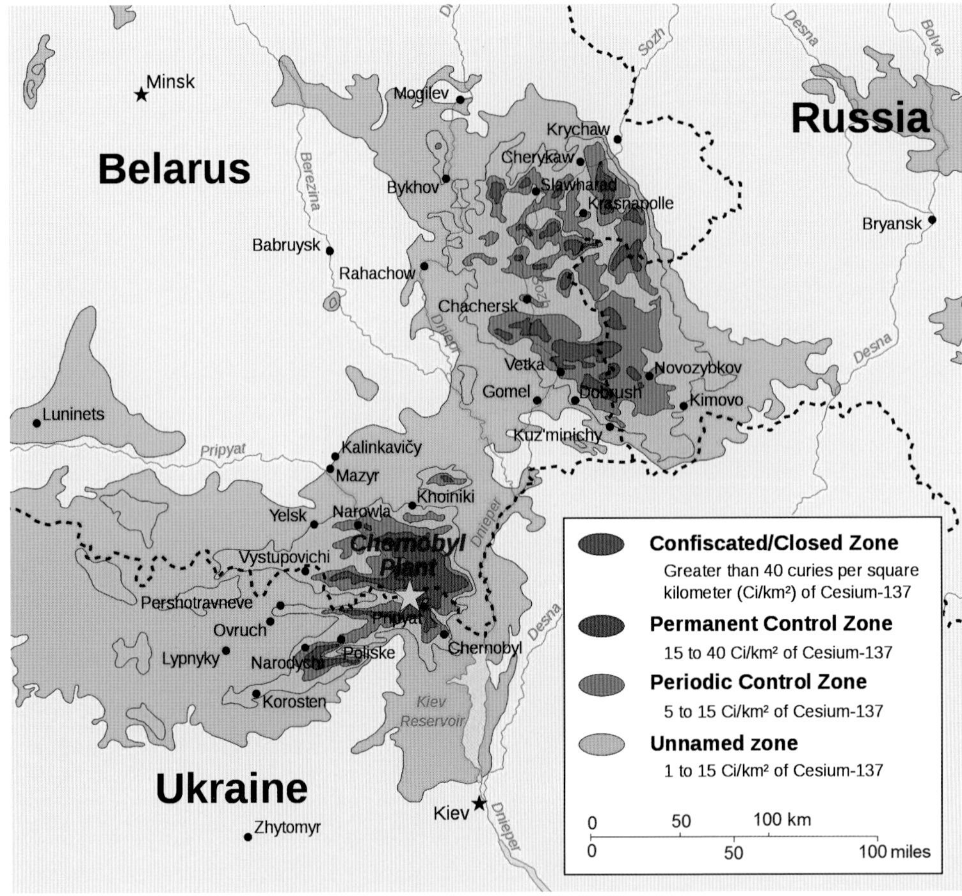

Plate 18: Chernobyl: A quarter century later. On April 26, 1986, as a combined result of design deficiencies and human error, reactor number four of the Chernobyl Nuclear Power Plant near the town of Pripyat, Ukraine, former Soviet Union, exploded. Radioactive material scattered over a wide area in Europe. In the vicinity of the Chernobyl Plant, the radiation level was so high that a 2,800-km^2 area in Belarus, Ukraine, and Russia, including three cities and more than one hundred villages, must be permanently abandoned. The accident crippled the economy of the former Soviet Union. Up to recently, the Chernobyl accident was the only one rated level 7 (a major accident) on the International Nuclear Event Scale (INES). On March 11, 2011, a tsunami damaged several nuclear reactors at the Fukushima Daiichi Nuclear Power Plant in Japan. Large quantity of radioactive materials was leaked. On April 11, 2011, the Japanese authorities raised the event level of the Fukushima accident to level 7, and predicted that the total radiation fallout will be comparable or even higher than that of Chernobyl. To prevent future nuclear accidents, a higher safety standards will drive up the cost of nuclear power. Source of the map: CIA Factbook.

6.6.2 Coefficient of Performance

In the industry, a dimensionless number called the *coefficient of performance* (COP) is used to characterize the performance of the heat pump and refrigerator. For the heat pump, it is defined as

$$\text{COP} = \frac{Q_{\text{H}}}{W}, \tag{6.65}$$

and for the refrigerator, it is

$$\text{COP} = \frac{Q_{\text{L}}}{W}. \tag{6.66}$$

The theoretical limit looks extremely attractive: The conversion ratio from mechanical work to heat can be as high as 10. Practically, a conventional vapor-compression heat pump or refrigerator can achieve a COP of 3–4. Nevertheless, this is still highly desirable. Replacing an electrical heating system with a ground-source heat pump can achieve an energy saving up to 70%.

6.6.3 Vapor-Compression Heat Pump and Refrigerator

The commercial geothermal systems use the vapor compression cycle for both heat pumping and refrigeration. Most are reversible, which can be used for room heating and cooling by using a reversing valve. Figure 6.6 shows the cooling cycle. It is very similar to an ordinary central air-conditioning system based on the principle of the vapor compression cycle. A liquid with a low boiling point serves as the working medium, called the *refrigerant.* Since the early twentieth century, ammonia (NH_3), also called R12, was the favorite refrigerant. Because of its toxicity, in air-conditioning systems, it was replaced by chlorodifluoromethane ($CHClF_2$), or R22. In spite of its flammability, propane (C_3H_8), or R290, is recently used in place of R22 because of its zero ozone depletion potential .

The central piece of a vapor compression refrigeration system is the *compressor*, (1 in Fig. 6.6). This is the point at which mechanical power is input. The compression process is approximately adiabatic. If the initial pressure is P_1 and the final pressure is P_2, according to Eq. 6.54,

$$P_1 V_1^{\gamma} = P_2 V_2^{\gamma}, \tag{6.67}$$

where γ is the ratio of specific heats (see Eq. 6.50), and the γ value of a typical refrigerant, R-22, is 1.26,

$$\gamma \equiv \frac{C_p}{C_v} = 1 + \frac{R}{C_v}. \tag{6.68}$$

Combining with the equation of state of an ideal gas, we find that the temperature is changed according to

$$\frac{T_2}{T_1} = \left(\frac{P_2}{P_1}\right)^{\frac{\gamma-1}{\gamma}} = \left(\frac{P_2}{P_1}\right)^{0.2857}. \tag{6.69}$$

If the gas is initially at room temperature (273 K), compressing it by a factor of 3, the temperature is raised to $3^{0.2857} \times 273 = 1.3687 \times 273 = 373$ K, which is about 100°C.

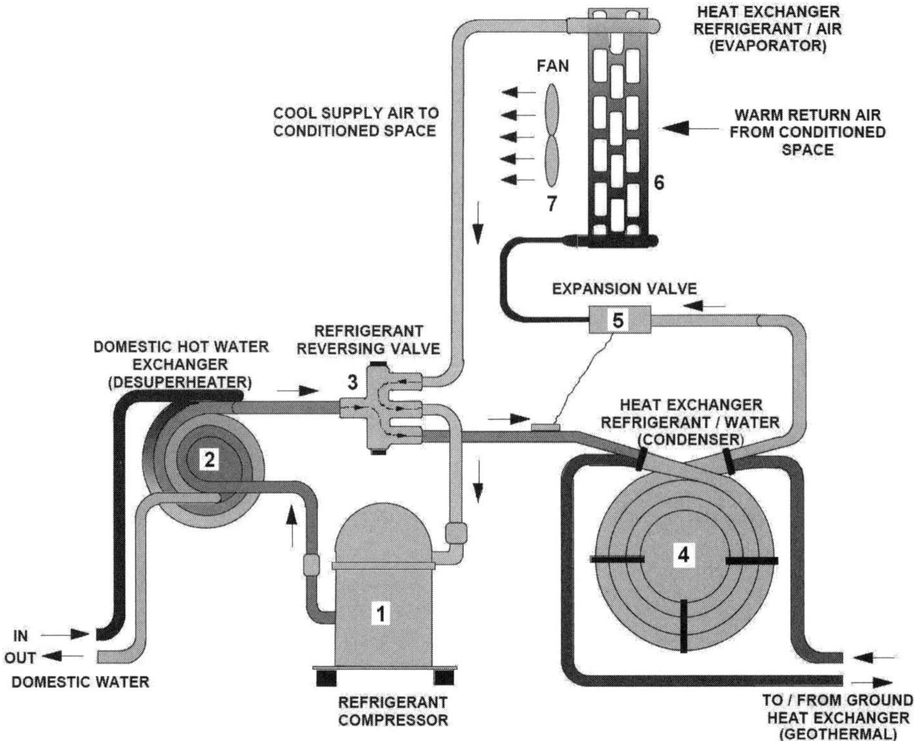

Figure 6.6 Ground-source heat pump: cooling mode. The refrigerant is compressed by a compressor (1) to become superheated. Through the heat exchanger (2), it first heats the domestic water. Then the still-hot refrigerant goes to the heat exchanger (4) to be cooled to underground temperature and condenses into liquid. Through an *expansion valve* (5) it becomes vapor. The evaporation process absorbs large quantity of heat from the space through the heat exchanger (6). The space is thus cooled. After Ref. [53].

Therefore, the refrigerant is a superheated gas. The heat thus generated is first utilized to heat the domestic water through the heat exchanger (2). Then, through reversing valve (3), the still hot refrigerant comes through the heat exchanger to contact with the water from the ground and is cooled to the underground temperature, for example, $T_0 = 15°\text{C}$, or 288 K. Because of the high pressure, the refrigerant condenses into liquid. Therefore, such a heat exchanger is called a *condenser*. The liquid refrigerant at T_0 and still under high pressure is letting through an *expansion valve* (5) to become vapor. The evaporation process absorbs large quantities of heat and the vapor temperature becomes very low, typically many degrees below 0°C. The cold vapor goes int the heat exchanger (6), a cold radiator. A fan blows warm air from the room onto the zigzag tubes and fins of the cold radiator (6). The cooled air then flows into the space to be conditioned. The warmed-up vapor of refrigerant is again sucked into the compressor (1) through the reversing valve (3). And the process goes on.

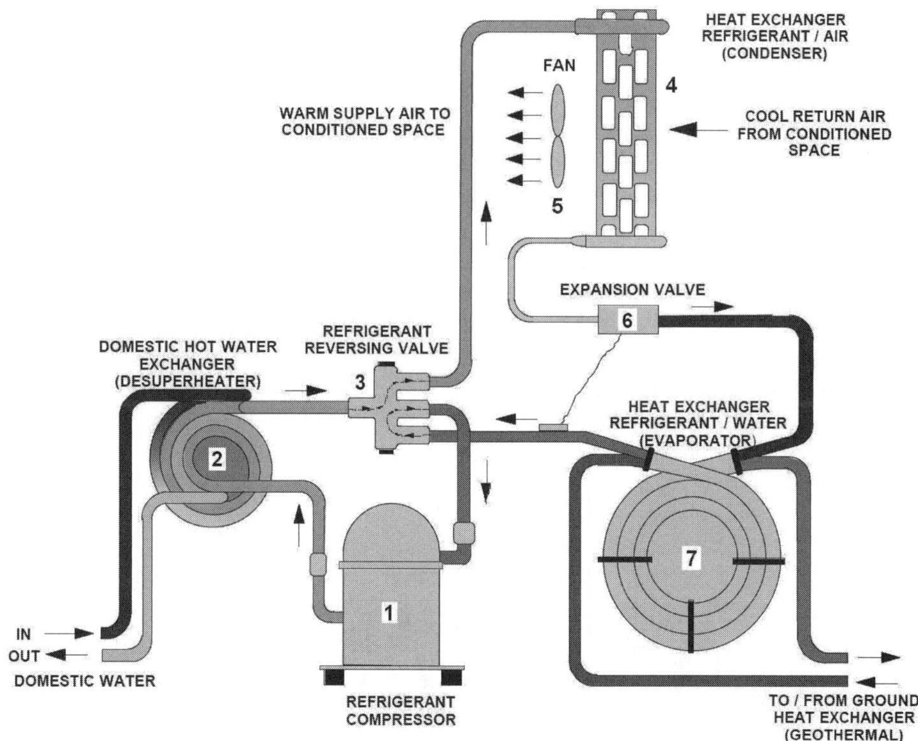

Figure 6.7 Ground source heat pump: heating mode. The refrigerant is compressed by a compressor (1) to become superheated. Through the heat exchanger (2), it first heats the domestic water. Then the still-hot pressurized gas-phase refrigerant goes to the radiator (4) to warm the space and become liquid. The liquid refrigerant goes through an expansion valve (5) to become supercooled vapor. The evaporation process absorbs large quantities of heat from the groundwater through the heat exchanger (6). After Ref. [53].

If domestic water heating is not required, the heat exchanger (2) can be eliminated. However, the fringe benefit of having this unit is obvious.

By turning the reversing valve (3) to another position, the system becomes a heat pump to be used in the winter; see Fig 6.7. Again, the compressor (1) squeezes the refrigerant into a superheated gas and transfers part of the heat to domestic water via the heat exchanger (2). Then, through the reversing valve (3), it goes into the heat exchanger (4), which is a heat radiator and refrigerant condenser. A fan (4) forces warm air into the space to be conditioned. At that time, the refrigerant is liquefied by the cool air. The liquid refrigerant goes through the expansion valve (6) and becomes supercooled vapor, often below the freezing point of water. By going through the heat exchanger (7) to contact the water at ground temperature, the refrigerant vapor picks up heat from the ground and the temperature recovers to T_0.

6.6.4 Ground Heat Exchanger

A ground heat exchanger can have many different configurations. Fig. 6.8 shows several of them. The requirement for a good ground heat exchanger is to access a large volume of soil, groundwater, or bedrock. Figs. 6.8(a), (b) show closed-loop heat exchangers. The vertical well does not require a large area, and so can be used in highly populated area. It is no surprise that most of the geothermal systems in New York City are vertical-well systems. However, drilling even a 100-m well is costly. For suburban or rural buildings, it is sufficient to bury a few hundred meters of copper coil in the ground a few meters deep. In places where groundwater is easily accessible, an open-loop system is even less expensive. By drilling two wells into the groundwater (Fig. 6.8(c)) or to access a pond (Fig. 6.8(d)), effective heat exchange could be established. In the United States, 46% are vertical closed-loop systems, 38% are horizontal closed-loop systems and 15% are open-loop systems.

Figure 6.9 shows the details of a vertical well at the Headquarters of the American Institute of Architects (AIA), Greenwich Village, New york City. It is a 1250-ft deep,

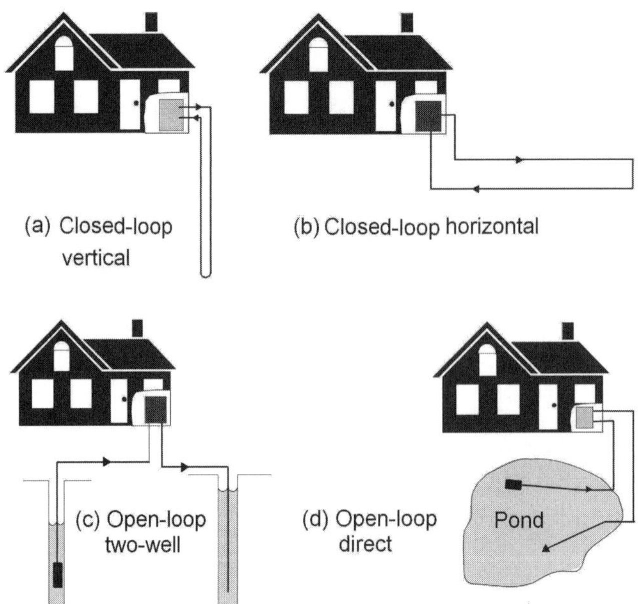

Figure 6.8 Heat-exchange configurations for ground-source heat pumps. The ground heat exchanger can have many different configurations. (a) and (b) are closed-loop heat exchangers. The vertical well in (a) does not require a large area, which means it can be used in a highly populated area. For suburban or rural buildings, the horizontal heat exchanger in (b) is much less expensive that the deep well. In places where groundwater is easily accessible, the open-loop system in (c) and (d) are even less expensive.

6-in-diameter well drilled into the sidewalk in front of the building. The underground water level is 40–50 ft down from the ground. A metal casing is placed to protect the well from collapsing due to the pressure of the soil. The lower part of the well is in the bedrock, which is strong enough to maintain the shape for probably hundreds of years. A 6-in-diameter polyvinyl chloride (PVC) tube, the so-called shroud straw, extends almost to the bottom of the well. Near the bottom, the shroud straw has 120 1-in-diameter holes. The water from the deep well is extracted by a submersible pump to the heat exchange unit; see Fig. 6.7. The water from the heat exchanger returns to the well through a return pipe and runs outside the shroud straw. Therefore, the water runs all the way through the full depth of the vertical well and thus reaches thermal equilibrium with the ground bedrock, where the temperature is always 12°C. Figure 6.10 is a photo of the ground connection to the geothermal well in the basement of the AIA.

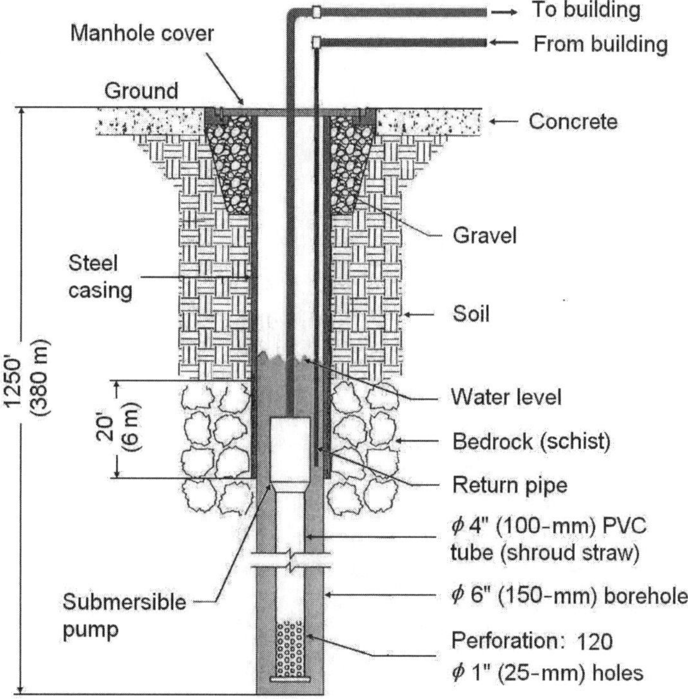

Figure 6.9 Vertical well in a heat pump system. Detail of the vertical well of a heat pump system at AIA. The depth is 1250 ft. About 20 ft below the underground water level, a submersible pump extracts water to the heat exchange unit. A 6-in-diameter PVC tube, the so-called shroud straw, extends almost to the bottom of the well. The water from the heat exchanger returns to the well through a return pipe and runs outside the shroud straw. Courtesy of American Institute of Architects.

Figure 6.10 Ground connection of geothermal well. Both incoming and outgoing pipes are shown. Photo taken by the author. Courtesy of American Institute of Architects.

The AIA ground source geothermal system was the first in New York City, but now there are many other systems in the City. One of the largest is the Knox Hall system at Columbia University; see Fig. 6.11. A 50,000-ft^2 office space, seminar rooms, and

Figure 6.11 Knox Hall of Columbia University. Knox Hall of Columbia University, a century-old historical building, is located at 606 West 122nd Street, Manhattan. During a 16-month-long renovation, ended on Februay 2010, one of the largest shallow geothermal energy facilities in New York City was installed. It has four 1800-ft (547 m) geothermal wells and heat pumps for both heating and cooling. The only visible marks of the project on the exterior of this historical building are four covered manholes on the sidewalk. Courtesy of Columbia University.

Figure 6.12 Heat pumps in the basement of Knox Hall. A photo taken by the author in the basement of Knox Hall, Columbia University. An array of heat pumps are shown on the left-hand side. The right-hand side is ground-well connections. Deep in the ground, the temperature is always at 57°F (14°C). In the winter, the heat pumps extract heat from the ground to warm the rooms. In the summer, the heat pumps extract heat from the rooms and dump it into the ground. The investment can be recovered in about 7 years from saving electricity. Courtesy of Columbia University.

classrooms for several departments of Columbia University, it has four 1800-ft (547-m) deep wells for both heating and cooling of the building for a 95-ton heating and cooling load. Their use contributes to a 23% reduction in energy consumption (over a conventional system). The four wells were drilled on the sidewalk, which has no interference with the entire building. After installation, the only visible marks are the four manhole covers on the sidewalk. Figure 6.12 shows the heat pumps inside the basement of Knox Hall.

Problems

6.1. The COP for heat pumps and refrigerators achieved in the industry is about 50% of the Carnot limit. Make estimates for the following cases:

1. For refrigerator to achieve $-10°$C in the freezer area with an external temperature of 25°C
2. For the same refrigerator to achieve -10°C in the freezer area but using the geothermal heat reservoir as the external source, at 10°C

6.2. A solar heat collector of 2 m^2 using the running water through underground pipes at 15°C to generate hot water at 55°C. If the efficiency of the solar collector is 80%, on average, how much hot water can this system generate per day?

1. In a region of 5 h of daily insolation
2. In a region of 3 h of daily insolation

Hint: The standard insolation is 1 kW/m^2.

Chapter 7

Quantum Transitions

As described in Chapter 2, sunlight embodies a wave–particle duality: it is both an electromagnetic wave and a stream of particles of indivisible energy packages, the photons. The utilization of solar energy can be divided into two categories. The first category is related to and can be explained by its wave nature, such as generating heat and the antireflection coatings for solar cells. The second category is related to the quantum nature of light. The light quanta, or photons, excite atoms, molecules, and solids from the ground state to an excited state, then convert part of the energy into electric power or chemical energy, such as various types of solar cells and photosynthesis.

In this chapter, we present the conceptual basis of the effect of sunlight as a stream of photons, that is, quantum mechanics. It includes the quantum states of atomic systems, its energy levels and wavefunctions, and transitions due to interactions with radiation.

7.1 Basic Concepts of Quantum Mechanics

Quantum mechanics is a branch of physics which deals with the phenomena at the atomic scale. It is a conceptual and mathematical description of how nuclei and electrons come together to form condensed matter, how the system evolves over time, and how the system interacts with radiation. The main difficulty of learning quantum mechanics is because the scale of atomic phenomena is below 1 nm, and quite different from the phenomena at the macroscopic scale. For example, Newtonian mechanics is a conceptual and mathematical description of macroscopic phenomena. Using a telescope, the motion of celestial bodies can be observed and measured directly for the verification of Newtonian mechanics. On the other hand, using optical microscopes, micrometer-scale objects, such as cells in biology, can be observed, verified, and understood directly. Only recently have the phenomena at the atomic scale, the entities of quantum mechanics, become directly observable using *scanning tunneling microscopy* (STM) and *atomic force microscopy* (AFM).

7.1.1 Quantum States: Energy Levels and Wavefunctions

In Newtonian mechanics, there are well-defined concepts thato describe the physical phenomena: position, velocity, acceleration, momentum, angular momentum, mass, and force. Newton's laws are the natural laws behind these phenomena. The motion of the planets in the solar system and the motion of the moons of the planets were the first and still among the best demonstrations of Newtonian mechanics.

The structure of an atom is similar to a miniature solar system: a heavy, positively charged proton surrounded by a number of light, negatively charged electrons. However, classical mechanics is totally inadequate for describing the atomic systems. It fails even to explain the fact that the atomic system can exist as a stable entity. Just as Newtonian mechanics is used to describe experimental observations of the motions of celestial bodies and macroscopic objects on Earth, quantum mechanics is used to describe the motions of objects at the atomic scale. Instead of the concepts of position, velocity, acceleration, momentum, angular momentum, mass, and force and so on, the basic concept of quantum mechanics is the *state*. According to Dirac, a quantum state is defined as one of the *various possible motions of the particles or bodies consistent with the laws of force* and is represented by either a *bra vector*, $\langle|$, or a *ket vector*, $|\rangle$. For confined systems, such as atoms and molecules, the states are discrete and can be labeled by discrete quantum numbers, such as

$$\langle n|, \quad |n\rangle : \quad n = 0,\, 1,\, 2,\, ...; \tag{7.1}$$

Quantum states in infinite space are labeled by continuous variables. For example, a plane-wave state can be labeled by a wave vector $\mathbf{k}$,

$$\langle \mathbf{k}|, \quad |\mathbf{k}\rangle : \quad \mathbf{k} = (k_x,\, k_y,\, k_z). \tag{7.2}$$

The states can be represented either by a vector with complex elements of infinite rank or by a complex function of the space coordinates $\psi(\mathbf{r})$, the *wavefunction*. For the case of a vector, the elements or the values of a bra are the complex conjugates of the elements or the values of a ket. For the case of a wavefunction, the bra is represented by the complex conjugate of that of the ket wavefunction, $\psi^*(\mathbf{r})$.

The terms *bra* and *ket* are taken from the word *bracket*. The expression of a complete bracket represents the inner product of a vector,

$$\langle a|b\rangle \equiv \sum_{n=0}^{\infty} a_n^*\, b_n, \tag{7.3}$$

where a_n^* is the complex conjugate of a_n, or the space integral of the wavefunctions,

$$\langle \psi|\chi\rangle \equiv \int \psi^*(\mathbf{r})\chi(\mathbf{r})d^3(\mathbf{r}). \tag{7.4}$$

7.1.2 Dynamic Variables and Equation of Motion

The dynamic variables are expressed as operators to the states. If a state is represented by a vector, then the operator is represented by a matrix. For clarity, we use the convention that an operator is marked by a hat on a letter. For example,

$$|b\rangle = \hat{V}|a\rangle \tag{7.5}$$

means

$$b_m = \sum_n V_{mn}\, a_n. \tag{7.6}$$

In order for the values of a dynamic variable are real, the matrix must be *Hermitian*, which means that the matrix must have the properties

$$V_{mn} = V^*_{nm}. \tag{7.7}$$

Consequently, the diagonal elements of the operator must be real. If an operator is not Hermitian, one can define an *adjoint operator* $\hat{A}^\dagger$ such that

$$A^\dagger_{mn} = A^*_{nm}. \tag{7.8}$$

It is easy to show that both $\hat{A}^\dagger + \hat{A}$ and $i(\hat{A}^\dagger - \hat{A})$ are Hermitian.

According to the creators (Heisenberg, Schrödinger and Dirac), quantum mechanics is formulated from the canonical form of classical mechanics with coordinate q, its conjugate momentum p, and the Hamiltonian function of q and p, $H(p,q)$. The quantum mechanics of the system is constructed by considering the dynamic variables as operators which satisfy the commutation relation

$$\hat{q}\hat{p} - \hat{p}\hat{q} = i\hbar, \tag{7.9}$$

where $\hbar$ is Dirac's constant, equal to the Planck constant h divided by 2π. For example, in the wavefunction representation, the momentum $\hat{p}_x$ is

$$\hat{p}_x|\Psi\rangle = -i\hbar\frac{\partial}{\partial x}|\Psi\rangle. \tag{7.10}$$

Then the commutation relation (Eq. 7.9), can be readily verified. A quantum state is in general time dependent, the evolution of which follows the *equation of motion*

$$i\hbar\frac{\partial}{\partial t}|\Psi\rangle = \hat{H}|\Psi\rangle, \tag{7.11}$$

where $\hat{H}$ is the Hamiltonian of the system as a function of the coordinate $\hat{q}$ and momentum $\hat{p}$. For a closed system, the Hamiltonian cannot contain time explicitly. In this case, Eq. 7.11 has a series of *stationary state solutions*,

$$|\Psi\rangle = e^{-iE_n t/\hbar}|\psi\rangle. \tag{7.12}$$

Here we use a capital letter to denote a state with time variation and a lowercase letter to denote the time-independent part of the state. The stationary quantum state of a system $|\psi\rangle$ is the solution of the Schrödinger's equation

$$\hat{H}|\psi\rangle = E|\psi\rangle. \tag{7.13}$$

7.1.3 One-Dimensional Potential Well

An instructive problem in quantum mechanics is an electron in a one-dimensional potential well; see Fig. 7.1. The Hamiltonian is

$$\hat{H} = \frac{\hat{p}_x^2}{2m_e} + V(x), \tag{7.14}$$

where

$$\begin{aligned} V(x) &= 0, \quad 0 < x < L, \\ V(x) &= \infty, \quad x < 0 \text{ and } x > L. \end{aligned} \tag{7.15}$$

In terms of coordinate x, Schrödinger's equation 7.13 becomes

$$\left[-\frac{\hbar^2}{2m_e}\frac{d^2}{dx^2} + V(x) \right] \psi(x) = E\psi(x). \tag{7.16}$$

Inside the box, $0 < x < L$, $V(x) = 0$. Equation 7.16 is reduced to

$$-\frac{\hbar^2}{2m_e}\frac{d^2\psi(x)}{dx^2} = E\psi(x). \tag{7.17}$$

At $x = 0$ and $x = L$, the wavefunction should vanish because the electron cannot penetrate an infinitely high potential barrier, which implies two boundary conditions

$$\psi(0) = 0, \tag{7.18}$$

$$\psi(L) = 0. \tag{7.19}$$

The solution satisfying the first boundary condition (Eq. 7.18) is

$$\psi(x) = C\sin(kx), \tag{7.20}$$

where C is a constant, and

$$k = \frac{\sqrt{2m_e E}}{\hbar}. \tag{7.21}$$

The second boundary condition (Eq. 7.19) requires

$$\psi(L) = C\sin(kL) = 0, \tag{7.22}$$

which implies

$$k = \frac{n\pi}{L}, \tag{7.23}$$

where n is an integer. From Eq. 7.21, we obtain the energy levels

$$E = \frac{n^2\pi^2\hbar^2}{m_e L^2}. \tag{7.24}$$

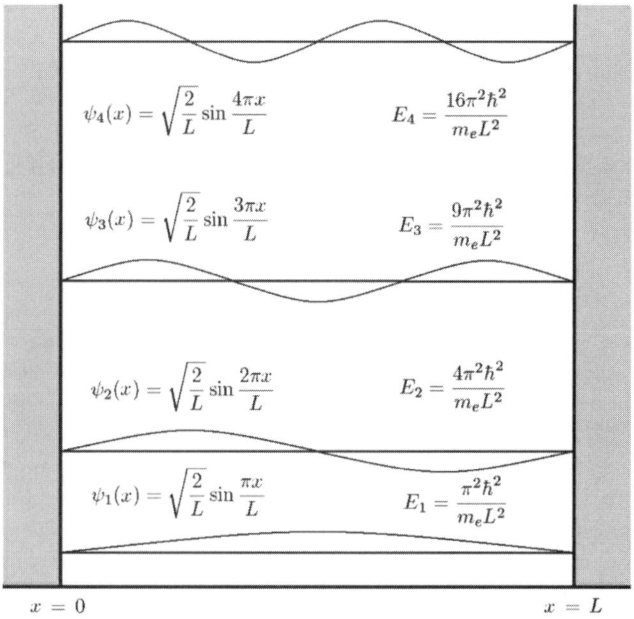

Figure 7.1 One-dimensional potential well. Energy levels and wavefunctions of an electron in a one-dimensional potential well.

The constant C is determined by the *normalization condition*

$$\int_0^L |\psi(x)|^2 \, dx = \int_0^L C^2 \sin^2(kx) dx = 1, \tag{7.25}$$

because the average value of $\sin^2(x)$ is $1/2$, which implies

$$C = \sqrt{\frac{2}{L}}. \tag{7.26}$$

By rewriting Eq. 7.21, one finds that the energy level E is a quadratic function of wavevector k,

$$E = \frac{\hbar^2 k^2}{2m_e}. \tag{7.27}$$

If mass is an unknown parameter, it can be obtained from the coefficient of that quadratic function,

$$\frac{1}{m_e} = \frac{1}{\hbar^2} \frac{\partial^2 E}{\partial k^2}, \tag{7.28}$$

which is an essential relation in semiconductor physics to determine the *effective mass* of an energy band.

7.1.4 Hydrogen Atom

The hydrogen atom, a miniature solar system composed of a positively changed proton and a negatively charged electron, is the first test case of quantum mechanics. To date, it is still the best system in nature for the illustration and understanding of quantum mechanics. By considering the proton immobile, the quantum state of a hydrogen atom is the solution of the equation,

$$\left[\frac{\hat{\mathbf{p}}^2}{2m_e} - \frac{\kappa}{r}\right]|\psi\rangle = E|\psi\rangle, \tag{7.29}$$

where $\kappa = q^2/4\pi\epsilon_0 = 2.30 \times 10^{-28}\ \mathrm{J \cdot m}$ (see Chapter 2) m_e is the mass of electron, and r is the distance from the proton to the electron. The mathematical form of the Hamiltonian in quantum mechanics is identical to the Hamiltonian in classical mechanics.

It is interesting to note that, before Schrödinger invented wave mechanics in 1927, the equation of motion of hydrogen in matrix form was resolved by Pauli in 1926, see Appendix C. Schrödinger's solution in terms of differential equations is more popular, and can be found in elementary quantum mechanics textbooks. Pauli's treatment is more compact and conceptually transparent.

The quantum states of the hydrogen atom can be described by a set of *quantum numbers*. The most widely used set of quantum numbers includes the principal quantum number n, the angular quantum number l, and the magnetic quantum number m. The energy value of the hydrogen atom depends on only the principal quantum number,

$$\hat{H}|n, l, m\rangle = E_n|n, l, m\rangle, \tag{7.30}$$

where

$$E_n = -\frac{\kappa^2\, m_e}{2n^2\hbar^2}. \tag{7.31}$$

The energy of the ground state is $E_1 = -2.17 \times 10^{-18}\ \mathrm{J} = -13.53\ \mathrm{eV}$. In a coordinate representation, the quantum states are represented by *wavefunctions*. For example, the wavefunction of the quantum state with the lowest energy level of a hydrogen atom is

$$\langle \mathbf{r}|1, 0, 0\rangle = \psi_{100}(r) = \frac{1}{\sqrt{\pi a_0^3}}\, e^{-r/a_0}, \tag{7.32}$$

where a_0 is the Bohr radius,

$$a_0 = \frac{\hbar^2}{m_e \kappa} \cong 0.0529\,\mathrm{nm} = 52.9\,\mathrm{pm}. \tag{7.33}$$

Although the wavefunction is often a complex function, and not directly observable, the probability of the charge density proportional to

$$\rho(r) = |\psi_{100}(r)|^2 = \frac{1}{\pi a_0^3}\, e^{-2r/a_0} \tag{7.34}$$

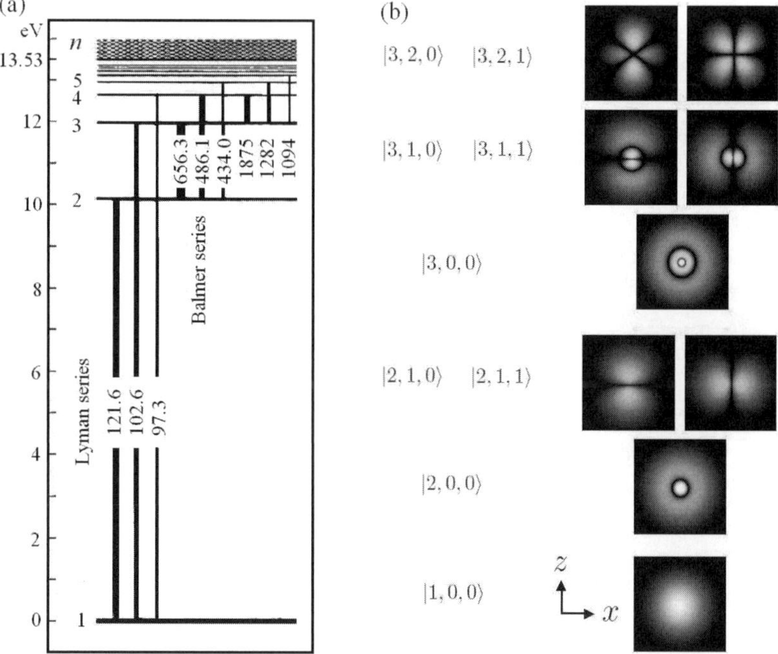

Figure 7.2 Quantum states of hydrogen atom. (a) The energy levels of the stationary states of the hydrogen atom. The lowest energy level of the hydrogen atom, the ground level, is 13.53 eV below the vacuum level. A series of energy levels lie between the vacuum level and the ground level. (b) The probability distributions of the stationary states of the hydrogen atom, determined by the wavefunctions. The numbers in the Dirac kets are, from left to right, the principal quantum number n, the angular-momentum quantum number l, and the magnetic quantum number m.

is observable. In fact, both the energy level and the change density can be directly observed using the scanning tunneling microscope. By definition, the probability is normalized,

$$\langle 1,0,0|1,0,0\rangle = \int \rho(r)\,d^3r = 4\pi \int_0^{\infty} |\psi_{100}(r)|^2\,r^2\,dr = 1. \tag{7.35}$$

Figure 7.2 shows the stationary states of the hydrogen atom. Figure 7.2(a) shows the energy levels. The lowest energy level of the hydrogen atom, the ground level, is 13.53 eV below the vacuum level. A series of energy levels lie between the vacuum level and the ground level. Radiative transitions, that is, absorption or emission, can occur between those levels. Figure 7.2(b) shows the charge density distributions of the stationary states of the hydrogen atom.

The quantum state of hydrogen can be excited to a state with a higher energy level by absorbing a photon, or de-excited to a state with a lower energy value by emitting a photon.

7.2 Many-Electron Systems

For atomic systems in nature, in terms of quantum mechanics, only a handful have analytic solutions: the hydrogen atom in both nonrelativistic and relativistic forms, and the hydrogen molecular ion, which also has only one electron. For systems with many electrons, an approximation is necessary. Because the nuclei are much heavier than electrons, in searching for the solutions of the Schrödinger equation for a system with N electrons, the coordinates of the nuclei $\mathbf{r}_I$ are taken as predetermined fixed values. The kinetic energy terms of the Hamiltonian T only contains the coordinates of the electrons,

$$T = -\frac{\hbar^2}{2m_e}\sum_{j=1}^{j=N}\nabla_j^2. \tag{7.36}$$

The potential energy includes the following three terms: the attractive potential between the electrons and the nuclei,

$$V_1 = -\sum_{I=1}^{I=N}\sum_{i=1}^{i=N}\frac{Z_I\kappa}{|\mathbf{r}_I - \mathbf{r}_i|}; \tag{7.37}$$

the repulsive potential among the electrons,

$$V_2 = \sum_{j=1}^{j=N}\sum_{i=1}^{i=N}\frac{\kappa}{|\mathbf{r}_j - \mathbf{r}_i|}; \tag{7.38}$$

and the repulsive potential among the nuclei,

$$V_3 = \sum_{I=1}^{I=N}\sum_{J=1}^{J=N}\frac{Z_I Z_J\kappa}{|\mathbf{r}_I - \mathbf{r}_J|}. \tag{7.39}$$

The Schrödinger equation is

$$[T + V_1 + V_2 + V_3]\,|\psi\rangle = E|\psi\rangle, \tag{7.40}$$

where $|\psi\rangle$ depends on the coordinates of the electrons.

7.2.1 Single-Electron Approximation

Despite the complexity of the problem, a simple and very successful concept, *single-electron approximation*, has been applied extensively in condensed-matter physics for many decades. An earlier version of this approach is the Hartree–Fock method or the self-consistent field (SCF) method. A later version, density-functional theory (DFT), makes the numerical computation more efficient and in general more accurate. For both methods, the problem of the many-electron system can be reduced to a number of single-electron problems, each electron moving in the mean potential ϕ formed by the nuclei and the rest of the electrons. Due to advances in computers and computational algorithms, quantum chemistry has reached a status where the predictions are reliable and useful.

In the case of a molecule, the solution of the SCF problem generates a series of stationary states, each with its own wavefunction and energy eigenvalue, which can be labeled by a positive integer n,

$$\left[-\frac{\hbar^2}{2m_e}\nabla + q\phi\right]|n\rangle = E_n|n\rangle. \tag{7.41}$$

Suppose the molecule is made of nuclei with order number Z_I. The total positive charge is

$$N = \sum_{I=1}^{\text{nuclei}} Z_I. \tag{7.42}$$

Charge neutrality requires that for the normal state of the molecule there must be N electrons. following the *Pauli exclusion principle*, the electrons fill up the states one by one, starting with the state with the lowest energy value. Once the N-th state is filled, the molecule becomes neutral.

For solid states, the number of nuclei is effectively infinite, and the labeling of the single-electron states needs a continuous variable, or a combination of an integer and a continuous variable. Because the lattice has translational symmetry, a wavevector $\mathbf{k}$ is a good quantum number.

The solutions to the resulting nonlinear equations behave as if each electron is subjected to the mean field generated by all other particles. Those equations can only be resolved numerically.

7.2.2 Direct Observation of Quantum States

Until the 1980s, the quantum states of atomic systems were considered merely as mathematical conveniences to explain macroscopic physical quantities. After the invention of the scanning tunneling microscope by Binnig and Rohrer in 1981, the quantum states of atomic systems, including the molecular orbitals, became directly observable [18].

Figure 7.3 shows a schematic of a scanning tunneling microscope. A probe tip, usually made of W or Pt–Ir alloy, is attached to a *piezodrive,* which consists of three mutually perpendicular piezoelectric transducers: $x-$ piezo, $y-$ piezo, and z-piezo. Upon applying a voltage, a piezoelectric transducer expands or contracts. By applying a sawtooth voltage on the x-piezo and a voltage ramp on the y-piezo, the tip scans on the xy-plane. Using the coarse positioner and the z-piezo, the tip and the sample are brought to within a fraction of a nanometer from each other. The electron wavefunctions in the tip overlap electron wavefunctions in the sample surface. A finite *tunneling conductance* is generated. By applying a bias voltage between the tip and the sample, a *tunneling current* is generated.

The tunneling current is converted to a voltage by the current amplifier, which is then compared with a reference value. The difference is amplified to drive the z-piezo.

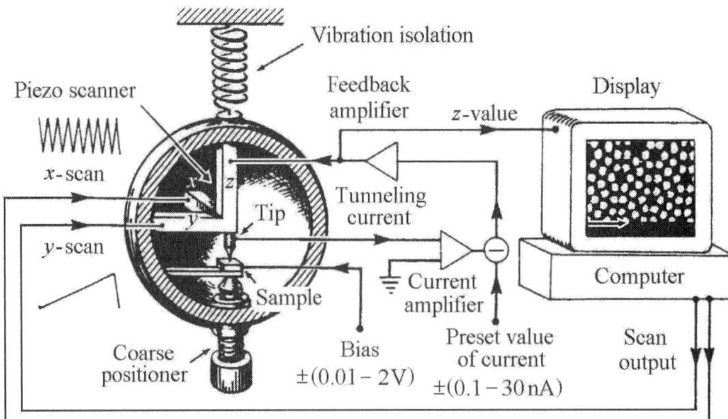

Figure 7.3 Scanning tunneling microscope. The scanning waveforms, on the x- and y-piezos, make the tip raster scan on the sample surface. A bias voltage is applied between the sample and the tip to induce a tunneling current. The z-piezo is controlled by a feedback system to maintain the tunneling current constant. The voltage on the z-piezo represents the local height of the topography. To ensure stable operation, vibration isolation is essential [18].

The phase of the amplifier is chosen to provide a negative feedback: if the absolute value of the tunneling current is larger than the reference value, then the voltage applied to the z-piezo tends to withdraw the tip from the sample surface, and *vice versa*. Therefore, an equilibrium z-position is established. As the tip scans over the xy-plane, a two-dimensional array of equilibrium z-positions, representing a contour plot of the equal tunneling-current surface, is obtained, displayed, and stored in the computer memory.

Therefore, the STM image can display simultaneously the energy level (by varying the bias voltage) and the wavefunction (by displaying the electron charge density).

7.2.3 Quantum States of Molecules: HOMO and LUMO

Consider a molecule with N positive charges; see Eq. 7.42. The Pauli exclusion principle requires that each state (with different spin as different states) can only be occupied by one electron. By adding electrons one after another starting from the lowest state, with N electrons, the molecule becomes neutral. The state or molecular orbital with highest energy level of a neutral molecule is the *highest occupied molecular orbital* (HOMO). The state with energy level above that HOMO is unoccupied for a neutral molecule, and is called the *lowest unoccupied molecular orbital* (LUMO). For organic molecules having a connection to a piece of conductor, HOMO is below the Fermi level. Otherwise the electron in that state can escape to form a positive molecular ion. On the other hand, the LUMO is above the Fermi level. Otherwise an electron from the connected conductor can occupy it to form a negative molecular ion.

Using STM, the electron charge density and the energy level of the HOMO or LUMO can be directly observed. Figure 7.4 illustrates the tunneling current and tunneling

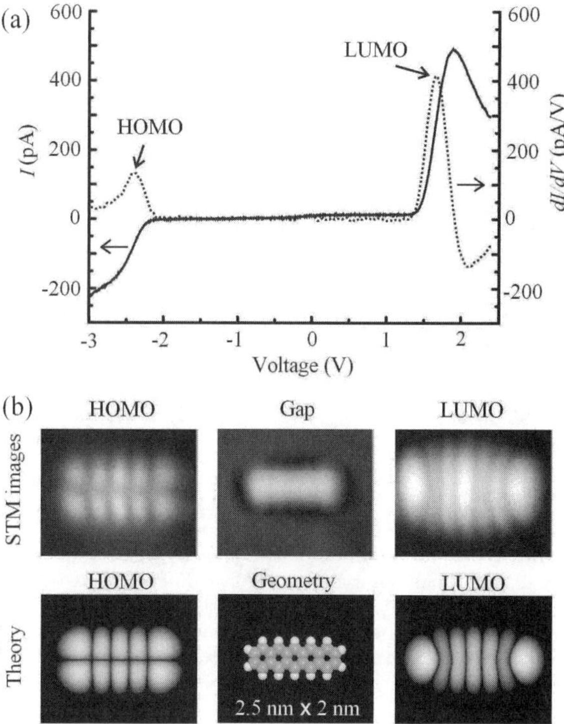

Figure 7.4 Experimental observation of HOMO and LUMO. Using a scanning tunneling microscope, the energy levels and the probability density distributions of those states can be determined by direct experimental measurements. (a) Tunneling current and tunneling conductivity curves. It shows a peak at -2.4 eV below the vacuum level, corresponding to HOMO, and a peak at +1.5 eV above the vacuum level, corresponding to LUMO. (b) Experimental observations of the charge density contours of pentacene at -2.4 and +1.5 eV, corresponding to the HOMO and LUMO, respectively. The theoretical results from numerical computations are also shown. Adapted from Ref. [18].

conductivity curves, where the peak at -2.4 eV below the Fermi level corresponding to HOMO, and the peak at +1.5 eV above the Fermi level corresponding to LUMO. Figure 7.4(b) shows experimental observations of the charge density contours of pentacene at -2.4 and +1.5 eV, corresponding to the HOMO and LUMO, respectively. The theoretical results from numerical computations are also shown.

7.2.4 Quantum States of a Nanocrystal

The scanning tunneling microscope is capable of not only mapping the electronic states, but also constructing artificial atomic structures not found in nature. Figure 7.5 shows the electronic states observed on a nanocrystal made of a linear chain of 15 copper atoms with spacing 225 pm assembled by STM. Electron density distributions at various bias

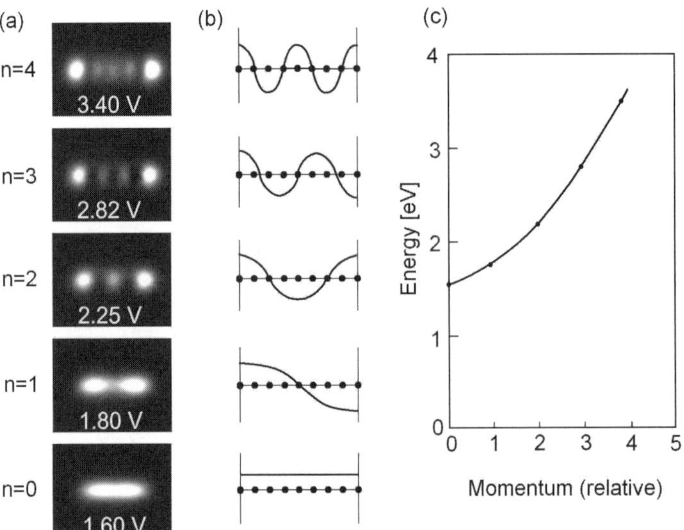

Figure 7.5 Quantum states of a nanocrystal. (a) A nanocrystal made of 9 copper atoms and the electron density distributions observed by STM at different bias voltages. Each bias voltage corresponds to the energy level of an electronic state of the nanocrystal. (b) Wavefunctions of the electronic states, which give rise to the observed electron density distributions. Each one has a different wavevector. (c) Relation between the energy levels and the wavevectors of the electronic states, showing a quadratic relation. Adapted from Ref. [18].

voltages are observed.

As shown, the wavefunction, exhibited as charge density distributions, as a function of the energy level, controlled by the bias voltage, can be directly observed. As shown in Section 7.1.3, the momemtum of the electron wavefunction is related to the number of nodes in the charge density distribution. Experimentally, a quadratic dependence on wavevector k is observed,

$$E = E_0 + \text{const.} \times k^2. \tag{7.43}$$

Similar to the theoretical argument in Eq. 7.27, an *effective mass* ca be obtained by Eq. 7.28,

$$\frac{1}{m_e} = \frac{1}{\hbar^2}\frac{\partial^2 E}{\partial k^2}. \tag{7.44}$$

7.3 The Golden Rule

In previous sections, the stationary states of atoms, molecules, and solid systems were described. For the utilization of solar energy, an understanding of the interaction of radiation with the quantum systems that cause transitions among various quantum states is of paramount importance. A quantum system can absorb a solar photon and be excited from a state with a lower energy value to a state with a higher energy value or decay from a state with a higher energy value to a state with a lower energy value by emitting a photon or giving up the energy to the surrounding atoms.

A simple, sketchy theory was provided by Einstein in 1916, as shown in Section 2.5. However, Einstein did not provide an explicit expression of those coefficients for a concrete system. That problem was resolved by Dirac in 1927 in a beautiful and comprehensive treatment of time-dependent quantum mechanical perturbation theory [22]. In that paper, a first-order time-dependent perturbation theory provided an explicit expression for the absorption and emission coefficients of radiation, thus also the rate of transition among quantum states. The resulting formula is so important that in 1950, Enrico Fermi coined a term Golden Rule for it in his book *Nuclear Physics* [29].

Dirac's 1927 paper deals with a number of issues concerning the interaction of radiation with atomic systems [22]. It is profound and difficult to read. Here, we provide an easy-to-follow derivation of the Golden Rule. As we will see, the derivation process provides a good understanding of the process of the interaction of radiation with atomic systems.

The derivation process also provides the quantum-mechanical foundation of the principle of detailed balance: a symmetry between absorption and emission.

7.3.1 Time-Dependent Perturbation by Periodic Disturbance

Consider a quantum system with a time-independent Hamiltonian $\hat{H}_0$. It could be a molecule, or a piece of semiconductor. In the absence of external disturbance, the equation of motion of a state $|\Psi\rangle$ is

$$i\hbar\frac{\partial}{\partial t}|\Psi\rangle = \hat{H}_0|\Psi\rangle, \tag{7.45}$$

where the state $|\Psi\rangle$ is understood to be in general time dependent. Because the Hamiltonian $\hat{H}_0$ is time independent, the system has a series of stationary states, labeled by an integer n,

$$\hat{H}_0|n\rangle = E_n|n\rangle. \tag{7.46}$$

A general solution of Eq. 7.45 can be written as

$$|\Psi\rangle = \sum_{n=0}^{\infty} c_n e^{-iE_n t/\hbar}|n\rangle, \tag{7.47}$$

where c_n are constant coefficients. That solution can be verified by direct substitution into Eq. 7.45.

Assuming that for $t \le 0$ the system is in an initial state with index i,

$$|\Psi\rangle = e^{-iE_i t/\hbar}|i\rangle, \qquad t = 0, \tag{7.48}$$

where the normalization condition implies $c_i = 1$.

The interaction of the system with radiation is described by a periodic disturbance

$$\hat{V}_{\text{rad}}(t) = \hat{V}e^{i\omega t} + \hat{V}^{\dagger}e^{-i\omega t}, \tag{7.49}$$

which is turned on for $t > 0$. It is easy show that the disturbance $\hat{V}_{\text{rad}}(t)$ is Hermitian. The first term represents an outgoing wave, and the second term represents an incoming wave. Consider the incoming term first. The total Hamiltonian becomes

$$\hat{H} = \hat{H}_0 + \hat{V}^{\dagger}e^{-i\omega t}. \tag{7.50}$$

The equation of motion becomes

$$i\hbar\frac{\partial}{\partial t}|\Psi\rangle = \left[\hat{H}_0 + \hat{V}^{\dagger}e^{-i\omega t}\right]|\Psi\rangle. \tag{7.51}$$

We look for a trial solution of Eq. 7.51 of the form

$$|\Psi\rangle = e^{-iE_i t/\hbar}|i\rangle + \sum_{n\neq i}^{\infty} c_n(t)e^{-iE_n t/\hbar}|n\rangle, \tag{7.52}$$

where the coefficients $c_n(t)$ are small but time dependent and are to be determined by the equation of motion 7.51. At $t = 0$, the disturbance is not yet effective, and therefore, $c_n(0) = 0$. Substituting Eq. 7.52 into Eq. 7.51, we have three types of terms. Because the disturbance $\hat{V}^{\dagger}$ is small, the first nonvanishing terms with $\hat{V}^{\dagger}$ are

$$\sum_{n\neq i}^{\infty}\frac{dc_n(t)}{dt}e^{-iE_n t/\hbar}|n\rangle = \hat{V}^{\dagger}|i\rangle e^{-iE_i t/\hbar - i\omega t}. \tag{7.53}$$

Now we look for the probability of transition to a final state, $|f\rangle$. Multiplying both sides with a bra of that final state, $\langle f|$, using the orthogonality condition $\langle f|n\rangle = \delta_{fn}$, one obtains

$$i\hbar\frac{dc_f(t)}{dt} = \langle f|\hat{V}^{\dagger}|i\rangle e^{-i(E_f - E_i)t/\hbar - i\omega t}. \tag{7.54}$$

The coefficient $c_f(t)$ can be obtained by direct integration using the initial condition $c_f(0) = 0$,

$$c_f(t) = \langle f|\hat{V}^{\dagger}|i\rangle\,\frac{e^{(E_f - E_i - \hbar\omega)t/\hbar} - 1}{E_f - E_i - \hbar\omega}. \tag{7.55}$$

The probability of finding a final state $|f\rangle$ at time t is

$$p_{fi} \equiv |c_f(t)|^2 = \frac{2\pi}{\hbar}t\,|\langle f|\hat{V}^{\dagger}|i\rangle|^2\left[\frac{\sin^2\left[(E_f - E_i - \hbar\omega)(t/2\hbar)\right]}{\pi\left[E_f - E_i - \hbar\omega\right]^2(t/2\hbar)}\right]. \tag{7.56}$$

Denote $E_f - E_i - \hbar\omega = u$ and $t/2\hbar = a$; the function in square brackets has a sharp peak near $u = 0$, as shown in Fig. 7.6. The area under it is 1:

$$\int_{-\infty}^{\infty}\frac{\sin^2 au}{\pi a u^2}du = 1. \tag{7.57}$$

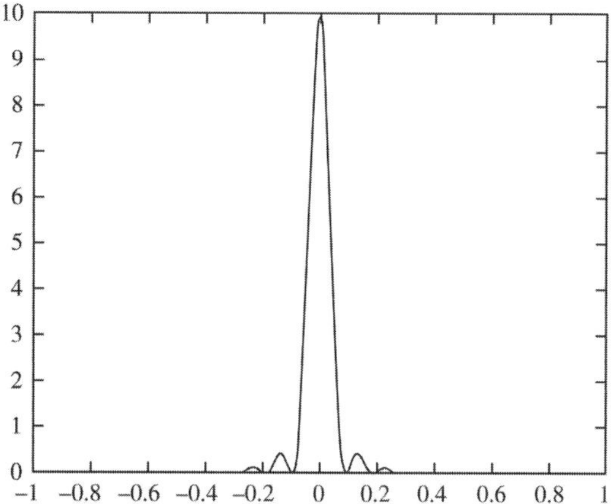

Figure 7.6 Condition of energy conservation. The integrand in Eq. 7.58 approaches a delta function when a approaches infinity. If the time of the experiment is not too short, the *condition of energy conservation* is valid.

As $a \to \infty$, it approaches a delta function,

$$\lim_{a\to\infty} \frac{\sin^2 au}{\pi a u^2} = \delta(u). \tag{7.58}$$

Therefore, when $t \to \infty$, the function in square brackets in Eq. 7.56 approaches a delta-function $\delta(E_f - E_i - \hbar\omega)$. Also, from Eq. 7.56, the probability of $|f\rangle$ is proportional to t, and therefore, the transition rate is

$$R_{fi} \equiv \frac{p_{fi}}{t} = \frac{2\pi}{\hbar} |\langle f|\hat{V}^\dagger|i\rangle|^2 \delta(E_f - E_i - \hbar\omega). \tag{7.59}$$

Equation 7.59 is the Golden Rule for atomic systems with discrete energy levels. The Bohr frequency condition,

$$\hbar\omega = E_f - E_i, \tag{7.60}$$

comes naturally. The radiation field can only transfer an energy quantum of $\hbar\omega$ to the atomic system, which is the essence of Einstein's theory of photons and provides an explanation of the line spectra, especially the Ritz combination principle.

7.3.2 Golden Rule for Continuous Spectrum

For systems with a continuous energy spectrum, the parameter f becomes a continuous variable. The number of states ΔN in an energy interval ΔE_f is given by the *density of states* of the final state, $\Delta N = \rho(E_f)\Delta E_f$. The transition rate R_{fi} should be integrated over a range of states,

$$
\begin{aligned}
R_{fi} &= \frac{2\pi}{\hbar} \lim_{t\to\infty} \int_{-\infty}^{\infty} |\langle f|\hat{V}^{\dagger}|i\rangle|^2 \left[\frac{\sin^2\left[(E_f - E_i - \hbar\omega)\,(t/2\hbar)\right]}{\pi\left[E_f - E_i - \hbar\omega\right]^2\,(t/2\hbar)} \right] \rho(E_f)\,dE_f \\
&= \frac{2\pi}{\hbar} \int_{-\infty}^{\infty} |\langle f|\hat{V}^{\dagger}|i\rangle|^2 \delta(E_f - E_i - \hbar\omega)\,\rho(E_f) dE_f \\
&= \frac{2\pi}{\hbar} |\langle f|\hat{V}^{\dagger}|i\rangle|^2 \rho(E_i + \hbar\omega).
\end{aligned} \tag{7.61}
$$

Similarly, the Bohr frequency condition $E_f = E_i + \hbar\omega$ comes naturally. Equation 7.61 is the Golden Rule for the continuous energy spectrum for the case of electronic excitation of an atomic system.

7.3.3 Principle of Detailed Balance

The above treatment is for the absorption of a photon from the radiation field. By changing the sign of exponential in Eq. 7.50,

$$
\hat{H} = \hat{H}_0 + \lambda \hat{V} e^{i\omega t}, \tag{7.62}
$$

and repeating the treatment, Eq. 7.59 becomes

$$
R_{fi} = \frac{2\pi}{\hbar} |\langle f|\hat{V}|i\rangle|^2 \delta(E_f - E_i + \hbar\omega), \tag{7.63}
$$

where the Bohr frequency condition becomes

$$
\hbar\omega = E_i - E_f, \tag{7.64}
$$

which means that the atomic system decays from a state of higher energy to a state of lower energy by emitting a photon to the radiation field. For a continuous spectrum, the corresponding Golden Rule is

$$
R_{fi} = \frac{2\pi}{\hbar} |\langle f|\hat{V}|i\rangle|^2 \rho(E_i - \hbar\omega). \tag{7.65}
$$

By exchanging the initial state and the final state, it is clear that the matrix element is identical in both absorption and emission processes,

$$
|\langle i|\hat{V}|f\rangle|^2 = |\langle i|\hat{V}^{\dagger}|f\rangle|^2. \tag{7.66}
$$

The symmetry of absorption and emission is the foundation of the *principle of detailed balance*, which plays a crucial role in the understanding of the efficiency limit of solar cells.

7.4 Interactions with Photons

The interaction of an atomic system with photons can be treated as a special case of the Golden Rule by using an explicit form of the interaction term $\hat{V}$. In Section 2.1.6, the classical Hamiltonian of an electron in electric and magnetic fields is Eq. 2.32,

$$H = \frac{1}{2m_e}\left(\mathbf{p} - q\mathbf{A}\right)^2 + q\phi, \tag{7.67}$$

where $\mathbf{A}$ is the vector potential of the electromagnetic field and ϕ is the electric (scalar) potential. The Hamiltonian in quantum mechanics has exactly the same form, but the dynamic variables become operators,

$$\hat{H} = \frac{1}{2m_e}\left(\hat{\mathbf{p}} - q\mathbf{A}\right)^2 + q\phi. \tag{7.68}$$

To treat the problems of the interaction between the solar radiation field and atomic systems, two simplifications can be made: First, the spatial variation of the radiation field, represented by the vector potential $\mathbf{A}$, is usually over 1 μm, much slower than the variation of the wavefunctions, which is less than 1 nm. Thus it can be treated as a constant over coordinates and consequently commutes with $\mathbf{p}$. Second, the square of the vector potential, $\mathbf{A}^2$, is much smaller than the other terms and thus can be neglected. Equation 7.68 becomes

$$\hat{H} = -\frac{\hbar^2}{2m_e}\nabla^2 + q\phi - \frac{q}{m_e}\left(\mathbf{A} \bullet \hat{\mathbf{p}}\right). \tag{7.69}$$

Here the first two terms represent the Hamiltonian for the atomic system, $\hat{H}_0$, and the last term is the perturbation potential $\hat{V}$.

A radiation field, a stream of sunlight, can be well represented by a plane wave. Take z as the direction of travel. The disturbance term in the total Hamiltonian is

$$\hat{V} = -\frac{q}{m_e}\,\mathbf{A}_0 \bullet \dot{\mathbf{p}}\, e^{i(k_z z - \omega t)}. \tag{7.70}$$

Equation 7.70 can be written in a more convenience form. Using Eqs. 2.33 and 2.8 and recalling that the term $e\mathbf{A}$ is a small perturbation, the spatial variation (on the order of 1 μm) is much slower than the variation of the atomic wavefunction, thus can be considered as a constant, and the temporal variation of the disturbance potential follows the $e^{-i\omega t}$ factor,

$$\begin{aligned}\hat{V} &= -i\omega q\mathbf{A}_0 \bullet \dot{\mathbf{r}}\, e^{-i\omega t} \\ &= \mathbf{E}_0 \bullet q\mathbf{r}\, e^{-i\omega t},\end{aligned} \tag{7.71}$$

which has an intuitive explanation. Because the spatial and temporal variations of the electric field of the radiation are much slower than that of the atomic states, the radiation field can be treated as uniform and static: The interaction of the radiation field is the electric field intensity of the light, $\mathbf{E}$, acting on the dipole moment of the atomic system, $q\mathbf{r}$. Hence, it is often called the dipole approximation.

Problems

7.1. Using the definition of Hermitian operators (Eq. 7.8), show that for an arbitrary matrix $\hat{A}$, both $\hat{A}^\dagger + \hat{A}$ and $i(\hat{A}^\dagger - \hat{A})$ are Hermitian.

7.2. A running wave in the x-direction with wavevector k_x can be represented by a function

$$\psi(x) = C\, e^{k_x x - \omega t}. \tag{7.72}$$

Using de Broglie's relation

$$p_x = \hbar k_x, \tag{7.73}$$

prove that the momentum operator is

$$\hat{p}_x \psi(x) = -i\hbar \frac{\partial}{\partial x} \psi(x). \tag{7.74}$$

7.3. Using the coordinate representation of the state in one dimension, the wavefunction $|\psi\rangle = \psi(x)$, and the definition of momentum

$$\hat{p_x} |\psi\rangle = -i\hbar \frac{\partial}{\partial x} |\psi\rangle, \tag{7.75}$$

prove the commutation relation, Eq. 7.9,

$$\hat{x}\hat{p}_x - \hat{p}_x\hat{x} = i\hbar. \tag{7.76}$$

7.4. The definition of angular momentum in quantum mechanics is

$$\hat{\mathbf{m}} = \hat{\mathbf{r}} \times \hat{\mathbf{p}}. \tag{7.77}$$

Prove that the angular momentum operator is Hermitian.

7.5. For the case of a string of atoms, the equation is the same as for an electron in a one-dimensional box (Eq. 7.17). At the two ends of the atom chain, the wavefunction should reach a maximum. The boundary conditions are

$$\psi'(0) = 0, \tag{7.78}$$

$$\psi'(L) = 0. \tag{7.79}$$

Write the wavefunctions satisfying those boundary conditions.

7.6. Using the same conditions in Problem 7.5, obtain the energy levels of that system.

7.7. Using the same conditions in Problem 7.5, obtain the normalized wavefunctions of that system.

7.8. Using the same conditions in Problem 7.5, obtain the wavefunctions with time variation.

Chapter 8

pn-Junctions

To date, most solar cells are made of semiconductors. A semiconductor is characterized by a relatively narrow energy gap, typically a fraction of an electron volt to a few electron volts. Electrons can be excited by a photon from the *valence band* to the *conduction band* and form an *electron–hole pair*. The electron–hole pair stores a substantial portion of the photon energy. After formation of the electron–hole pairs, the *pn-junction* separates the electrons and holes to generate external electric current. In this chapter, we will learn the basic physics of semiconductors and *pn*-junctions.

8.1 Semiconductors

8.1.1 Conductor, Semiconductor, and Insulator

When a large number of atoms come together to form a solid, the wavefunctions of atoms interact to form extended states which are similar to the one-dimensional crystal in Section 7.2.4. A number of *energy bands* are formed. In general, the number of states is infinite. According to the Pauli exclusion principle, each state can only be occupied

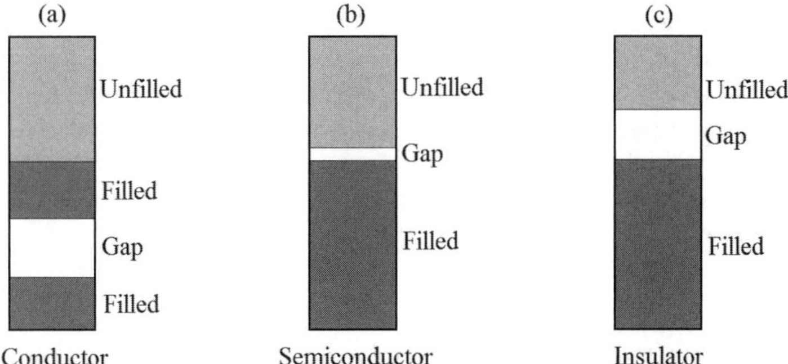

Figure 8.1 Conductor, semiconductor, and insulator. (a) For conductors, the highest occupied energy level is in the middle of an energy band. (b) For semiconductors, the highest occupied energy level matches the top of the valance band but the energy gap to the conduction band is small. (c) If the energy gap is big, the solid is an insulator.

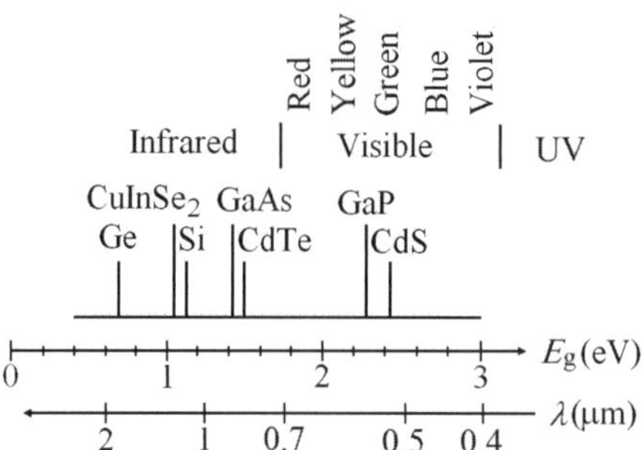

Figure 8.2 Band gaps of a number of semiconductors. The wavelength of light corresponding to the energy gap is also shown. Most of the semiconductors used in solar cells have an energy gap corresponding to the photons of near-infrared or visible light.

by one electron. Imagine electrons arte added to a system one by one, stating from the state with lowest energy. At one point, the number of electrons equals the number of protons in the system and the system becomes neutral.

According to the relative position of the energy bands and the highest occupied energy level, there are three different cases, as shown in Fig. 8.1.

In Fig. 8.1(a), the highest occupied energy level is in the middle of an energy band. The electrons can move to the unoccupied parts of the energy band. This type of material is called a *conductor* or a *metal.*

If the highest occupied energy level matches the top of an energy band, which is called the *valence band,* marked as E_v, and the distance to the next energy band is large, the electrons are not easily excited to the higher band. This type of materials is called an *insulator*; see Fig. 8.1 (c).

An important case between those two is the *semiconductor*, where the gap between the top of the valance band E_V and the bottom of the next energy band E_c is small such that when the temperature is not too low electrons can be excited to the next energy band, the *conduction band.* Typically the energy gap is less than a few electron volts. Once the electrons are excited to the conduction band, some conduction can take place. See Fig. 8.1(b).

Figure 8.2 shows the band gaps of a number of important semiconductors related to solar cells.

8.1.2 Electrons and Holes

At low temperature, pure semiconductors have almost no mobile electrons, and the conductivity is very low. By raising the temperature, electrons in the valance band can be excited to the conduction band; see Fig. 8.3. Therefore, a semiconductor has an important property: Conductivity depends critically on temperature — the higher the temperature, the higher the conductivity.

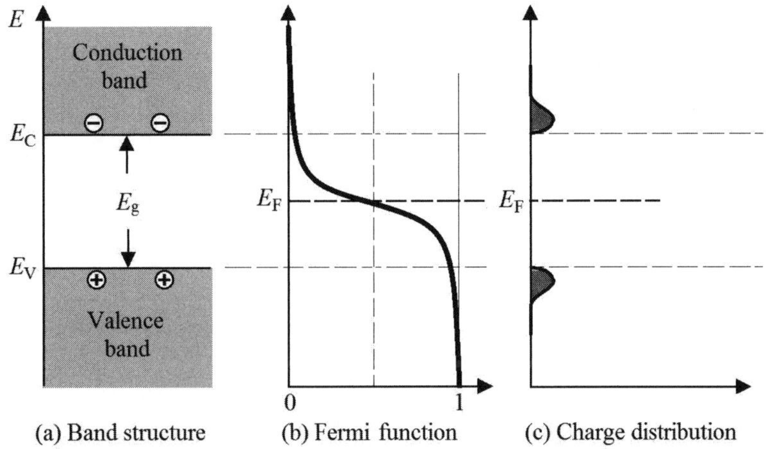

(a) Band structure (b) Fermi function (c) Charge distribution

Figure 8.3 Intrinsic semiconductors: Free electrons and holes. Thermal excitation could raise electrons from the valence band to the conduction band, to make electron–hole pairs. Charge neutrality requires that the concentration of electrons equals the concentration of holes.

According to Fermi–Dirac statistics (see Appendix D), at temperature T, the concentration of electrons n_0 at the bottom of the conduction band is

$$n_0 = N_c\, f(E_c), \tag{8.1}$$

where N_c is the effective density of states of the conduction band, a quantity determined by the property of the semiconductor, and $f(E_c)$ is the Fermi function (Eq. D.22 in Appendix D), the distribution function of electrons at absolute temperature T. At room temperature, $k_B T \approx 0.026$ eV, the value of $E_c - E_F$ is about 1 eV, and the factor 1 in the Fermi function can be neglected. To high accuracy, we have

$$f(E_c) = \frac{1}{e^{(E_c - E_F)/k_B T} + 1} \approx e^{-(E_c - E_F)/k_B T}. \tag{8.2}$$

Therefore, the concentration of electrons in the conduction band is

$$n_0 = N_c\, e^{-(E_c - E_F)/k_B T}. \tag{8.3}$$

In the valance band, there is a deficiency of electrons from the saturated situation. The deficiency of electrons in the valence band forms the mobile carriers, the *holes*. Similarly, the concentration of holes, p_0, is given as

$$p_0 = N_v[1 - f(E_v)] \approx N_v\, e^{-(E_F - E_v)/k_B T}, \tag{8.4}$$

where N_v is the effective density of states in the valence band and E_v is the energy level of the top of the valence band.

It is an interesting and important fact that the product $n_0\, p_0$ is *independent of the Fermi level.* Actually, by combining Eqs. 8.3 and refeq08110, we obtain

$$\begin{aligned} n_0\, p_0 &= \left[N_c\, e^{-(E_c-E_F)/k_\mathrm{B}T}\right] \left[N_v\, e^{-(E_F-E_v)/k_\mathrm{B}T}\right] \\ &= N_c\, N_v\, e^{-(E_c-E_v)/k_\mathrm{B}T} = N_c\, N_v\, e^{-E_g/k_\mathrm{B}T}. \end{aligned} \tag{8.5}$$

For intrinsic semiconductors, or semiconductors without impurities, charge neutrality requires that $n_0 = p_0$. Therefore, an *intrinsic carrier concentration* n_i can be defined,

$$n_i = \sqrt{N_c\, N_v}\, e^{-E_g/2k_\mathrm{B}T}, \tag{8.6}$$

with the general property

$$n_0\, p_0 = n_i^2, \tag{8.7}$$

which is valid even for semiconductors with impurities.

8.1.3 p-Type and n-Type Semiconductors

Semiconductors have an even more important property: their conductivity critically depends on the type and concentration of *impurities.* According to the position of the energy level of the atoms in the band gap of a semiconductor, there are two major types of impurities.

The energy level of *donor* atoms is just below the bottom of the conduction band. The impurity atom can easily be ionized to contribute an electron to the conduction

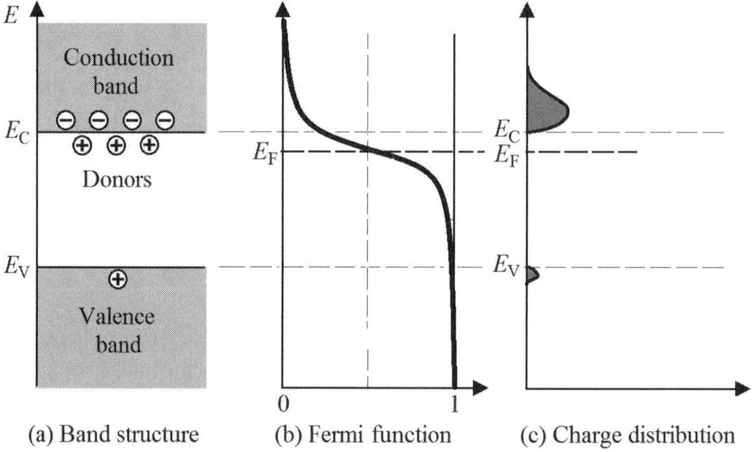

Figure 8.4 ***n*-type semiconductor.** The donor atoms release electrons into the conduction band. The Fermi level shifts toward the conduction band. The concentration of free electrons approximately equals the concentration of donor atoms.

band. For silicon and germanium, atoms from group V of the periodic table (N, P, As and Sb) are effective donors. The Fermi distribution is still valid, but the Fermi level has shifted toward the conduction band, as shown in Fig. 8.4. Assume the concentration of donor atoms is N_D. If the temperature is moderately high, all donor atoms could be ionized. The concentration of free electrons in an n-type semiconductor, n_n, approximately equals the concentration of donor atoms,

$$n_n = N_D. \tag{8.8}$$

On the other hand, the energy level of *acceptor* atoms is just above the top of the valence band. An electron in the valence band can easily be trapped by the acceptor atoms and leave a hole in the valence band. For silicon and germanium, atoms from Group IIIA (B, Al, Ga, and In) are effective acceptors. The Fermi distribution is still valid, but the Fermi level has shifted toward the valence band, as shown in Fig. 8.5. Assuming the concentration of acceptor atoms is N_A. If the temperature is moderately high, all acceptor atoms become negative ions. The concentration of holes in a p-type semiconductor, p_p, approximately equals the acceptor concentration,

$$p_p = N_A. \tag{8.9}$$

In both cases, the product of the concentrations of free electrons and holes equals the square of the intrinsic carrier concentration,

$$n_n\, p_n = p_p\, n_p = n_i^2, \tag{8.10}$$

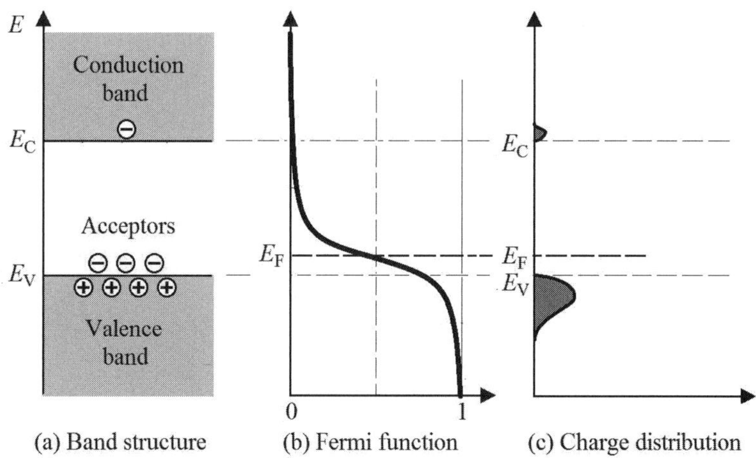

Figure 8.5 The p-type semiconductor. The acceptor atoms grab electrons from the valence band to create holes. The Fermi level shifts toward the valence band. The concentration of holes approximately equals the concentration of acceptor atoms.

where p_n is the hole concentration in an n-type semiconductor and n_p is the free-electron concentration in a p-type semiconductor. Each is a *minority-carrier concentration.*

8.2 Formation of a *pn*-Junction

When a p-type semiconductor and an n-type semiconductor are brought together, a *built-in potential* is established. Because the Fermi level of a p-type semiconductor is close to the top of the valence band and the Fermi-level of an n-type semiconductor is close to the bottom of the conduction band, there is a difference between the Fermi levels of the two sides. When the two pieces are combined to form a single system, the Fermi levels must be aligned. As a result, the energy levels of the two sides must undergo a shift with a potential V_0. Letting E_{cp} be the energy level of the bottom of the conduction band for the p-type semiconductor versus the Fermi level and E_{cn} that for the n-type semiconductor, the built-in potential is

$$qV_0 = E_{cp} - E_{cn}; \tag{8.11}$$

see Fig. 8.6. The establishment of a built-in potential near the boundary of a *pn*-junction can be understood from another point of view, the flow of carriers. Because in the n-region the hole concentration is very low, the holes diffuse from the p-region to the n-region. After a number of holes move to the n-region, an electrical field is formed to drive the holes back to the p-region. At equilibrium, the net current $J_p(x)$ must be zero,

$$J_p(x) = q\left[\mu_p p(x) E_x(x) - D_p \frac{dp(x)}{dx}\right] = 0, \tag{8.12}$$

where μ_p is the mobility of the holes, $p(x)$ is the concentration of holes as a function of x, $E_x(x)$ is the x-component of electric field intensity as a function of x, and D_p is the diffusion coefficient of the holes. Using Einstein's relation,

$$\frac{D_p}{\mu_p} = \frac{k_B T}{q}, \tag{8.13}$$

and the relation between the potential $V(x)$ and electric field intensity, $E_x(x) = -dV(x)/dx$, Eq. 8.12 becomes

$$-\frac{q}{k_B T}\frac{dV(x)}{dx} = \frac{1}{p(x)}\frac{dp(x)}{dx}. \tag{8.14}$$

Integrating Eq. 8.14 over the entire transition region yields

$$-\frac{q}{k_B T}(V_n - V_p) = \ln\frac{p_n}{p_p}. \tag{8.15}$$

Because $V_n - V_p = V_0$, Eq. 8.15 can be rewritten as

$$p_n = p_p \exp\left(\frac{-qV_0}{k_B T}\right). \tag{8.16}$$

Similarly, because in the *p*-region the free-electron concentration is very low, the free electrons diffuse from the *n*-region to the *p*-region. After a number of free electrons move to the *p*-region, an electrical field is formed to drive the free electrons back to the *n*-region. At equilibrium, the net current of free electrons must be zero. A similar equation is found,

$$n_p = n_n \exp\left(\frac{-qV_0}{k_B T}\right). \tag{8.17}$$

The meaning of Eqs. 8.16 and 8.17 is as follows: p_n is the concentration of holes in the *n*-region of the *pn*-junction and p_p is the concentration of holes in the *p*-region of the *pn*-junction. Because of the potential established by the space charge, V_0, the former is reduced by a factor of $\exp(-qV_0/k_B T)$ from the latter. The situation for the concentration of free electrons is similar.

To make a mental picture, we will make an order-of-magnitude estimate of the two equations. A typical value of the built-in potential is $V_0 \approx 0.75$ eV. At room temperature, $k_B T \approx 0.026$ eV. The factor $\exp(-0.75/0.026) \approx 10^{-12.5}$. Therefore, the absolute values are very small. For obvious reasons, both p_n and n_p are called *minority carriers*.

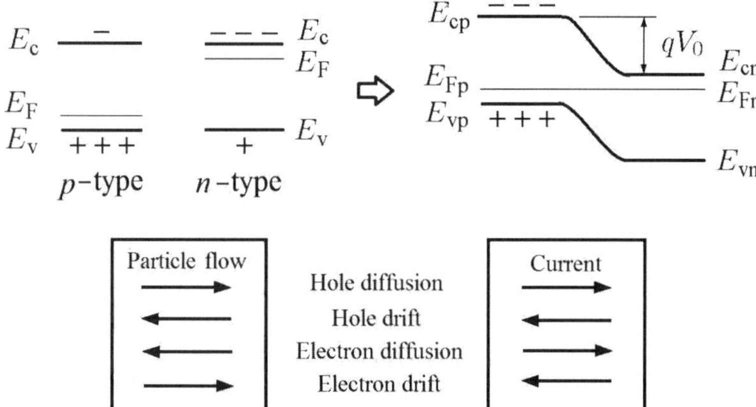

Figure 8.6 Formation of a *pn*-junction. By bringing a piece of *p*-type semiconductor and a piece of *n*-type semiconductor together to form a junction, the Fermi levels must be aligned. This would happen naturally as follows: The holes in the *p*-side moves to the *n*-side, and the free electrons in the *n*-side moves to the *p*-side, to form a double charged layer, until an equilibrium is established. The dynamic equilibrium is established by a balance of *drift current* driven by the electrical field, and a *diffusion current* driven by the gradient of carrier density.

Equations 8.16 and 8.17 are essential in being able to understand the current-voltage behavior of the *pn*-junction and the derivation of the diode equation.

To better understand the *pn*-junction, we need to establish a mathematical model for the space charge and the potential curve. A very effective and fairly accurate model is based on the *depletion approximation*; see Fig. 8.7. Under such an approximation, in the *p*-region near the junction boundary there is a layer of thickness x_p where all the holes are removed and the charge density ρ_p is determined by the density of the acceptors N_A, which are negatively charged,

$$\rho_p = -qN_A. \tag{8.18}$$

The electrostatic potential ϕ in this region is given by Poisson's equation,

$$\frac{d^2\phi}{dx^2} = \frac{1}{\epsilon}qN_A, \tag{8.19}$$

where ϵ is the dielectric constant or permittivity of the semiconductor and is the product of the permittivity of a vacuum, ϵ_0, and the relative dielectric constant ϵ_r of the semiconductor. The permittivity of a vacuum is given as $\epsilon_0 = 8.85 \times 10^{-14}$ F/cm. For example, for silicon, $\epsilon_r = 11.8$, $\epsilon \approx 1.04 \times 10^{-12}$ F/cm. Alternatively, Eq. 8.19 can be

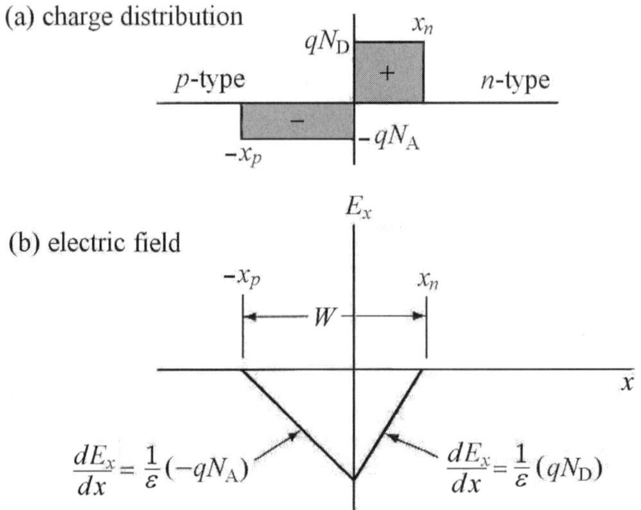

Figure 8.7 Charge and field distributions. The standard model of the *pn*-junction is based on the *depletion approximation*. In the *p*-region near the junction boundary, all holes are depleted, leaving the negatively charged acceptor ions to make the space charge. In the *n*-region near the junction boundary, all free electrons are depleted, leaving the positively charged donor ions to make the space charge. The electrostatic problem can be solved analytically to provide an easy-to-handle mathematical model which is sufficiently accurate for most problems in semiconductor devices, especially solar cells.

expressed in terms of electric field intensity,

$$\frac{dE_x}{dx} = -\frac{1}{\epsilon} q N_A. \tag{8.20}$$

Similarly, there is a slab of thickness x_n where all the free electrons are removed, and the charge density ρ_n is determined by the density of the donors, N_D, which are positively charged,

$$\rho_n = q N_D. \tag{8.21}$$

Poisson's equation gives

$$\frac{d^2\phi}{dz^2} = -\frac{1}{\epsilon} q N_D. \tag{8.22}$$

and the corresponding equation for electric field intensity is

$$\frac{dE_x}{dx} = \frac{1}{\epsilon} q N_D. \tag{8.23}$$

The boundary conditions for Eqs. 8.19, 8.20, and 8.22 are as follows. First, the charge neutrality of the entire transition region requires that

$$N_A x_p = N_D x_n. \tag{8.24}$$

Second, outside the transition region, the electric field should be zero:

$$E_x = 0 \quad \text{for} \quad x \leq -x_p \quad \text{and } x \geq x_n. \tag{8.25}$$

Third, the electrostatic potential should match the values at the boundaries of the transition region:

$$\begin{aligned} \phi &= 0, \quad \text{at } x = -x_p, \\ \phi &= V_0, \quad \text{at } x = x_n. \end{aligned} \tag{8.26}$$

The solutions of Eqs 8.20 and 8.23 with boundary condition Eq. 8.25 are

$$\begin{aligned} E_x &= -\frac{q N_A}{\epsilon}(x + x_p), \quad \text{for } -x_p \leq x < 0; \\ E_x &= \frac{q N_A}{\epsilon}(x - x_n), \quad \text{for } 0 < x \leq x_n. \end{aligned} \tag{8.27}$$

Using boundary conditions 8.26 and the definition of the width of the transition region W,

$$W = x_p + x_n, \tag{8.28}$$

the following relation is obtained:

$$V_0 = \frac{q}{2\epsilon} \frac{N_A N_D}{N_A + N_D} W^2. \tag{8.29}$$

The width of the transition region as a function of V_0 is

$$W = \sqrt{\frac{2\epsilon V_0}{q}\left(\frac{1}{N_A} + \frac{1}{N_D}\right)}. \tag{8.30}$$

Most solar cells are manufactured from a lightly doped p-type silicon wafer as the *base*, typically 100 – 300 μm thick, doped with boron of density $N_A \approx 1 \times 10^{16}\,\text{cm}^{-3}$ having resistivity $\rho \approx 1\,\Omega \cdot \text{cm}$. The n-type emitter is created by doping heavily on one side with phosphorus, with density $N_D \approx 1 \times 10^{19}\,\text{cm}^{-3}$, having resistivity $\rho \approx 10^{-3}\,\Omega \cdot \text{cm}$. For the case of $N_D \gg N_A$, Eqs 8.29 and 8.30 are simplified to

$$V_0 = \frac{q}{2\epsilon} N_A W^2 \tag{8.31}$$

$$W = \sqrt{\frac{2\epsilon V_0}{q N_A}}. \tag{8.32}$$

From Eq. 8.32, we obtain the capacitance of the pn-junction,

$$C \equiv \frac{\epsilon}{W} = \sqrt{\frac{q\epsilon N_A}{2V_0}} \;\left(\text{F/cm}^2\right). \tag{8.33}$$

8.3 Analysis of *pn*-Junctions

As we saw in Section 8.2, especially Fig. 8.6, in the absence of external applied voltage, there is no current running through a pn-junction, because the diffusion current and the drift current cancel each other for both holes and free electrons. By applying an external voltage on a pn-junction, the equilibrium is broken and a net current is generated.

Qualitatively, the mechanism can be explained as follows; see Fig. 8.8. At equilibrium, as shown in Fig. 8.8(a), for both electrons and holes, there is a concentration gradient which gives rise to diffusion and an electrical field pointing to $-x$-direction which drives the carriers in an opposite direction. The net current is zero. By applying a positive bias voltage, namely, connecting the positive terminal of a battery to the p-side and the negative terminal to the n-side, as shown in Fig. 8.8(b), the external potential pushes the holes to the n-side and the free electrons to the p-side. The potential barrier is reduced. Diffusion currents of both holes and free electrons are increased. The drift current, depending on the available carriers, are unchanged. The net current is nonzero. On the other hand, by applying a reversed bias, as shown in Fig. 8.8(c), the holes are pushed further back into the p-region and the free electrons are pushed further back into the n-region. The diffusion current is further reduced. The drift currents are unchanged and become the dominant factor. Eventually the current reaches a saturated value determined by the drift currents.

8.3.1 Effect of Bias Voltage

To account for the current quantitatively, we should further develop the concepts in Section 8.2. By applying a forward bias, as shown in Fig. 8.8(b), the potential difference across a pn-junction becomes $V_0 - V$. The electron concentration in the p-region, n_p, changes:

$$n_p \longrightarrow n_n \exp\left(\frac{-q(V_0 - V)}{k_B T}\right). \tag{8.34}$$

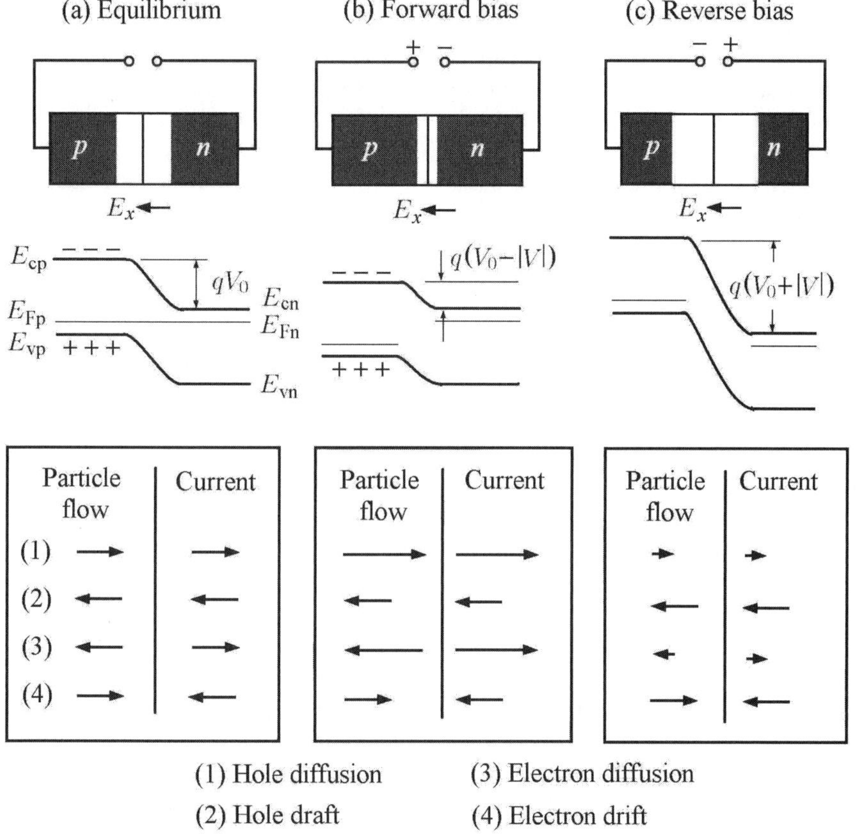

Figure 8.8 Effect of bias in a pn-junction. (a) At equilibrium, without external bias, the diffusion current and the drift current cancel each other. (b) A positive bias voltage pushes the holes to the n-side and the free electrons to the p-side. The potential barrier is reduced. Both diffusion currents of holes and free electrons are increased. The drift currents, depending on the available carriers, are unchanged. The net current is nonzero. (c) By applying a reversed bias, the holes are pushed further back into the p-region and the free electrons are pushed further back into the n-region. The diffusion current is reduced. Only the drift current persists.

Comparing with Eq. 8.16, one finds an *excess free-electron concentration at the border of the neutral p-region,*

$$\delta n_p(x=0) = n_p \left[\exp\left(\frac{qV}{k_B T}\right) - 1\right]. \tag{8.35}$$

Similarly, the external forward bias voltage V generates an *excess hole concentration at the border of the neutral n-region,*

$$\delta p_n(x=0) = p_n \left[\exp\left(\frac{qV}{k_B T}\right) - 1\right]. \tag{8.36}$$

The excess carrier concentrations generate an excess diffusion current, which is the main part of the forward-bias current of a diode.

To obtain an explicit expression of the current as a function of the bias voltage, we notice first that even with a substantial forward bias, for example, 0.5 V, the excess minority carriers δp_n and δn_p are still much smaller than the majority carrier concentrations p_p and n_n. For example, using V=0.5 V, from Eq. 8.34, $\delta p_n \approx \exp(-250/26) \times p_p \approx 10^{-5.8} \times p_p$. Therefore, the concentration of majority carriers can be treated as a constant even with a substantial bias voltage.

8.3.2 Lifetime of Excess Minority Carriers

Diffusion of excess minority carriers is the origin of junction current. However, there is a competing process which limits the junction current. The excess minority carriers are surrounded by a sea of majority carriers which are constantly courting for recombination. Because the concentration of majority carriers, p_p or n_n, is several orders of magnitude greater than the concentration of excess minority carriers, even with recombination, p_p or n_n is virtually a constant. The rate of decay of excess minority carriers is thus proportional to its concentration, which can be characterized by a *lifetime.* The combined effect of diffusion and lifetime of the excess minority carriers can be summarized in the following equations. For free electrons,

$$\frac{\partial \delta n_p(x,t)}{\partial t} = -\frac{\delta n_p(x,t)}{\tau_n} + D_n \frac{\partial^2 \delta n_p(x,t)}{\partial x^2}, \tag{8.37}$$

where D_n is the diffusion coefficient, and τ_n is the lifetime of free electrons. For holes,

$$\frac{\partial \delta p_n(x,t)}{\partial t} = -\frac{\delta p_n(x,t)}{\tau_p} + D_p \frac{\partial^2 \delta p_n(x,t)}{\partial x^2}, \tag{8.38}$$

where D_p is the diffusion coefficient and τ_p is the lifetime of holes.

8.3.3 Junction Current

At equilibrium, the concentration of carriers is independent of time. For example, Eq. 8.37 becomes

$$D_n \frac{d^2 \delta n_p(x)}{dx^2} = \frac{\delta n_p(x)}{\tau_n}. \tag{8.39}$$

It is equivalent to two first-order differential equations in the same sense that $x^2 = 4$ is equivalent to $x = 2$ and $x = -2$,

$$\frac{d\delta n_p(x)}{dx} = \frac{1}{\sqrt{\tau_n D_n}} \delta n_p(x) \tag{8.40}$$

$$\frac{d\delta n_p(x)}{dx} = -\frac{1}{\sqrt{\tau_n D_n}} \delta n_p(x), \tag{8.41}$$

which represent decays to the $+x$ and $-x$ directions, respectively. To account for junction current, for each side of the junction, only one of the two is needed. The diffusion current of electrons is

$$I_n = qD_n \frac{d\delta n_p(x)}{dx} = q\sqrt{\frac{D_n}{\tau_n}} \delta n_p(x). \tag{8.42}$$

At $x = 0$, using Eq. 8.35, the junction current of electrons is

$$I_n(x_p = 0) = -q\sqrt{\frac{D_n}{\tau_n}}\, n_p \left[\exp\left(\frac{qV}{k_B T}\right) - 1\right]. \tag{8.43}$$

Similarly, for holes,

$$I_p(x_n = 0) = q\sqrt{\frac{D_p}{\tau_p}}\, p_n \left[\exp\left(\frac{qV}{k_B T}\right) - 1\right]. \tag{8.44}$$

The total junction current is

$$I = q\left(\sqrt{\frac{D_n}{\tau_n}}\, n_p + \sqrt{\frac{D_p}{\tau_p}}\, p_n\right) \left[\exp\left(\frac{qV}{k_B T}\right) - 1\right]. \tag{8.45}$$

Furthermore, using the approximate relations

$$p_n = \frac{n_i^2}{N_D}, \quad n_p = \frac{n_i^2}{N_A}, \tag{8.46}$$

Eq. 8.45 can be reduced to

$$I = qn_i^2 \left(\frac{1}{N_A}\sqrt{\frac{D_n}{\tau_n}} + \frac{1}{N_D}\sqrt{\frac{D_p}{\tau_p}}\right) \left[\exp\left(\frac{qV}{k_B T}\right) - 1\right]. \tag{8.47}$$

8.3.4 Shockley Equation

Denoting a constant

$$I_0 \equiv qn_i^2 \left(\frac{1}{N_A}\sqrt{\frac{D_n}{\tau_n}} + \frac{1}{N_D}\sqrt{\frac{D_p}{\tau_p}}\right), \tag{8.48}$$

Eq. 8.47 is simplified to the well-known form of the *diode equation*, also known as the *Shockley equation*,

$$I = I_0 \left(e^{qV/k_B T} - 1 \right). \tag{8.49}$$

By applying a large reversed bias voltage on the diode, the exponential term vanishes. The current $I = -I_0$. Therefore, the constant I_0 is the *reverse saturation current density*. A question that remains to be answered is whether in the above derivation of the diode equation only the current at the border of the space-charge region (the transition region) and the neutral region is counted. What about the current deep in the neutral regions? The simple answer is, the electric current must be continuous. As the minority carriers are recombined with the majority carriers, there is a continuous current of majority carriers to replenish the lost ones. Because the concentration of majority carriers is orders of magnitude greater than that of the minority carriers, the disturbance caused by such a current is very minor.

Figure 8.9 shows the current–voltage behavior of a *pn*-junction. According to the diode equation 8.49, in the first quadrant with a forward bias at room temperature, the current is growing exponentially, about one order of magnitude per 60 mV. In the third quadrant, as the applied voltage exceeds 100 mV the current reaches a saturated value, which is dominated by the drift current. The two sides are highly asymmetric, which makes the *pn*-junction the most widely used rectifier.

The quality of a rectifier critically depends on the magnitude of the reverse saturation current density I_0. As shown in Eq. 8.48, the decisive factor is the minority-carrier lifetimes τ_n and τ_p. The longer the lifetime, the smaller the reverse saturation current density, and more improved the quality of the rectifier. This fact is equally important

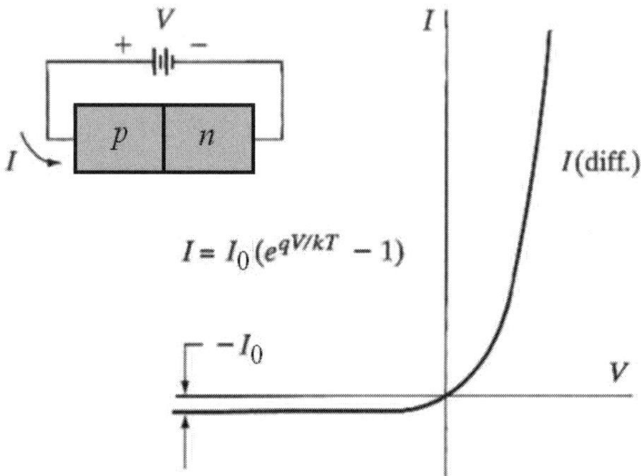

Figure 8.9 Current-voltage behavior of a *pn*-junction. With a forward bias, at room temperature, the current is growing exponentially, about one order of magnitude per 60 millivolts. In the third quadrant, as the applied reverse voltage exceeds 100 millivolts, the current reaches a saturate value, which is dominated by the drift current.

for solar cells. As we will show in Chapter 9, the most significant limiting factor of solar cell efficiency is minority-carrier lifetime. Therefore, in the design and manufacture of solar cells, efforts should be devoted to reducing the rate for minority-carrier recombination and thus prolong the minority-carrier lifetime.

Problems

8.1. A typical silicon solar cell has the following parameters: The *p*-type material has acceptor concentration $N_A = 1 \times 10^{16}\text{cm}^{-3}$, hole diffusion coefficient $D_p = 40\ \text{cm}^2/\text{s}$, and lifetime $\tau_p = 5\ \mu\text{s}$. The *n*-type material has donor concentration $N_D = 10^{19}\text{cm}^{-3}$, electron diffusion coefficient $D_n = 40\ \text{cm}^2/\text{s}$, and lifetime $\tau_n = 1\ \mu\text{s}$. The permittivity of free space is $\epsilon_0 = 8.85 \times 10^{-14}$ F/cm, and the relative permittivity of silicon is $\epsilon_r = 11.8$. The intrinsic carrier concentration in silicon is $n_i = 1.5 \times 10^{10}\ \text{cm}^{-3}$. Assuming the built-in potential is $qV_0 = 0.75$ V, calculate the following:

1. Width of transition layer, W

Hint: The donor concentration is too high, thus the *n*-layer is very thin. The width of transition layer is effectively that of the *p*-layer. The dielectric constant (permittivity) of silicon is the product of the permittivity of free space and the relative permittivity of silicon.

2. Diffusion lengths for holes and electrons
3. Saturation current density I_0
4. The pn-junction capacitance of the 10-cm × 10-cm solar cell.

8.2. Show that excess minority carriers decay with x by

$$\delta p_n(x) = \delta p_n(0) e^{-x/L_p} \tag{8.50}$$

and

$$\delta n_p(x) = \delta n_p(0) e^{-x/L_n}, \tag{8.51}$$

where the *diffusion lengths* are defined as

$$L_p = \sqrt{D_p \tau_p} \tag{8.52}$$

$$L_n = \sqrt{D_n \tau_n}. \tag{8.53}$$

8.3. Using diffusion length as a parameter, show that the junction current is

$$I = q\left(\frac{D_n p_n}{L_n} + \frac{D_p n_p}{L_p}\right)\left(e^{qV/k_BT} - 1\right). \tag{8.54}$$

8.4. The mobility of holes in silicon is $\mu_p = 480\ \text{cm}^2/\text{V·s}$ at room temperature. What is its diffusion coefficient D_p?

Hint: use Einstein's relation.

Chapter 9

Semiconductor Solar Cells

The photovoltaic effect, the direct generation of electric power by light in a solid material, was discovered by British scientists William Grylls Adams and his student Richard Evens Day in the 1870s using selenium. A few years later, Charles Fritt of New York constructed the first photovoltaic module for generating power from sunlight. However, the efficiency of the selenium solar cells was less than 0.5%, which meant it would not generate sufficient energy economically.

An important breakthrough was made in the 1950s by Gerald Pearson, Darryl Chapin, and Calvin Fuller at Bell Labs. Using silicon, they demonstrated a solar cell of efficiency 5.7%, ten times greater than that of the selenium solar cell; see Section 1.4. Solar cells first found applications in space. The efficiency of silicon cells has been improved to about 24% in the early 2000s, very close to the theoretical limit of 28%.

To date, semiconductor solar cells account for roughly 90% of the market share. Especially, silicon solar cells account for more than 80% of the solar cell market. Thin-film solar cells, especially those based on CIGS (copper–indium–gallium–selenide) and CdTe-CdS, are second to silicon solar cells in market share. Organic solar cells, which will be presented in Chapter 10, is a promising emerging technology.

9.1 Basic Concepts

The solar cell is a solid-state device which converts sunlight, as a stream of photons, into electrical power. Figure 1.22(b) shows the structure of a typical silicon solar cell. The base is a piece of *p*–type silicon, lightly doped with boron, a fraction of a millimeter thick. A highly doped *n*–type silicon, with a thickness of a fraction of one micrometer was generated by doping with phosphorus of much higher concentration. Because of the built-in potential of the *pn*-junction, electrons migrate to the *n*–type region, and generate electric power similar to an electrochemical battery.

According to the theory of quantum transitions presented in Chapter 7, radiation, as a stream of photons, interacts with a semiconductor in two ways; see Fig. 9.1. A photon with energy greater than the gap energy of the semiconductor material can be absorbed and create an electron–hole pair. An electron–hole pair can recombine and emit a photon of energy roughly equal to the energy gap of the semiconductor. According to the principle of detailed balance (see section 7.3.3), the probabilities of

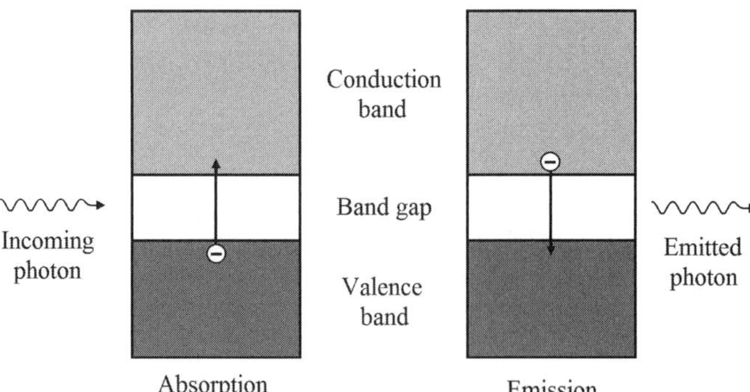

Figure 9.1 Interaction of radiation with semiconductors. According to the theory of quantum transitions, a photon with energy greater than the gap energy of the semiconductor material can be absorbed and create an electron–hole pair. An electron–hole pair can recombine and emit a photon of energy roughly equal to the energy gap of the semiconductor.

the two processes should equal. This fact has a significant consequence to the efficiency of solar cells; see Section 9.2.3.

Because the potential energy of the electron–hole pair equals the value of the energy band, the best material should have a band gap close to the center of the solar spectrum. Another factor that affects the efficiency of solar cells is the type of the energy gap. Depending on the relative positions of the top of the valence band and

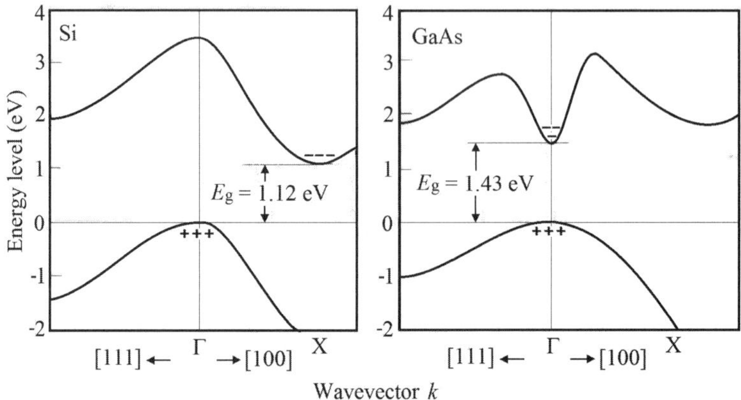

Figure 9.2 Direct and indirect semiconductors. Depending on the relative positions of the top of the valence band and the bottom of the conduction band in the wavevector space, the energy gap of a semiconductor can be direct or indirect. Direct semiconductors have a much higher absorption coefficient than that of indirect semiconductors. As shown, Si is an indirect semiconductor and GaAs is a direct semiconductor.

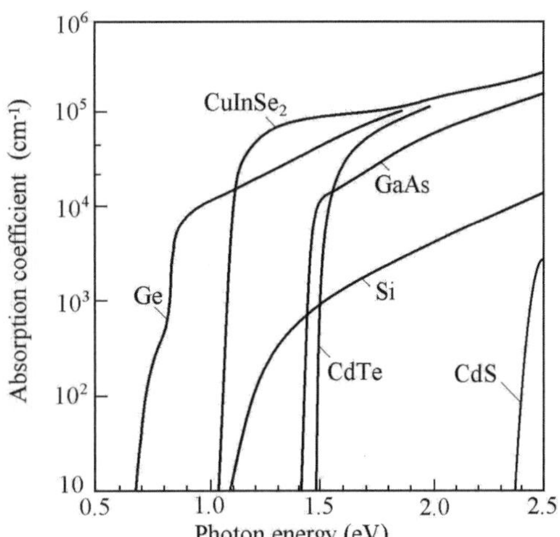

Figure 9.3 Absorption spectra of semiconductors commonly used for solar cells. The most commonly used material for solar cells, silicon, is an indirect semiconductor, and has a relatively low absorption coefficient, typically $10^3\,\text{cm}^{-1}$. A thickness of 0.01 cm is required for efficient absorption. Direct semiconductors, such as GaAs, $CuInSe_2$ and CdTe, have absorption coefficients ranging from 10^4 to $10^5\,\text{cm}^{-1}$. A thickness of a few micrometers is sufficient for an almost complete absorption.

the bottom of the conduction band in the wavevector space, the energy gap of a semiconductor can be direct or indirect; see Fig. 9.2. For semiconductors with a direct gap, such as GaAs, $CuInSe_2$, and CdTe, a photon can directly excite an electron from the valence band to the conduction band; the absorption coefficient is high, typically greater than $1\times10^4\,\text{cm}^{-1}$. For semiconductors with an indirect gap, such as Ge and Si, the top of valence band and the bottom of the conduction band are not aligned in the wavevector space, and the excitation must be mediated by a phonon, in other words, by lattice vibration. Therefore, the absorption coefficient is low, typically smaller than $1\times10^3\text{cm}^{-1}$. A thicker substrate is required. Figure 9.3 shows the absorption spectra of commonly used semiconductor materials for solar cells. Table 9.1 gives the properties of frequently used solar cell materials.

Table 9.1: Properties of Common Solar-Cell Materials.

Material	Ge	$CuInSe_2$	Si	GaAs	CdTe
Type	Indirect	Direct	Indirect	Direct	Direct
Band gap (eV)	0.67	1.04	1.11	1.43	1.49
Absorption edge (μm)	1.85	1.19	1.12	0.87	0.83
Absorption coef. (cm^{-1})	5.0×10^4	1.0×10^5	1.0×10^3	1.5×10^4	3.0×10^4

9.1.1 Generation of Electric Power

Figure 9.4 shows a quantitative explanation of how the electric power is generated by a solar cell. As shown in Fig. 9.4(a), a photon generates a electron–hole pair in the p–type region. Because of the built-in electric field, which points towards the p–type region, the electrons drift into the n–type region, whereas the holes stay in the p–type region. By connecting the two terminals in Fig. 9.4 together, almost all the electrons generated by the photons can migrate to the n–type region, and complete the circuit. The *short-circuit current* is the current of the electrons generated by sunlight. As shown in Fig. 9.4(b), if the two regions are not connected externally, the charges accumulated in the two regions generate a potential across the junction capacitance. The potential becomes a forward voltage for the diode. The thickness of the transition region is reduced, and a forward diode current is created. When the forward diode current equals the drift current of the electrons generated by the photons, an equilibrium is established. The voltage on the two terminals is the *open-circuit voltage* of the solar cell under illumination.

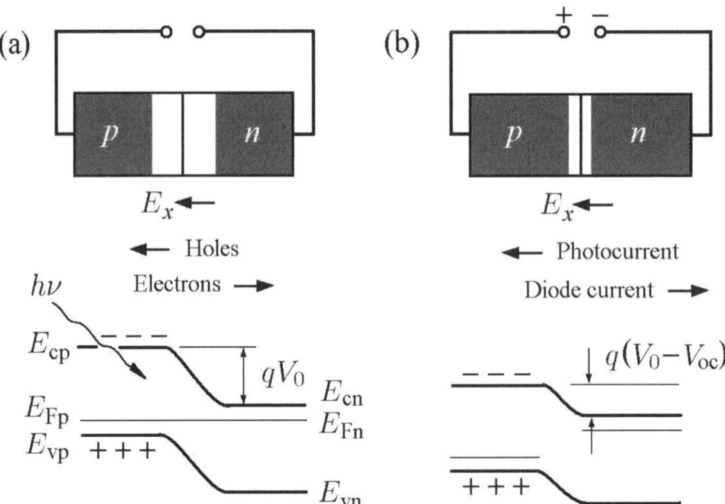

Figure 9.4 Separation of holes and electrons in a solar cell. (a) The incident photons generate electron–hole pairs. The built-in electrical field in the transition region is pointing toward the p–type region. The electrons, negatively charged, is dragged by the field into the n–type region. By connecting the two terminals, an electrical current is generated. The *short-circuit current* is determined by the rate of electron–hole pair generation from the radiation. (b) If the terminals are not connected, the electrons migrate to the n–type region are accumulated and builds a voltage across the junction capacitance. The direction of the voltage is the same as the forward bias voltage of the diode, which generates a current to compensate the electron current. At equilibrium, an *open-circuit voltage* is established.

9.1.2 Solar Cell Equation

A solar cell can be represented by a current source connected in parallel with a *pn*-junction diode; see Fig. 9.5. The current source is the photocurrent generated by incoming sunlight, defined by Eq. 9.27. the diode equation is changed to

$$I = I_0 \left(e^{qV/k_BT} - 1 \right) - I_{sc}. \tag{9.1}$$

which is the fundamental equation of solar cells, in a format consistent with the diode equation. However, in Eq. 9.1, while the voltage is always positive, the current is always negative. This is understandable because the diode is a passive device which consumes energy. By considering the solar cell as a battery, the direction of current should be reversed. Therefore, a better form of the solar cell equation is

$$I = I_{sc} - I_0 \left(e^{qV/k_BT} - 1 \right), \tag{9.2}$$

where both voltage and current are always positive; see Fig. 9.5.

The open-circuit voltage is the voltage when the current is zero, defined by the condition

$$I_0 \left(e^{qV_{oc}/k_BT} - 1 \right) = I_{sc}. \tag{9.3}$$

Consequently,

$$V_{oc} = \frac{k_BT}{q} \ln \left(\frac{I_{sc}}{I_0} - 1 \right). \tag{9.4}$$

Because I_{sc} is always much bigger than I_0, Eq. 9.3 can be simplified to

$$V_{oc} = \frac{k_BT}{q} \ln \frac{I_{sc}}{I_0}. \tag{9.5}$$

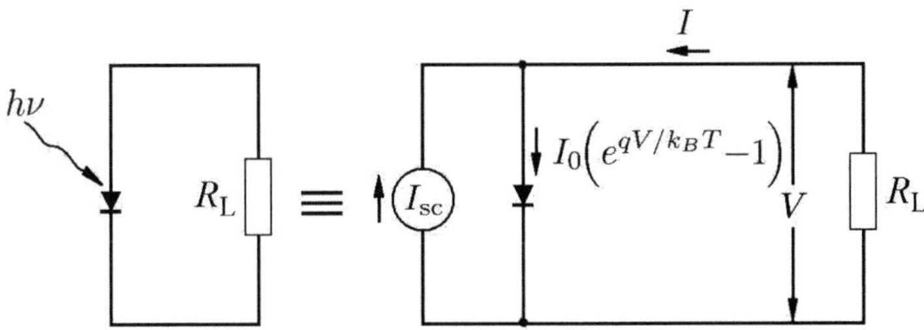

Figure 9.5 Equivalent circuit of solar cell. A solar cell can be represented by a current source connected in parallel with a *pn*-junction diode. The current source is the photocurrent generated by incoming sunlight.

9.1.3 Maximum Power and Fill Factor

The output power of a solar cell is determined by the product of voltage and current, $P = IV$. As discussed in Section 1.4.2, it is always smaller than the product of the short-circuit current I_{sc} and the open-circuit voltage V_{oc}, see Fig. 1.25. The rated power of a solar cell is the maximum power output with an influx of photons of one sun, or 1 kW/m^2, under favorable impedance-matching conditions. In general, the condition of maximum power is

$$dP = IdV + VdI = 0, \tag{9.6}$$

in other words,

$$\frac{dI}{dV} = -\frac{I}{V}. \tag{9.7}$$

According to the solar cell equation 9.2, the output power as a function of the output voltage V is

$$P = IV = \left[I_{sc} - I_0\left(e^{qV/k_BT} - 1\right)\right]V. \tag{9.8}$$

From Fig. 1.25, we observe that the voltage of maximum power is only slightly smaller than the open-circuit voltage. Introduce a voltage offset v, we write

$$V = V_{oc} - v. \tag{9.9}$$

Using Eq. 9.5, Eq. 9.8 can be simplified to

$$P \approx I_{sc}\left(V_{oc} - v\right)\left[1 - e^{-qv/k_BT}\right]. \tag{9.10}$$

Taking the derivative of P with respect to v, the condition of maximum power is

$$e^{qv/k_BT} = 1 + \frac{qV_{oc}}{k_BT}. \tag{9.11}$$

Again, using Eq. 9.5, Eq. 9.10 becomes

$$e^{qv/k_BT} = 1 + \ln\frac{I_{sc}}{I_0}. \tag{9.12}$$

Because $I_{sc} \gg I_0$, we find

$$v = \frac{k_BT}{q}\ln\ln\frac{I_{sc}}{I_0}. \tag{9.13}$$

Therefore, the voltage at maximum power is

$$V_{mp} = V_{oc} - v = V_{oc}\left(1 - \frac{\ln\ln(I_{sc}/I_0)}{\ln(I_{sc}/I_0)}\right), \tag{9.14}$$

and the current at maximum power is

$$I_{mp} = I_{sc}\left(1 - e^{-qv/k_BT}\right) = I_{sc}\left(1 - \frac{1}{\ln(I_{sc}/I_0)}\right). \tag{9.15}$$

After some simplification, maximum power is

$$P_{\text{mp}} = I_{\text{mp}}V_{\text{mp}} = V_{\text{oc}}I_{\text{sc}}\left(1 - \frac{1+\ln\ln(I_{\text{sc}}/I_0)}{\ln(I_{\text{sc}}/I_0)}\right). \tag{9.16}$$

The fill factor η_f,[1] defined as

$$\eta_f \equiv \frac{I_{\text{mp}}V_{\text{mp}}}{V_{\text{oc}}I_{\text{sc}}}, \tag{9.17}$$

is

$$\eta_f = 1 - \frac{1+\ln\ln(I_{\text{sc}}/I_0)}{\ln(I_{\text{sc}}/I_0)}. \tag{9.18}$$

Typically, $\eta_f = 0.8 - 0.85$.

9.2 The Shockley–Queisser Limit

In 1961, William Shockley and Hans Queisser made a thorough analysis of pn-junction solar cell, and established an upper limit for the efficiency of single-junction photovoltaic cells as a consequence of the principle of detailed balance[77]. The efficiency is defined as the ratio of power delivered to a matching load versus the incident radiation power to the solar cell. Three parameters are involved: The temperature of the Sun, $T_\odot$; the temperature of the cell, T_c; and the energy gap of the semiconductor, E_g. Actually, efficiency only depends on two dimensionless ratios:

$$x_s = \frac{E_g}{k_{\text{B}}T_\odot}, \tag{9.19}$$

$$x_c = \frac{E_g}{k_{\text{B}}T_c}. \tag{9.20}$$

Typically, $k_{\text{B}}T_\odot = 0.5$ eV, $k_{\text{B}}T_c$– 0.025 eV, and E_g is on the order of 1–2 eV. Therefore, the typical order of magnitude is $x_s \approx 2 - -4$, and $x_c \approx 40 - -80$.

The analysis of Shockley and Queisser was based on the following assumptions:

1. A single p–n-junction
2. One electron–hole pair excited per incoming photon
3. Thermal relaxation of the electron–hole pair energy in excess of the band gap
4. Illumination with unconcentrated sunlight

The above assumptions are well satisfied by the great majority of conventional solar cells, and the limit is well verified by experiments, unless one or more assumptions are explicitly removed, for example, using concentrated sunlight or using tandem solar cells. In these cases, the arguments of the Shockley–Queisser theory are still valid.

[1] In the literature, the notation FF is often used to represent the fill factor, a factor of efficiency in the treatment of solar cells by Shockley and Queisser. To maintain consistency of notations, we use η_f instead.

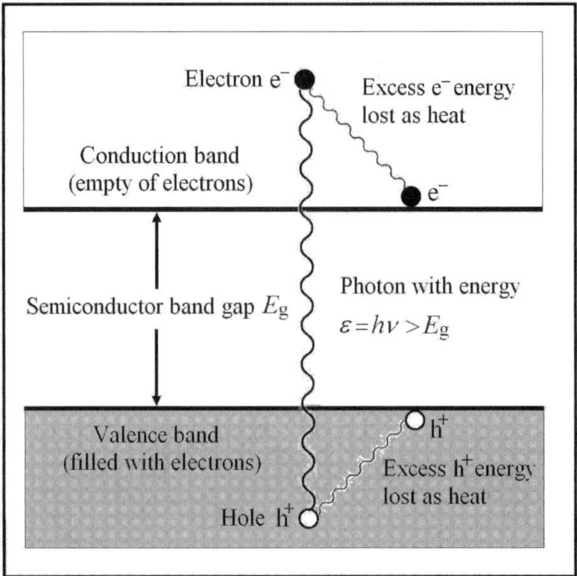

Figure 9.6 Generation of an electron–hole pair. A photon of energy greater than the band gap of the semiconductor can excite an electron from the valence band to the conduction band. The electron–hole pair energy in excess of the band gap dissipated into thermal energy of the electrons quickly (with a time scale of 10^{-11} s). The useful part of the photon energy for converting to electrical energy equals the band gap.

9.2.1 Ultimate Efficiency

Shockley and Queisser first considered the effect of band gap [77]. Assuming that solar radiation is proportional to blackbody radiation of temperature $T_\odot$, the power spectrum is (see Eq. 2.96),

$$u(\epsilon, T_\odot)\, d\epsilon = \frac{2\pi q^4}{c^2\, h^3} \frac{\epsilon^3}{\exp(\epsilon/k_B T_\odot) - 1}\, d\epsilon. \tag{9.21}$$

Equation 9.21 represents the radiation power density at the surface of the Sun. At the location of Earth, the spectral power density is diluted by a factor f defined by Eq. 2.103,

$$f = \left(\frac{r_\odot}{A_\odot}\right)^2 = \frac{\left[6.96 \times 10^8\right]^2}{\left[1.5 \times 10^{11}\right]^2} = 2.15 \times 10^{-5}. \tag{9.22}$$

Shockley and Queisser evaluated the ratio of the power of electron–hole pairs and the incident radiation power by assuming that the absorptivity of the semiconductor for photons with energy less than E_g is zero and those greater than E_g is unity. This simplification is also used to evaluate the rate of radiative combination of the electron–hole pair thus generated; see section 9.2.3. For photons with $\epsilon > E_g$, the electron-pair energy is quickly thermalized to E_g; see Fig. 9.6. By replacing one of variables ϵ in Eq. 9.21 is by E_g, the power of thus generated electron–hole pairs is

$$P_{ep} = \frac{2\pi q^4 E_g f}{c^2\, h^3} \int_{E_g}^{\infty} \frac{\epsilon^2\, d\epsilon}{\exp(\epsilon/k_\mathrm{B} T_\odot) - 1} = \frac{2\pi q^4 f (k_\mathrm{B} T_\odot)^4}{c^2\, h^3}\, x_s \int_{x_s}^{\infty} \frac{x^2\, dx}{e^x - 1}. \tag{9.23}$$

On the other hand, the incident radiation power is

$$P_s = \frac{2\pi q^4 f}{c^2\, h^3} \int_0^{\infty} \frac{\epsilon^3}{\exp(\epsilon/k_\mathrm{B} T_\odot) - 1} d\epsilon = \frac{2\pi q^4 (k_\mathrm{B} T_\odot)^4 f}{c^2\, h^3} \frac{\pi^4}{15}. \tag{9.24}$$

The efficiency as a function of the dimensionless variable x_s is

$$\eta_u(x_s) = \frac{P_{ep}}{P_s} = \frac{15}{\pi^4}\, x_s \int_{x_s}^{\infty} \frac{x^2\, dx}{e^x - 1}. \tag{9.25}$$

The integral in Eq. 9.25 can be evaluated using a simple numerical integration program or the quickly converging expansion in Problem 9.1. The result is shown in Fig. 9.7. A

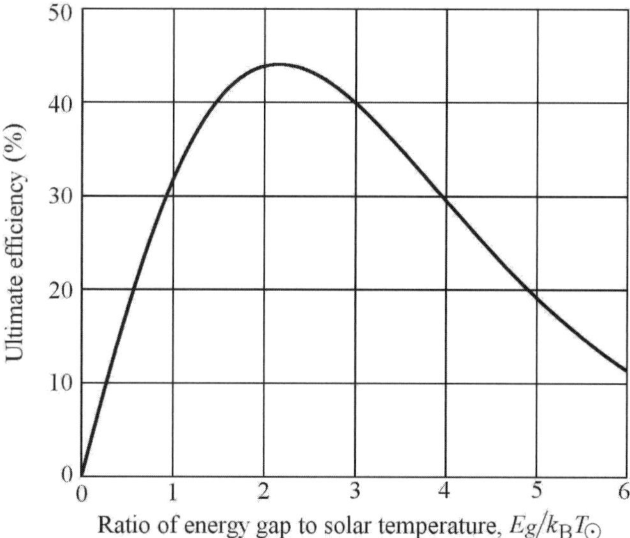

Figure 9.7 Ultimate efficiency of solar cells. Only photons with energy greater than the band gap can be absorbed, and the electron–hole pair energy in excess of the band gap is dissipated into thermal energy of the electrons, which make an ultimate limit to the efficiency of solar cells. The maximum efficiency is 0.44, at $E_g = 2.2\, k_B T_\odot$. For $T_\odot = 5800$ K, $k_B T_\odot \approx 0.5$ eV, the optimum band gap is 1.1 eV.

qualitative explanation of the result is as follows. If the band gap is small, the range of photon absorption is large. However, most of the photon energy is dissipated as heat; see Fig. 9.6. For a large band gap, the range of spectral absorption is reduced. Therefore, it should have a maximum somewhere. The position and the value of the maximum efficiency can be obtained using a numerical program, and the results are

$$E_{g\,\max} = 2.2\,k_B\,T_{\odot}; \qquad \eta_{u\,\max} = 0.44. \tag{9.26}$$

Taking $T_{\odot} = 5800$ K, $k_B T_{\odot} \approx 0.5$ eV. The optimum band gap is 1.1 eV. Shockley and Queisser called it the *ultimate efficiency*.

9.2.2 Role of Recombination Time

The ultimate efficiency determines the maximum open-circuit current of a solar cell. If the solar radiation power received by a solar cell is P_S, the power of the electron–hole pairs generated by the solar radiation is $\eta_u P_S$. It corresponds to the maximum short-circuit current of the solar cell,

$$I_{\text{sc}} = \frac{q}{E_g}\,\eta_u P_S. \tag{9.27}$$

The open-circuit voltage at the terminals of the solar cell is determined by the diode equation 9.4. Combining Eq. 9.4 with Eq. 9.27, we have the nominal power, defined as the product of the short-circuit current and the open-circuit voltage,

$$P_{\text{no}} = I_{\text{sc}} V_{\text{oc}} = \frac{\eta_u(x_s)}{x_c}\,\ln\left(\frac{I_{\text{sc}}}{I_0} - 1\right) P_S. \tag{9.28}$$

Clearly, the *reverse saturation current* of a *pn*-junction, I_0, is the limiting factor, determined by Eq 8.48,

$$I_0 = qn_i^2\left(\frac{1}{N_A}\sqrt{\frac{D_n}{\tau_n}} + \frac{1}{N_D}\sqrt{\frac{D_p}{\tau_p}}\right). \tag{9.29}$$

The reverse saturation current, Eq. 9.29, can be estimated using actual data from semiconductors. A general observation is as follows. The higher the reverse saturation current, the smaller the open-circuit voltage. By looking at Eq. 9.29, it is clear that the determining factor is the *recombination time*, τ_n and τ_p. Once an electron–hole pair is generated by absorbing a photon, the pair has a tendency to recombine by generating radiation or giving up the energy to the lattice. Shockley and Queisser found a fundamental limit due to radiative recombination of electron–hole pairs based on a detailed-balance argument.

9.2.3 Detailed-Balance Treatment

In the steady state, the electron–hole pairs in a solar cell are undergoing two major processes: the generation of the pairs by solar radiation with a rate F_s and various recombination processes. The generation rate F_s is calculated by assuming that the spectral density of sunlight is a blackbody radiation with temperature $T_\odot$ but diluted because of the distance between the Sun and Earth by a factor $f = 2.15 \times 10^{-5}$, defined by Eq. 9.22. The number of electron–hole pairs generated per unit area per unit time on a surface perpendicular to the sunlight is expressed by the power of the electron–hole pair (Eq. 9.23),

$$F_s = \frac{P_{ep}}{E_g} = \frac{2\pi q^4 f}{c^2 h^3} \int_{E_g}^{\infty} \frac{\epsilon^2}{\exp(\epsilon/k_B T_\odot) - 1} d\epsilon. \tag{9.30}$$

Many factors that contribute to recombination, such as those related to defects or surfaces, can be reduced or avoided. The radiative recombination F_c, however, is a process which sets a fundamental limit on the efficiency of solar cells. To evaluate the rate of radiative recombination of the electron–hole pairs, Shockley and Queisser considered the equilibrium of the solar cell with the environment at the cell temperature T_c, typically 300 K, without sunlight. The principle of detailed balance requires that the generation rate must equal the recombination rate. This principle, manifested as Kirchhoff's law, states that emissivity equals absorptivity at any given wavelength. With respect to the simplified model of Shockley and Queisser [77], for photons with energy greater than the energy gap of the semiconductor, the emissivity is 1; otherwise it is 0 (see Fig. 9.8). The rate of radiative electron–hole recombination can be calculated using an integral similar to Eq. 9.30 but at the environment temperature T_c,

$$F_{c0} = \frac{2\pi q^4}{c^2 h^3} \int_{E_g}^{\infty} \frac{\epsilon^2}{\exp(\epsilon/k_B T_c) - 1} d\epsilon. \tag{9.31}$$

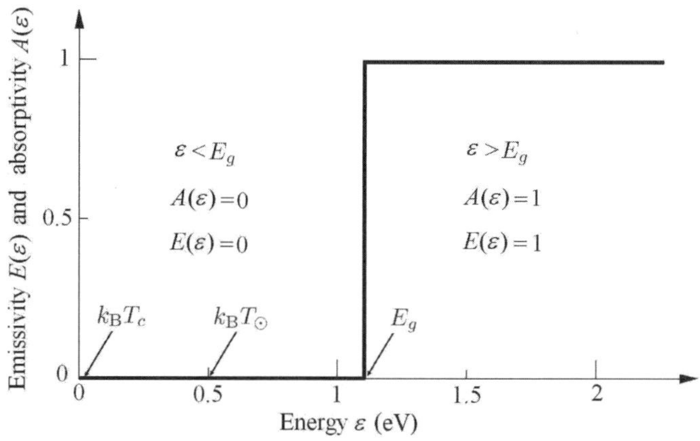

Figure 9.8 A simplified optical model of semiconductors. To evaluate the effect of radiative recombination, Shockley and Queisser used a simplified optical model of semiconductors. For photons with energy greater than the band gap of the semiconductor, the absorptivity is 1, and hence the emissivity is also 1. Otherwise the emissivity is 0.

Typically, the parameter x_c in Eq. 9.20 equals $E_g/k_B T_c \approx 40 - 80$, and thus the term 1 in the denominator of the integrand of Eq. 9.31 can be neglected. To high accuracy, the result is

$$F_{c0} = \frac{2\pi q^4 (k_B T_c)^3}{c^2\, h^3} \int_{x_c}^{\infty} x^2\, e^{-x}\, dx = \frac{2\pi q^4 (k_B T_c)^3}{c^2\, h^3} e^{-x_c} \left[x_c^2 + 2x_c + 2\right]. \tag{9.32}$$

The above result is valid when the semiconductor is at equilibrium. With sunlight, excess minority carriers are generated. For example, if the bulk of a solar cell is made of p–type semiconductor, according to Eq. 8.34, with an external voltage V, the electron concentration changes from its equilibrium value n_{p0} to

$$n_p = n_{p0} \exp\left(\frac{qV}{k_B T_c}\right). \tag{9.33}$$

Thus, the radiative recombination rate is increased to

$$F_c(V) = F_{c0} \exp\left(\frac{qV}{k_B T_c}\right). \tag{9.34}$$

At a steady state, the rate of electron–hole pair generation must equal the rate of radiative recombination plus the rate of electron consumption due to the current I drawn by the external circuit,

$$F_s = F_c(V) + \frac{I}{q}. \tag{9.35}$$

By shorting the terminals together, the voltage V is zero, and the *short-circuit current* is

$$I_{sc} = q\,(F_s - F_{c0}). \tag{9.36}$$

Using Eq. 9.34, the current on the external load, I, is given as

$$I = I_{sc} + qF_{c0}\left[1 - e^{qV/k_B T_c}\right]. \tag{9.37}$$

Defining the reverse saturation current I_0 as

$$I_0 = qF_{c0}, \tag{9.38}$$

Eq. 9.37 becomes

$$I = I_{sc} + I_0\left[1 - e^{qV/k_B T_c}\right]. \tag{9.39}$$

The open-circuit voltage can be obtained by setting $I = 0$,

$$V_{oc} = \frac{k_B T_c}{q} \ln\left(\frac{I_{sc}}{I_0} - 1\right). \tag{9.40}$$

Because $F_s \gg F_{c0}$, Eq. 9.40 can be simplified to

$$V_{oc} = \frac{k_B T_c}{q} \ln\left(\frac{F_s}{F_{c0}} - 1\right). \tag{9.41}$$

9.2.4 Nominal Efficiency

Shockley and Queisser defined the *nominal efficiency* as

$$\eta_n = \frac{V_{\rm oc} I_{\rm sc}}{Ps}. \tag{9.42}$$

Using Eqs. 9.41, 9.27, and 9.20, after some reduction, we find the expression

$$\eta_n = \eta_u \eta_d, \tag{9.43}$$

where the ultimate efficiency η_u is defined by Eq. 9.25, and the detailed balance efficiency η_d is defined as

$$\eta_d = \frac{qV_{\rm oc}}{E_g} = \frac{1}{x_c} \ln\left(\frac{F_s}{F_{c0}} - 1\right), \tag{9.44}$$

Using Eqs. 9.30 and 9.31, we obtain

$$\frac{F_s}{F_{c0}} = \frac{f\left(\dfrac{x_c}{x_s}\right)^3 \displaystyle\int_{x_s}^{\infty} \frac{x^2\,dx}{e^x - 1}}{e^{-x_c}\left[x_c^2 + 2x_c + 2\right]}. \tag{9.45}$$

As shown in Eq. 9.44, the detailed-balance efficiency η_d is the fundamental upper limit of the ratio of the open-circuit voltage and the band gap of the semiconductor in volts. It depends on the temperature of the cell. If the temperature of the solar cell is very low, $x_c \to \infty$, the detailed-balance efficiency becomes

$$\eta_d \to \frac{1}{x_c} \ln\left(C\, e^{x_c}\right) \to \frac{x_c + \ln C}{x_c} \to 1, \tag{9.46}$$

because the expression C varies much less than the exponential. Therefore, at very low cell temperatures, if there is no other recombination mechanism other than radiative recombination, the open-circuit voltage approaches the band gap energy in volts and the nominal efficiency approaches the ultimate efficiency.

The reverse saturation current in Eq. 9.38 is the absolute limit of the observed reverse saturation current (Eq. 9.29). In addition to radiative recombination, there are other types of recombination processes that increase the reverse saturation current and then reduce the efficiency of solar cells; see Section 9.3.

9.2.5 Shockley–Queisser Efficiency Limit

With a load of matched impedance, the output power of a solar cell can be maximized. As shown in Section 9.1.3, the maximum power is related to the nominal power by a form factor η_f approximately

$$\eta_{\rm SQ} = \eta_u\, \eta_d\, \eta_f \approx \eta_u\, \eta_d \left[1 - \frac{1 + \ln\ln(I_{\rm sc}/I_0)}{\ln(I_{\rm sc}/I_0)}\right]. \tag{9.47}$$

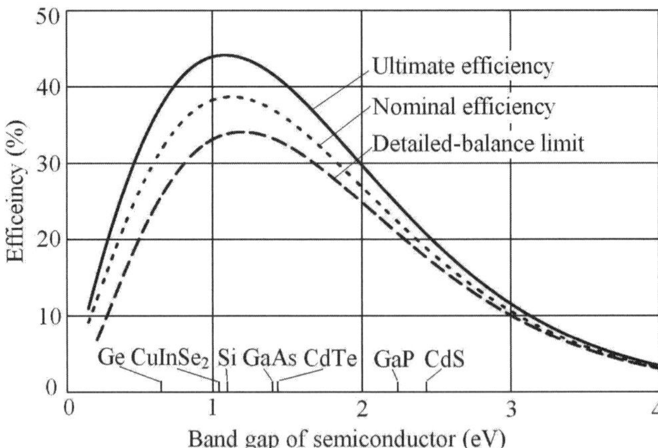

Figure 9.9 Efficiency limit of solar cells. Efficiency limit as determined by detailed balance. The solar radiation is approximated by blackbody radiation at 5800 K. The temperature of the solar cell and its surroundings is assumed as 300 K. The abscissa is the band gap of the semiconductor in electron volts. The values of several important solar cell materials are marked. The ultimate efficiency is determined by the absorption edge and thermalization of excited electrons. The nominal efficiency, which is always lower than the ultimate efficiency, is a result of radiative recombination of electrons and holes. While driving an external load at maximum power, the radiative recombination is further increased, and the efficiency is further reduced. The detailed-balance limit of efficiency is a fundamental limit for the conditions listed by Shockley and Queisser. After Ref. [77].

The detailed-balance efficiency limit, together with the ultimate efficiency and the nominal efficiency, for $T_\odot = 5800$ K and $T_c = 300$ K, is shown in Fig. 9.9. The abscissa is the band gap of the semiconductor in electron volts. Values of several important solar cell materials are indicated. The ultimate efficiency is determined by the absorption edge and thermalization of excited electrons. The nominal efficiency, which is always lower than the ultimate efficiency, is a result of radiative recombination of electrons and holes. While driving an external load at maximum power, the radiative recombination is further increased, and the efficiency is further reduced.

In the original paper [77], Shockley and Queisser also discussed several factors that affect the efficiency limit, such as both sides of a solar cell can radiate which only one side receives sunlight, the non-blackbody behavior of the cell, nonradiative electron–hole pair recombination processes, and the difference between the solar constant and the AM-1.5 solar radiation. The efficiency is further reduced by a few percentage points.

9.2.6 Efficiency Limit for AM1.5 Radiation

Shockley and Queisser assumed that solar radiation is a blackbody radiation [77]. The actually solar radiation spectrum received on Earth is influenced by scattering and the absorption of water vapor, carbon dioxide, and so on; thus the spectrum is different from blackbody radiation; see Section 5.2. How is the efficiency limit for actual solar

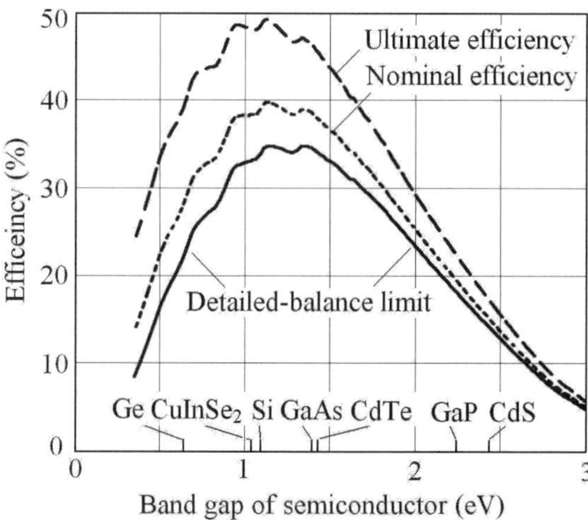

Figure 9.10 Efficiency limit of solar cells for AM1.5 solar radiation. Efficiency limit as determined by detailed balance for the AM1.5 solar radiation. The abscissa is the band gap of the semiconductor in electron volts. The values of several important solar cell materials are marked. Similar to Fig. 9.9, the ultimate efficiency, the nominal efficiency, and the detailed-balance limit of efficiency, the fundamental limit for the conditions listed by Shockley and Queisser, are shown.

radiation spectrum changed from that of blackbody radiation?

The problem can be resolved using the same method of Shockley and Queisser. Instead of using the blackbody radiation spectrum, the measured solar radiation, the standard AM1.5 spectrum (see Plate 1) is used to compute the integrals in Eqs. 9.25 and 9.45. Because the strongest absorption and scattering are in the mid- to far-infrared and ultraviolet regions, visible light is basically unchanged. Because the infrared radiation with photon energy lower than the energy gap of the semiconductor does not participate in the generation of the electron–hole pair, and ultraviolet photons would lose more energy in the thermalization process, efficiency with regard to total radiation power is slightly higher. The strong absorption peaks and valleys are smoothed out by the integration process. Therefore, the final result is qualitatively identical to that from a blackbody radiation approximation. Thus, the blackbody approximation is appropriate and useful.

9.3 Nonradiative Recombination Processes

The Shockley–Queisser limit is solely based on the thermodynamics of radiative recombination of electron–hole pairs. There are several other recombination mechanisms and factors that could limit the solar-cell efficiency. Some of the factors are intrinsic and others can be mitigated or avoided by better cell design and manufacturing.

As shown in Section 9.2.2, the most serious limiting factor to solar cell efficiency is the recombination rate of electrons and holes. According to the basic equation of solar cells (Eq. 9.5), the open-circuit voltage is determined as

$$V_{\rm oc} = \frac{k_B T}{q} \ln \frac{I_{\rm sc}}{I_0}. \tag{9.48}$$

The dark current I_0 is determined by Eq. 8.48. The solar cells are always made of a thicker, lightly doped substrate and a thinner, highly doped film. For example, the typical silicon solar cell is made of a p–n^+ junction. Therefore, only one of the two terms dominate. Thus, Eq. 8.48 can be simplified to

$$I_0 = \frac{q n_i^2}{N} \sqrt{\frac{D}{\tau}}. \tag{9.49}$$

The open-circuit voltage is

$$V_{\rm oc} = \text{const} + \frac{k_B T}{2q} \ln \tau. \tag{9.50}$$

Obviously, the greater the recombination time τ, the greater the open-circuit voltage $V_{\rm oc}$, and the efficiency will be greater.

If there are several recombination processes, the rate is additive, and thus the inverse of the recombination time is additive,

$$\frac{1}{\tau} = \frac{1}{\tau_1} + \frac{1}{\tau_2} + \ldots + \frac{1}{\tau_n}. \tag{9.51}$$

Each additional recombination process causes a reduction of the total recombination time, and thus a reduction of efficiency.

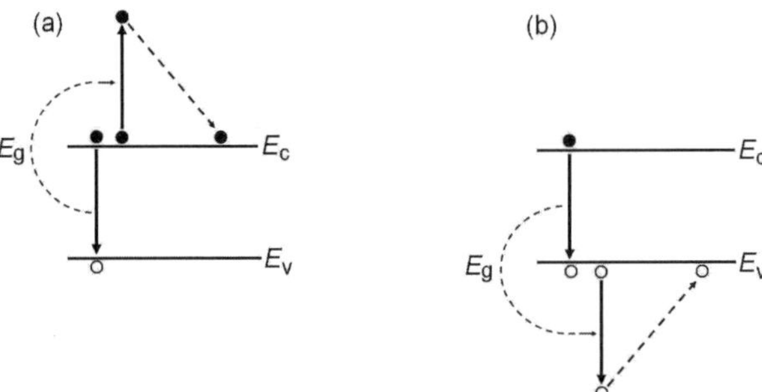

Figure 9.11 The Auger recombination process. The electron–hole pair can recombine and transfer the energy E_g into either a free electron near the conduction band edge E_c, (a), or free hole near the valance band edge, E_v, (b). Then the excited electron quickly loses its excess energy to the lattice as phonons.

Various types of nonradiative recombination processes are discussed in details in Chapter 7 of Jacques I. Pankove's book [65]. Their effect on the efficiency of solar cells is discussed in Chapter 3 of Martin A. Green's book [34] and subsequent papers [35, 83].

9.3.1 Auger Recombination

As shown in Section 9.2.3, after an electron–hole pair is created, both the free electron and free hole quickly transfer excess energy to the lattice as phonons and stay near the band edge. The electron–hole pair can recombine and emit a photon, as shown in Section 9.2.3. An alternative process, the Auger process, is to transfer the energy E_g into either a free electron near the conduction band edge, E_c, as shown in Fig. 9.11(a), or a free hole near the valance band edge, E_v, as shown in Fig. 9.11(b). Then the excited electron quickly loses its excess energy to the lattice as phonons.

Clearly, Auger recombination is an intrinsic process which cannot be eliminated by smart design. Detailed calculations and experiments have shown that, for good-quality crystalline silicon, it is the dominant recombination process besides radiative recombination, which would further reduce the theoretical efficiency from the Shockley–Queisser limit of about 32% to about 28% [35, 83].

9.3.2 Trap-State Recombination

As shown in Section 8.1.3, the impurities in a semiconductor create states in the energy gap. The gap states are effective intermediate media for a two-step recombination process; see Fig. 9.12(a). Clearly, the higher the concentration of impurities, the more the gap states, and thus the shorter the electron–hole pair lifetime. As a general guideline, high-purity semiconductor materials are preferred.

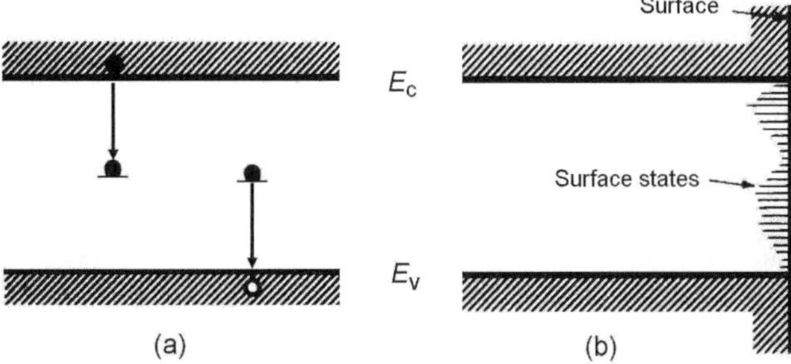

Figure 9.12 Two-step recombination processes. The electron–hole pair can recombine and transfer the energy E_g into either a free electron near the conduction band edge, E_c (a), or free hole near the valance band edge, E_v (b). Then the excited electron or hole quickly loses its excess energy to the lattice as phonons.

9.3.3 Surface-State Recombination

Surfaces of the semiconductor materials can represent a high concentration of defects, or surface states. Each surface state can become a mediator for two-step recombination; see Fig. 9.12(b). An experimentally verified effective method to reduce or eliminate surface states is *passivation*. For silicon, typical methods are either to create a Si–SiO_2 interface by oxidation, or to create a Si-Al_2iO_3 interface by depositing a thin layer of alumina. Both SiO_2 and Al_2iO_3 are insulators, and thus prevent the formation of direct metallic conductor on the surface. At both the top and back sides, only a small area is allowed to make ohmic contact. We will discuss this issue in section 9.5.2.

9.4 Antireflection Coatings

As discussed in Section 2.2.3, because all semiconductors have high refractive indices, according to the Fresnel formulas, reflection loss at the semiconductor–air interface is significant.

The solution, that is, antireflection coatings, were invented in early twenteenth century and has been applied to reduce the reflection of lenses in eyeglasses, cameras, telescopes, and microscopes. The concept of antireflection coatings is shown in Fig. 9.13. Without antireflection coatings, the reflection coefficient at the interface of two media with refractive indices n_1 and n_2 is determined by the Fresnel formula 2.78,

$$\mathcal{R} = \left(\frac{n_1 - n_2}{n_1 + n_2} \right)^2 . \tag{9.52}$$

For example, for silicon, where $n = 3.8$, $\mathcal{R} = 0.34$, which is very high. By coating the

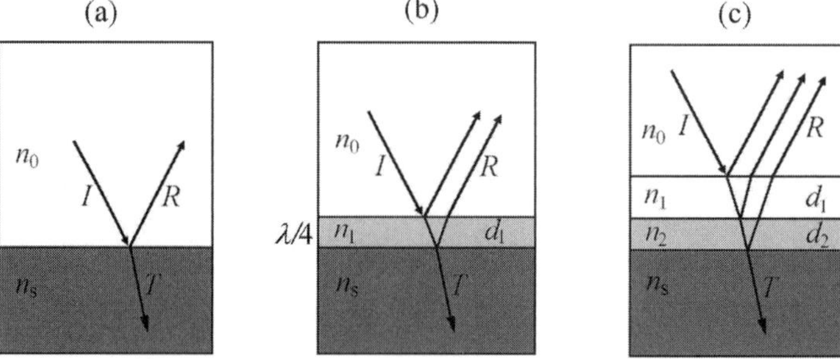

Figure 9.13 Antireflection coatings (a) At the interface of two dielectric media, the reflection coefficient is determined by the Fresnel formula. (b) By coating the interface with a film of dielectric material with an refractive index equal to the geometric mean of the indices of air and the bulk and a thickness equal to a quarter of the wavelength in that medium, the reflection can be eliminated. However, it only works for one wavelength. (c) Using multiple coatings, the wavelength range for almost zero reflection can be extended.

surface with a film of thickness equal to a quarter wavelength in that medium, the two reflected light waves should have a phase difference of 180°. If the intensities of the two reflected light waves are equal, complete cancellation can take place. The condition of cancellation is then

$$\left(\frac{n_1 - n_2}{n_1 + n_2}\right)^2 = \left(\frac{n_2 - n_3}{n_2 + n_3}\right)^2. \tag{9.53}$$

It can happen only when $n_1/n_2 = n_2/n_3$, or

$$n_2 = \sqrt{n_1 n_3}. \tag{9.54}$$

In other words, the reflection can be completely eliminated when the refractive index of the thin film is the geometric mean of the refractive indices of air and the bulk medium. Obviously, it only works for a single wavelength. As shown in Fig. 9.13(c), by using multiple coatings, two or more reflection minima can be created, and the wavelength range for almost zero reflection can be extended.

The above argument based on interference is intuitive but not accurate. Multiple reflections take place in the film. The simple interference argument does not work for multiple antireflection coatings. The standard treatment is based on a matrix method, as presented in the next section.

9.4.1 Matrix Method

A general treatment of the optics of stratified media can be found in Born and Wolf [13] and Macleod [57]. In both books, very general cases are discussed, and the mathematics is quite complicated. Here we present a simple treatment for the case of normal incidence. For antireflection coatings on solar cells, it represents most of the essential physics and is mathematically transparent. For advanced readers, it can serve as a bridge to understand more sophisticated treatments.

Consider an antireflection coating with s films each with refractive index n_j and thickness d_j; see Fig. 9.14. Consider the electromagnetic wave of one polarization with circular frequency ω; see Fig. 2.3. Because, for a vacuum, $E_x = cB_y$ (see Eq. 2.26), we combine them as a two-dimensional vector,

$$\mathbf{F}(z) = \begin{pmatrix} F_1(z) \\ F_2(z) \end{pmatrix} = \begin{pmatrix} E_x(z) \\ cB_y(z) \end{pmatrix}. \tag{9.55}$$

Because both E_x and B_y are continuous at a boundary of two layers of dielectrics, F_1 and F_2 are also continuous at a boundary of two dielectrics. Within the jth film, following Eqs. 2.41 – 2.44, Maxwell's equations for $F_1(z)$ and $F_2(z)$ are

$$\frac{dF_1(z)}{dz} = ik_0 F_2(z), \tag{9.56}$$

$$\frac{dF_2(z)}{dz} = in_j^2 k_0 F_1(z), \tag{9.57}$$

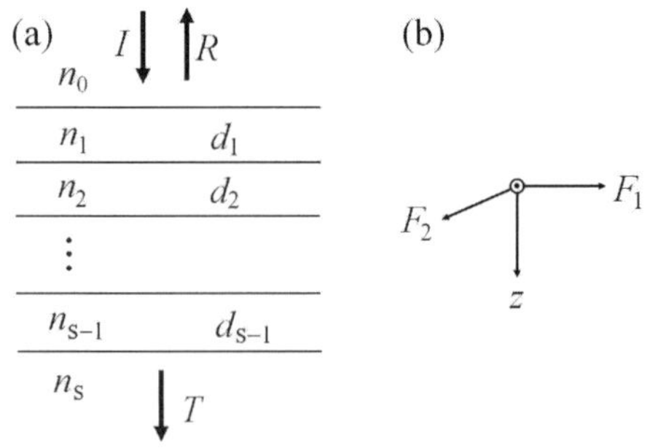

Figure 9.14 Matrix method for antireflection coatings. For each layer of antireflection coating, Maxwell's equations for a continuous medium is valid. Its effect can be represented by a 2 × 2 matrix. (a) The incident, reflected, and transmitted light can be represented by a two-dimensional vector for each. (b) The field vectors and the z-direction.

(a) n_0 I R; n_1 d_1; n_2 d_2; ⋮; n_{S-1} d_{S-1}; n_S T

(b) F_2 F_1 z

where $k_0 = \omega/c$ as the wavevector of the electromagnetic wave in a vacuum. Both F_1 and F_2 satisfy the following second-order differential equations:

$$\frac{d^2 F_1(z)}{dz^2} + n_j^2 k_0^2 F_1(z) = 0, \tag{9.58}$$

$$\frac{d^2 F_2(z)}{dz^2} + n_j^2 k_0^2 F_2(z) = 0, \tag{9.59}$$

There are two easily verifiable solutions: the forward-running wave,

$$\mathbf{F}(z) = \begin{pmatrix} 1 \\ n_j \end{pmatrix} F_0 \, e^{in_j k_0 z} \tag{9.60}$$

and the backward-running wave,

$$\mathbf{F}(z) = \begin{pmatrix} 1 \\ -n_j \end{pmatrix} F_0 \, e^{-in_j k_0 z}. \tag{9.61}$$

In general, the field is a linear combination of both waves. To obtain the general solution in a multilayer film system, we write down the solutions of Eqs. 9.56 and 9.57 with boundary conditions $F_1(0)$ and $F_2(0)$ at $z = 0$. The solutions can be easily verified as

$$F_1(z) = F_1(0) \cos k_0 n_j z + \frac{i}{n_j} F_2(0) \sin k_0 n_j z, \tag{9.62}$$

$$F_2(z) = in_j \, F_1(0) \sin k_0 n_j z + F_2(0) \cos k_0 n_j z. \tag{9.63}$$

Equations 9.62 and 9.63 can be written conveniently in matrix form:

$$\begin{pmatrix} F_1(z) \\ F_2(z) \end{pmatrix} = \begin{pmatrix} \cos k_0 n_j z & \dfrac{i}{n_j} \sin k_0 n_j z \\ in_j \sin k_0 n_j z & \cos k_0 n_j z \end{pmatrix} \begin{pmatrix} F_1(0) \\ F_2(0) \end{pmatrix}. \tag{9.64}$$

Introducing the two-by-two matrix

$$\mathsf{M}_j(z) = \begin{pmatrix} \cos k_0 n_j z & \dfrac{i}{n_j} \sin k_0 n_j z \\ in_j \sin k_0 n_j z & \cos k_0 n_j z \end{pmatrix}, \tag{9.65}$$

Eq. 9.64 can be written in concise form

$$\mathbf{F}(z) = \mathsf{M}_j(z)\,\mathbf{F}(0). \tag{9.66}$$

The matrix format has some interesting properties. By direct arithmetic, it is easy to prove that

$$\mathsf{M}_j(z_1 + z_2) = \mathsf{M}_j(z_1)\,\mathsf{M}_j(z_2), \tag{9.67}$$

and then

$$\mathbf{F}(z_1 + z_2) = \mathsf{M}_j(z_1 + z_2)\,\mathbf{F}(0). \tag{9.68}$$

Using the inverse matrix

$$\mathsf{M}_j^{-1}(z) = \begin{pmatrix} \cos k_0 n_j z & -\dfrac{i}{n_j}\sin k_0 n_j z \\ -in_j \sin k_0 n_j z & \cos k_0 n_j z \end{pmatrix}, \tag{9.69}$$

where it can be verified directly that $\mathsf{M}_j(z)\,\mathsf{M}_j^{-1}(z) = \mathsf{M}_j^{-1}(z)\,\mathsf{M}_j(z) = 1$, a reverse equation can be established,

$$\mathbf{F}(0) = \mathsf{M}_j^{-1}(z)\,\mathbf{F}(z). \tag{9.70}$$

Because at the boundaries the x- and y-components of E and B are continuous, for a series of films with thicknesses d_j,

$$\mathbf{F}(d) = \mathsf{M}_{s-1}(d_{s-1})\,\mathsf{M}_{s-2}(d_{s-2})\ldots\mathsf{M}_2(d_2)\,\mathsf{M}_1(d_1)\,\mathbf{F}(0), \tag{9.71}$$

where d is the total thickness of the antireflection films. The inverse expression is used more often,

$$\mathbf{F}(0) = \mathsf{M}_1^{-1}(d_1)\,\mathsf{M}_2^{-1}(d_2)\ldots\mathsf{M}_{s-2}^{-1}(d_{s-2})\,\mathsf{M}_{s-1}^{-1}(d_{s-1})\,\mathbf{F}(d). \tag{9.72}$$

In the following, we will treat the problems of single-layer antireflection (SLAR) coatings and double-layer antireflection (DLAR) coatings.

9.4.2 Single-Layer Antireflection Coating

The reflectance of the SLAR coating can be easily calculated using the matrix method. Because on the semiconductor side there is only transmitted light, the calculation starts with the field vector of transmitted light,

$$\mathbf{F}(0) = \mathsf{M}_1^{-1}(d_1)\,\mathbf{F}(d). \tag{9.73}$$

Since we are only interested in the ratio of the intensities of the incident light and the reflected light, the absolute magnitude and phase of the light waves are not important. Following Eq. 9.60, the transmitted light can be represented by

$$\mathbf{F}(d) = \begin{pmatrix} 1 \\ n_s \end{pmatrix}, \tag{9.74}$$

where n_s is the refractive index of the substrate. The field vector $\mathbf{F}(0)$ is a mixture of incident and reflected light waves. It can be separated by a projection matrix

$$\mathsf{P} = \begin{pmatrix} n_0 & -1 \\ n_0 & 1 \end{pmatrix}. \tag{9.75}$$

In fact, from Eqs. 9.60 and 9.61, at $z = 0$,

$$\mathsf{P}\,\mathbf{F}(0) = \mathsf{P}\left[\begin{pmatrix} 1 \\ n_0 \end{pmatrix} I + \begin{pmatrix} 1 \\ -n_0 \end{pmatrix} R\right] = 2n_0 \begin{pmatrix} R \\ I \end{pmatrix}. \tag{9.76}$$

Since we are interested in the ratio of R and I, the factor $2n_0$ has no effect. Combining Eqs. 9.69 and 9.73 – 9.75, we find

$$R = (n_0 - n_s)\cos\delta_1 - i\left(n_1 - \frac{n_0\, n_s}{n_1}\right)\sin\delta_1, \tag{9.77}$$

$$I = (n_0 + n_s)\cos\delta_1 - i\left(n_1 + \frac{n_0\, n_s}{n_1}\right)\sin\delta_1, \tag{9.78}$$

where δ_1 is the phase shift of the film,

$$\delta_1 = n_1 k_0 d_1. \tag{9.79}$$

The reflectivity is

$$\mathcal{R} = \left|\frac{R}{I}\right|^2 = \frac{n_1^2(n_0 - n_s)^2\cos^2\delta_1 + (n_1^2 - n_0\, n_s)^2\sin^2\delta_1}{n_1^2(n_0 + n_s)^2\cos^2\delta_1 + (n_1^2 + n_0\, n_s)^2\sin^2\delta_1}. \tag{9.80}$$

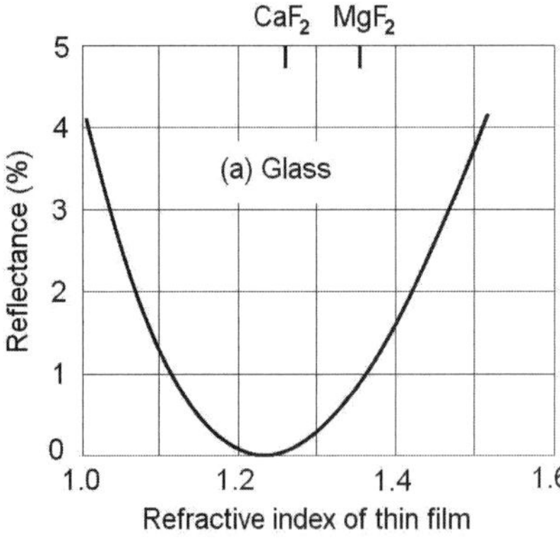

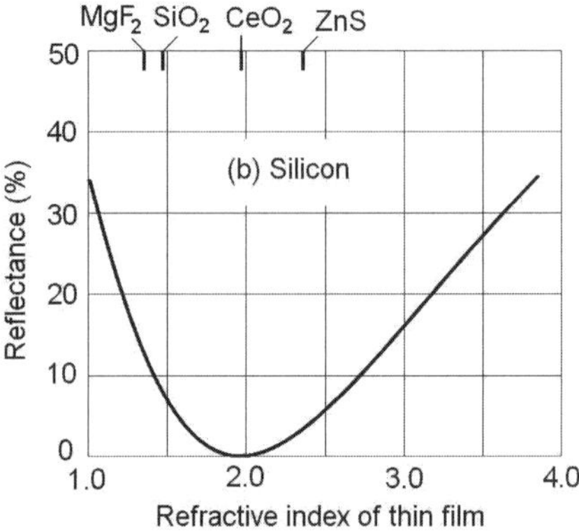

Figure 9.15 Choice of materials for SLAR coatings. The minimum reflectance for an SLAR coating is determined by the refractive index of the material; see Eq. 9.81. Two cases are shown. For glass, calcium fluoride and magnesium fluoride are the best choices. For silicon, cerium (ceric) oxide is the best choice.

Two special cases are worth noting. If the thickness of the coating is one quarter of the wavelength in that medium, that is, $\delta_1 = \pi/2$, then

$$\mathcal{R} = \left(\frac{n_1^2 - n_0\, n_s}{n_1^2 + n_0\, n_s}\right)^2 . \tag{9.81}$$

When $n_1^2 = n_0\, n_s$, the reflectivity is zero. This verifies the result based on the naïve interference argument resulting in Eq. 9.54.

If the thickness of the coating is one-half of the wavelength in that medium, that is, $\delta_1 = \pi$, then

$$\mathcal{R} = \left(\frac{n_0 - n_s}{n_0 + n_s}\right)^2 , \tag{9.82}$$

which coincides with the Fresnel formula 2.80, as if the antireflection coating does not exist.

9.4.3 Double-Layer Antireflection Coatings

The above treatment can be extended to double-layer antireflection (DLAR) coatings readily. The mathematical details are left as an exercise. The general result for the reflectance is

$$\mathcal{R} = \frac{R_1^2 + R_2^2}{I_1^2 + I_2^2} , \tag{9.83}$$

where

$$\begin{aligned}
R_1 &= (n_0 - n_s)\cos\delta_1\,\cos\delta_2 - \left(\frac{n_0 n_2}{n_1} - \frac{n_s n_1}{n_2}\right)\sin\delta_1\,\sin\delta_2,\\
R_2 &= \left(\frac{n_0 n_s}{n_2} - n_2\right)\cos\delta_1\,\sin\delta_2 + \left(\frac{n_0 n_s}{n_1} - n_1\right)\sin\delta_1\,\cos\delta_2,\\
I_1 &= (n_0 + n_s)\cos\delta_1\,\cos\delta_2 - \left(\frac{n_0 n_2}{n_1} + \frac{n_s n_1}{n_2}\right)\sin\delta_1\,\sin\delta_2,\\
I_2 &= \left(\frac{n_0 n_s}{n_2} + n_2\right)\cos\delta_1\,\sin\delta_2 + \left(\frac{n_0 n_s}{n_1} + n_1\right)\sin\delta_1\,\cos\delta_2,
\end{aligned} \tag{9.84}$$

and the phase shift of the jth film is given as

$$\delta_j = n_j k_0 d_j . \tag{9.85}$$

As a special case, if the thicknesses of both films are a quarter-wavelength, where $\delta_1 = \delta_2 = \pi/2$, all cosines are zero and all sines are unity. We have

$$\mathcal{R} = \left|\frac{R}{I}\right|^2 = \left(\frac{n_0 n_2^2 - n_s n_1^2}{n_0 n_2^2 + n_s n_1^2}\right)^2 . \tag{9.86}$$

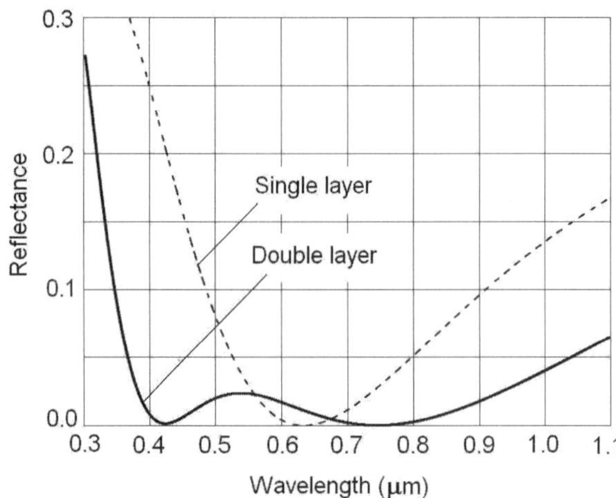

Figure 9.16 Wavelength range of antireflection coatings. Single-layer antireflection coating has one minimum-reflectance wavelength. Shown here is the reflectance for a 80-nm-thick CeO_2 SLAR film on silicon substrate. DLAR coatings can have two minimum-reflectance wavelengths. Shown here is the reflectance of a 101-nm ZnS film and a 56-nm MgF_2 film on a silicon substrate. The wavelength range of low reflectance is greatly increased. See Ref. [91].

The condition for zero reflectance is

$$n_0 n_2^2 = n_s n_1^2. \tag{9.87}$$

Therefore, the choice of materials can be broadened.

The major advantage of multilayer antireflection coatings is the wavelength range. As discussed in the previous section, even if the best material is chosen, for an SLAR coating, complete reflection cancellation can only happen for a single wavelength. For DLAR coatings, this can happen for two different wavelengths. This can be seen from the expression of reflectance $\mathcal{R}$. If the condition in Eq. 9.87 is satisfied, when $\cos\delta_1$ and $\cos\delta_2$ vanishes at two different wavelengths, reflection is eliminated at both wavelengths. Figure 9.16 shows two cases used in manufacturing. The dashed curve is the reflectance for an 80-nm-thick CeO_2 SLAR film on silicon. The solid curve is the reflectance of a DLAR cotings with a 101-nm-ZnS film and a 56-nm MgF_2 film on a silicon substrate, which provides two minimum-reflectance wavelengths. The wavelength range of low reflectance is greatly increased. See Ref. [91].

9.5 Crystalline Silicon Solar Cells

The first practical solar cell, invented in 1954, used crystalline silicon. To date, crystalline silicon solar cells still have 80 – 90% of market share. The material has many advantages:

1. Silicon, comprising 27% of Earth's crest, is the second most abundant element after oxygen.
2. Its band gap is almost optimum regarding the solar spectrum.
3. Chemically, silicon is very stable.
4. Silicon is nontoxic.
5. Because of the microelectronics industry, the production and processing of ultra-pure silicon are well developed.
6. After more than 50 years of research and development, the efficiency of silicon solar cells, 24.7% for research prototypes, is already close to its theoretical limit. The mass-produced modules, limited for cost reduction considerations, have reached 20% efficiency.

9.5.1 Production of Pure Silicon

The raw material for silicon, silica, is the most abundant mineral on Earth, including quartz, chalcedony, white sand, and numerous noncrystalline forms. The first step in silicon production is to reduce silica with coke to generate metallurgical-grade silicon,

$$\mathrm{SiO_2 + C \longrightarrow Si + CO_2}. \tag{9.88}$$

The silicon thus produced is typically 98% pure. For applications in solar cells, at least 99.9999% purity is required, so-called solar-grade silicon. However, many processes can generate silicon with impurity levels less than 10^{-9}, which is needed for high-efficiency solar cells.

Two processes are commonly used. In the Siemens process, high-purity silicon rods are exposed to trichlorosilane at 1150°C. The trichlorosilane gas decomposes and deposits additional silicon onto the rods, enlarging them:

$$\mathrm{2HSiCl_3 \longrightarrow Si + 2HCl + SiCl_4}. \tag{9.89}$$

Silicon produced from this and similar processes is called polycrystalline silicon. The byproduct, silicon tetrachloride, cannot be reused and becomes waste. The energy consumption in this process is also significant.

In 2006 Renewable Energy Corporation in Norway (REC) announced the construction of a plant based on fluidized-bed technology using silane, taking silicon tetrachloride as the starting point:

$$\begin{aligned} \mathrm{3SiCl_4 + Si + 2H_2} &\longrightarrow \mathrm{4HSiCl_3}, \\ \mathrm{4HSiCl_3} &\longrightarrow \mathrm{3SiCl_4 + SiH_4}, \\ \mathrm{SiH_4} &\longrightarrow \mathrm{Si + 2H_2}. \end{aligned} \tag{9.90}$$

The purification process takes place at the silane (SiH_4) stage. According to REC, the energy consumption of this new process is significantly reduced from the Siemens process. Also, using the almost free hydroelectric power in Norway, REC is expecting to reduce the cost of solar-grade pure silicon to less than $20 per kilogram.

To prepare for solar cell production, pure silicon can go through two alternative processes. For single-crystal silicon solar cells, the Czochraski process or the float-zone process is applied to generate single-crystal silicon ingots. Alternatively, the pure silicon can be melted in an oven to produce polycrystalline ingots.

9.5.2 Solar Cell Design and Processing

The efficiency of solar cells has gradually improved since the invention of the silicon solar cell in 1954. Recently, the 25% efficiency of laboratory prototypes of monocrystalline silicon solar cells has approached its theoretical limit; see Plate 5.

There are several silicon solar cell designs. Here we present one by the University of New South Wales in Australia, the passivated emitter, rear locally diffused (PERL) cells (Fig. 9.17), which has achieved up to 24.7% efficiency under standard global solar spectra; see Refs. [91] and [36].

The design of the PERL solar cell is shown in Fig. 9.17. In addition to using a rather thick wafer (370- or 400-μm) high-quality monocrystalline silicon, it has several features that enhance efficiency.

Passivation and Metal Contacts

To reduce surface recombination, both sides of the wafer are passivated with a layer of silicon dioxide. Because SiO_2 is an insulator, metal contacts must be made with small holes in the SiO_2 film. By locally diffusing boron to the rear contact areas, the effective

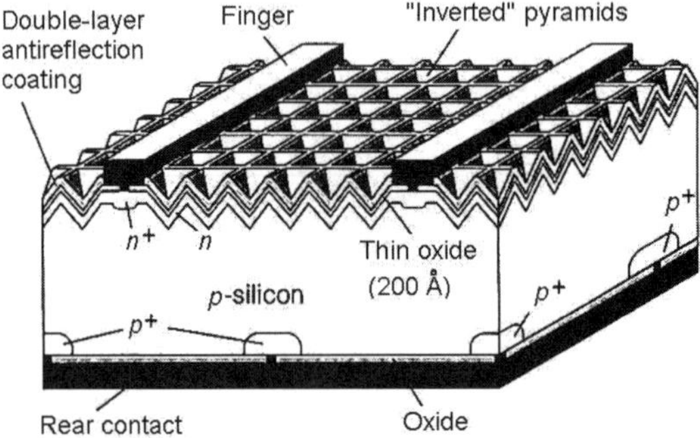

Figure 9.17 Typical high-efficiency silicon solar cell. The front side of the relatively thick silicon wafer is textured to capture light. A double-layer antireflection film is applied. The back side is passivated with silicon dioxide to reduce surface recombination. The rear contacts are enabled through highly doped $p+$ regions. After Ref. [91].

recombination rate is reduced by suppressing minority-carrier concentration in these regions. The doping process uses BBr_3 as the source dopant. The contact resistance is also markedly reduced. On the front side, heavy phosphorous doping is made using liquid PBr_3 as the dopant carrier. The width of the front metal contact is reduced to expand the

Textured Front Surface

The front surface is textured to become a two-dimensional array of inverted pyramids. The photons entering the substrate can be trapped by the textured top surface.

Double-Layer Antireflection Coating

As presented in Section 9.4, DLAR coatings can significantly improve overall efficiency. A ZnS and MgF_2 DLAR coating is evaporated onto the cells.

9.5.3 Module Fabrication

Single silicon solar cells are fragile and vulnerable to the elements. In all applications, silicon solar cells are framed and protected to become solar modules. Figure 9.18 shows a cross section of a typical solar module. To manufacture such a module, a piece of low-iron window glass, two sheets of ethylene vinyl acetate (EVA) film (with typical thickness of 0.5 mm), a rectangular array of solar cells, and a back plate, either a Mylar film or a sheet of metal, are stacked together in a heated press machine. The EVA is softened at about 150°C, then tightly bonded with the solar cells and other components. Finally, the glass module is framed by a metal structure with a protection gasket.

The monocrystalline solar module and polycrystalline solar module can be identified by the look. Figure 9.19 shows the two types of solar nodules. Figure 9.19(a) is a monocrystalline solar module. The solar cells are cut from a cylindrical single crystal

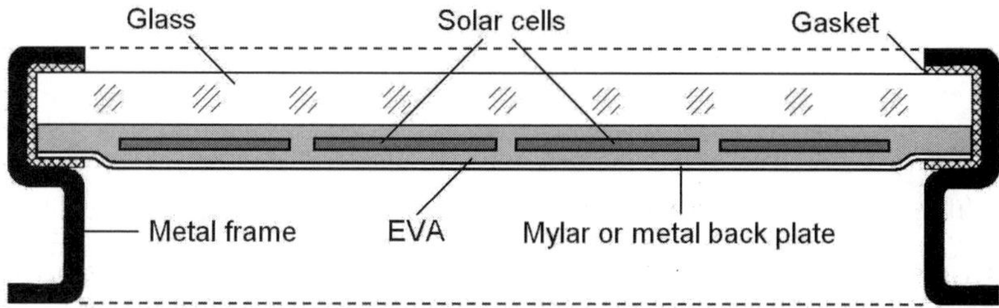

Figure 9.18 Cross section of typical solar module. A complete solar module is made of a piece of low-iron window glass, two sheets of EVA film, an array of solar cells, and a back plate. Those components are bonded together in a heated press.

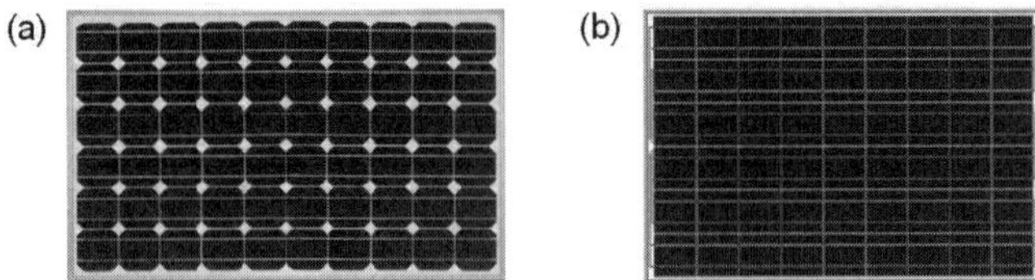

Figure 9.19 Monocrystalline solar module and polycrystalline solar module. (a) The monocrystalline solar cells are cut from a cylindrical single crystal. To save material and space, the solar cells are cut to an octagonal piece. There is always some wasted space because of the cut corners. (b) The polycrystalline solar cells are cut from a rectangular ingot. The solar cells are usually perfect squares. There is no wasted space.

to become an octagonal piece. Figure 9.19(b) is a module made of polycrystalline solar cells which are cut from a rectangular ingot.

9.6 Thin-Film Solar Cells

Silicon as a solar cell material has many advantages. However, it also has a disadvantage. As shown in Section 9.1, silicon is an indirect semiconductor. The absorption coefficient near its band edge is low. Therefore, a fairly thick substrate is required. The wafer is cut from a single crystal or an polycrystal ingot. The minimum thickness to maintain reasonable absorption and mechanical strength is 0.1 – 0.2 mm. The cost of the material and mechanical processing is substantial. The direct semiconductors often have an absorption coefficient one or two orders of magnitude higher than silicon; see Fig. 9.3. For those materials, a thickness of a few micrometers is sufficient. In addition to a high absorption coefficient near the band gap which is close to 1 eV, there are many other factors which determine the practicality of making a solar cell. To date, besides silicon, only two such materials have reached the status of mass production, namely, cadmium telluride (CdTe) and copper indium–gallium diselenade, Cu(InGe)Se_2, often called CIGS. However, the material cost for these is still high. Amorphous-silicon thin-film solar cells, in spite of their relatively low efficiency, are mass produced for applications where a low efficiency is tolerable; see Table 1.6.

9.6.1 CdTe Solar Cells

Due to its high absorption coefficient and ease in making a p–type material, cadmium telluride (CdTe) is currently the most popular material for thin-film solar cells [16, 19, 69]. Another advantage is its compatibility with CdS, a wide-band-gap semiconductor for which it is easy to generate an n–type film. Because the absorption edge of CdS is 2.4 eV, it is transparent to the bulk of solar radiation. The typical structure of a CdTe solar cell is shown in Fig. 9.20. As shown, the solar cell is sandwiched between two sheets of window glass. The solar cell is made of a 5-μm film of CdTe, covered with a 100-nm CdS thin film, to form a pn-junction. To the sunny side is a film of TCO, to

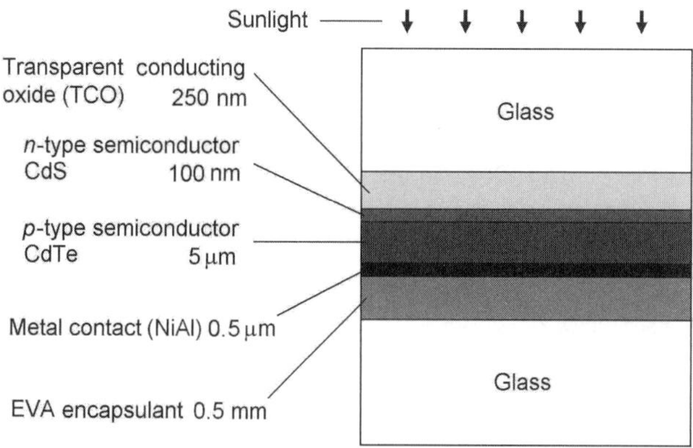

Figure 9.20 Typical structure of CdTe thin film solar cell. The *pn*-junction made of CdTe and CdS is sandwiched between two glass plates. See Refs. [16], [19], and [69].

allow radiation to come in and serves as a conductor. To the back side is a metal film for electric contact. A 0.5-mm EVA film is added for mechanical protection. The best efficiency of CdTe solar cells is 16.5%, and expected to reach 20%.

One question often asked is about the toxicity of cadmium. According to a recent study, because the quantity of cadmium is very small and is completely sealed by glass, the environment impact is negligible.

The largest manufacturer of CdTe solar cells is First Solar, with headquarters in Tempa, Arizona. For several years since 2002, First Solar was the largest manufacturer of solar cells in the world. Only in late 2011 it was dethorned by Suntech. In September 2009, First Solar signed a contract with China to built a 2 GW solar field in Ordos, Inner Mongolia. It is by far the largest solar field under development to date.

9.6.2 CIGS Solar Cells

The $CuInSe_2/CdS$ system was discovered in 1974 as a photovoltaic light detector [87]. In 1975, a solar cell was built with this materials which showed an efficiency comparable to the silicon solar cells at that time [76]. In 2000s, the efficiency of CIGS thin-film solar cells has reached 19.9%, comparable to polycrystalline silicon solar cells [19, 69, 78], see Plate 5. As shown in Fig. 9.3, the band gap of $CuInSe_2$ is very close to the optimum value, but its absorption coefficient is about 100 times greater than silicon. Therefore, even if the semiconductor film were as thin as 2 μm, more than 90% of the near-infrared and visible light would be absorbed.

The typical structure of a CIGS solar cell is shown in Fig. 9.21. Similar to the CdTe solar cells, a 50-nm *n*-type CdS film is used to form a *pn*-junction. Again, because the quantity of cadmium used here is very small, and sandwiched between two glass plates, the environment impact is negligible.

The CIGS solar cell can be produced with a wet process, without requiring a vac-

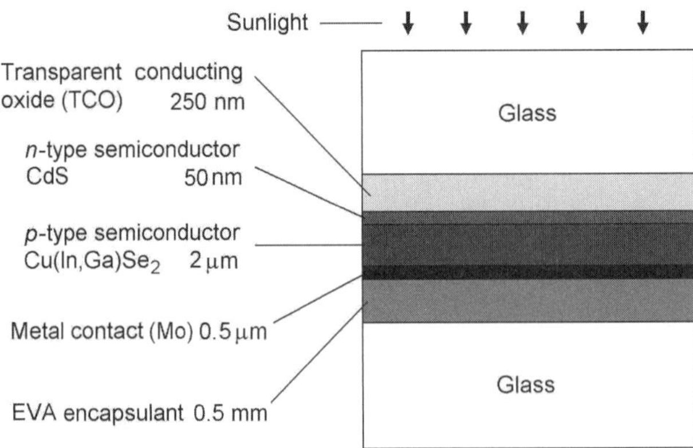

Figure 9.21 Typical structure of CIGS thin-film solar cell. A *pn*-junction of $Cu(InGa)Se_2$ and CdS is sandwiched between two glass plates. See Refs. [19], [69], [76], [78], and [87].

uum. Therefore, the manufacturing cost can be low. Another advantage of CIGS solar cells is that interconnections can be made on the same structure, similar to an integrated circuit, thus a higher voltage single cell, for example 12 V, can be made without requiring external connections. Figure 9.22 is an experimental 5-V CIGS solar cell, made from 10 individual CIGS solar cells in a single glass envelop. The boundaries between adjacent solar calls are apparent.

9.6.3 Amorphous Silicon Thin-Film Solar Cells

Because the cost of some key materials in CdTe and CIGS thin-film solar cells, notably tellurium and indium, is high, silicon thin-film solar cells, in spite of their low efficiency, have been mass produced for many years. For applications where efficiency is not crucial, such as hand-held calculators and utility-scale solar fields in deserts, silicon thin-film solar cells provide an advantage. Especially, silicon thin-film solar cells can

Figure 9.22 CIGS solar cell integrated circuit. An experimental solar cell made of 10 individual CIGS cells. The nominal voltage of each CIGS solar cell is 0.5 V. By connecting 10 individual CIGS solar cells in series internally, a 5-V solar cell is built. Because the connections are integrated, the device is compact and rugged. Photo taken by the author.

be manufactured on flexible substrates.

A major disadvantage of silicon is its low absorption coefficient. However, by substantially doping amorphous silicon with hydrogen, up to 10%, its absorption coefficient can be made as high as 10^5 cm^{-1}, with the band gap shifted from 1.1 eV to 1.75 eV, similar to that of CdTe. In the literature, this material is often abbreviated as a-Si:H [19]. Because of high defect density, the recombination rate in high. The efficiency of the best experimental a-Si:H solar cells is around 10%, and for the mass-produced solar cells, it is around 5%.

9.7 Tandem Solar Cells

As discussed in Section 9.2.1, over the entire solar spectrum, the highest efficiency comes from photons with energy just above the band gap of the semiconductor material. For photons with energy lower than the band gap, the semiconductor is transparent. There is no energy conversion. For photons with energy much higher than the band gap, the energy of the electron–hole pair quickly relaxes to that of the energy gap. The excess photon energy over the band gap is lost. Therefore, by stacking two or more solar cells in tandem, the efficiency can be much higher than the Shockley–Queisser limit.

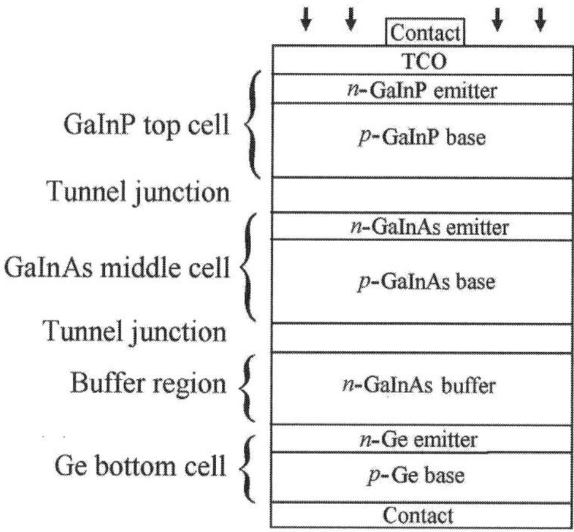

Figure 9.23 Multijunction tandem solar cell. Through the TCO layer, the sunlight falls on the top cell first. The semiconductor material of the top cell has a large band gap, in this case GaInP, 1.9 eV. It is transparent to photons of energy smaller than 1.9 eV. Photons with energy greater than 1.9 eV will generate an electron–hole pair with energy about 1.9 eV. The middle cell, made of GaInAs, has band gap 1.35 eV. Photos with energy between 1.35 and 1.9 eV will generate an electron–hole pair of about 1.35 eV. The bottom cell is made of Ge, with a band gap of 0.67 eV. For photons with energy greater than 0.67 eV and smaller than 1.35 eV, an electron–hole pair of about 0.67 eV is generated. The voltage is additive, and thus it can generate much more power than the single cells, often exceeding the Shockley–Queisser limit.

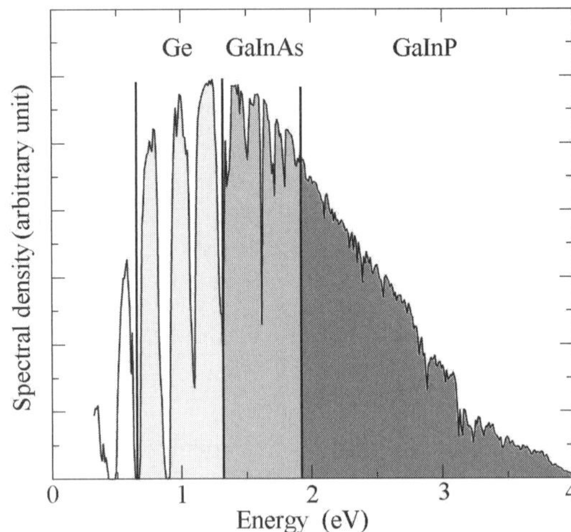

Figure 9.24 Working principle of multijunction tandem solar cells. The solar spectrum is divided into three sections. Each section generates a voltage. The tandem cell can be considered as a solar battery made of three cells. The voltage is additive, and thus it could generate much more power than the single cells, often exceeding the Shockley–Queisser limit.

Figure 9.23 shows the schematic of a three-junction tandem cell. The top cell is made of GaInP, with a band gap of 1.9 eV. The photons with energy greater than 1.9 eV will generate an electron–hole pair with energy about 1.9 eV. For photons with energy smaller than 1.9 eV, that GaInP layer is transparent. The middle cell, made of GaInAs, with a band gap of 1.35 eV, would absorb photons with energy between 1.35 and 1.9 eV, and generate an electron–hole pair of about 1.35 eV. The GaInAs film is transparent to photons with energy smaller than 1.35 eV. Those photons then go to a layer of Ge, with a band gap of 0.67 eV. For photons with energy greater than 0.67 eV and smaller than 1.35 eV, an electron–hole pair is generated in the bottom cell. The current should be continuous, but the voltage is additive. Therefore, for solar radiation with a rich spectrum, the tandem cell can generate much more power than the single cells, thus often exceeding the Shockley–Queisser limit. Recently, a group at Spectrolab has demonstrated a tandem solar cell of efficiency exceeding 40%. See [43].

Much of the material used in tandem cells is expensive. The major application of tandem solar cells is with concentrated solar radiation. By concentrating solar radiation 100 folds or more, the area of the solar cell is less than 1% without concentration. Economically it can be even better than crystaline silicon solar cells.

Problems

9.1. The ultimate efficiency of Shockley and Queisser can be easily computed by expanding the denominator of Eq. 9.25,

$$\begin{aligned}\eta_u(x_s) &= \frac{15}{\pi^4}\,x_s \int_{x_s}^{\infty} \frac{x^2\,dx}{e^x - 1} \\ &= \frac{15}{\pi^4} x_s \sum_{n=1}^{\infty} \int_{x_s}^{\infty} e^{-nx}\, x^2\, dx.\end{aligned} \tag{9.91}$$

Prove that

$$\eta_u(x_s) = \frac{15}{\pi^4} x_s \sum_{n=1}^{\infty} e^{-nx_s} \left[\frac{x_s^2}{n} + \frac{2x_s}{n^2} + \frac{2}{n^3} \right]. \tag{9.92}$$

9.2. If the temperature of the solar cell can be maintained at room temperature, what is the effect of concentration of the sunlight by 100 times on the efficiency of solar cells? For blackbody radiation, work out the enhancement for a silicon solar cell.
Hint: Instead of using the usual geometric factor $f = 2.15 \times 10^{-5}$, using a larger number, for example, $100f$ in Eq. 9.45.

9.3. The absorption coefficients of some frequently used solar cell materials at the center of the visible-light spectrum are listed in Table 9.1. To achieve a total absorption of 95%, what is the required thickness?

9.4. A typical solar cell is made from a 10-cm × 10-cm piece of single-crystal silicon. Reverse saturation current I_0 is 3.7×10^{-11} A/cm^2. For V=0, 0.05, 0.1, ... , 0.55, 0.6V, calculate the forward current for the solar cell at room temperature.

9.5. For a typical silicon solar cell, the acceptor concentration in the p-region is $N_A = 1 \times 10^{16}$ /cm^3, and the donor concentration in the n-region is $N_D = 1 \times 10^{19}$/cm^3. Assuming that the built-in potential is 0.5 V, calculate the pn-junction capacitance of the 10-cm × 10-cm solar cell. (The permittivity of free space is $\varepsilon_0 = 8.85 \times 10^{-14}$F/cm, and the relative permittivity of silicon is $\varepsilon_r = 11.8$.)

Hint: the donor concentration is too high, thus the thickness of the n-type region is negligible. Only the acceptor concentration is needed for calculation. The dielectric constant (permittivity) of silicon is the product of the permittivity of free space and the relative permittivity of silicon.

9.6. For an interface of air and glass with refractive index n, show that the transmittance of each interface is

$$\tau = \frac{4n}{(1+n)^2}. \tag{9.93}$$

Verify the validity of the relation by showing that if n=1, $\tau = 1$.

9.7. If a solar device is covered with N sheets of glass with refractive index n, show that the transmittance of the entire cover is

$$\tau = \frac{(4n)^{2N}}{(1+n)^{4N}}. \tag{9.94}$$

9.8. A solar cell of 100 square centimeter area has a reversed-bias dark current of 2×10^{-9} A, and a short-circuit current at one sun (1 kW/m^2) of 3.5 A. At room temperature, what is the open-circuit voltage? What is the optimum load resistance ($V_{\text{mp}}/I_{\text{mp}}$)? What is the maximum power output?

9.9. A typical silicon solar cell has the following parameters:

Band gap of silicon is 1.1 eV.

The p-type material has an acceptor concentration $N_A = 1 \times 10^{16}$ cm^3, with hole diffusion coefficient $D_p = 40$ cm^2/s and lifetime $\tau_p = 5\,\mu$s.

The n-type material has a donor concentration $N_D = 10^{19}$ cm^3, with electron diffusion coefficient $D_n = 40$ cm^2/s and lifetime $\tau_n = 1\mu$s.

The intrinsic carrier concentration in silicon is $n_i = 1.5 \times 10^{10}$ cm^3.

For a 1-cm × 1-cm solar cell, calculate the following:

1. Reverse saturation current I_0
2. Short-circuit current under one sun solar radiation
3. Open-circuit voltage at room temperature

Chapter 10

Solar Electrochemistry

In the previous chapter, we discussed semiconductor solar cells, where photons from the Sun generate electron–hole pairs, and then the energy in the electron–hole pair is converted into electric energy. Solar photochemistry follows a different route: photons from the Sun causes a molecule to go from its ground state to an excited state. The energy stored in the excited molecule can be converted into either electrical energy, or permanent chemical energy. The most important example of solar photochemistry is photosynthesis, the conversion of solar energy into chemical energy stored in organic material, such as glucose.

10.1 Physics of Photosynthesis

To date, most of the energy human society uses originates from photosynthesis. The energy stored in food and traditional fuels, such as firewood, hey, and vegetable and animal oil, comes directly or indirectly from photosynthesis. Fossil fuel is the remains of ancient organisms, that is, stored product of photosynthesis. As the energy source of all life on Earth, photosynthesis is a natural process that has evolved in the last one billion years through natural selection. The study of photosynthesis will inspire us to create high-efficiency systems to harvest solar energy. For more details on photosynthesis, see Blankenship [11] and Voet [86].

Photosynthesis is arguably the most important chemical reaction on Earth. As evidence, nine Nobel Prizes in Chemistry have awarded for research of photosynthesis:

1915: Richard Martin Wilstaetter, for chlorophyll purification and structure, carotenoids, anthocyanins.

1930: Hans Fischer, for haemin synthesis, chlorophyll chemistry.

1937: Paul Karrer, for carotenoid structure, flavins, vitamin B_2.

1938: Richard Kuhn, for carotenoids, vitamins.

1961: Melvin Calvin, for carbon dioxide assimilation.

1965: Robert Burns Woodward, for total synthesis of vitamin B_{12}, chlorophyll, and other natural products.

1978: Peter Mitchell, for oxidative and photosynthetic phosphorylation, chemiosmotic theory.

1988: Hartmut Michel, Robert Huber, and Johannes Deisenhofer, for X-ray structure of the bacterial photosynthetic reaction center.

1992: Rudolph Marcus, for electron transfer theory and application to the primary photosynthetic charge separation.

1997: Paul D. Boyer and John E. Walker, for elucidation of enzymatic mechanism underlying the synthesis of adenosine triphosphate (ATP).

A typical chemical equation for photosynthesis is the generation of glucose or fructose, $C_6H_{12}O_6$, from carbon dioxide, water, and energy from solar radiation:

$$6CO_2 + 6H_2O + 29.79\,\text{eV} \longrightarrow C_6H_{12}O_6 + 6O_2. \tag{10.1}$$

To understand the elementary process, reaction 10.1 is often written as

$$CO_2 + H_2O + 4.965\,\text{eV} \longrightarrow \frac{1}{6}(\text{glucose}) + O_2. \tag{10.2}$$

Because the typical energy of a photon from sunlight is 1–3 eV, the above process is inevitably a multiphoton process.

10.1.1 Chlorophyll

Although there have been many different types of photosynthetic processes, only very few survived the natural selection process to support life on Earth. Photosynthesis relies on chlorophyll, the green pigment on plants. The word chlorophyll is derived from two Greek words, *cholos* ("green") and *phyllon* ("leaf").

The most common chlorophyll in plants and algae is chlorophyll *a*. The chemical structure is shown in Fig. 10.1. It is a squarish planar molecule about 1 nm on each side. A Mg atom at the center of the molecule is coordinated with four nitrogen atoms. Each nitrogen atom is part of a pyrrole ring. At the external sites groups such as CH_3

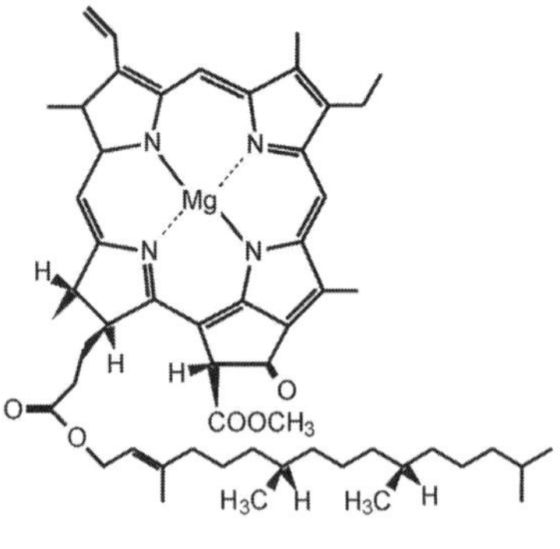

Figure 10.1 Chlorophyll. Chemical structure of the most common chlorophyll, chlorophyll *a*. At the center of the squarish molecule is a Mg atom. Each nitrogen atom is part of a pyrrole ring. A long hydrocarbon tail is attached through an oxygen site. At the external sites, different groups are bonded.

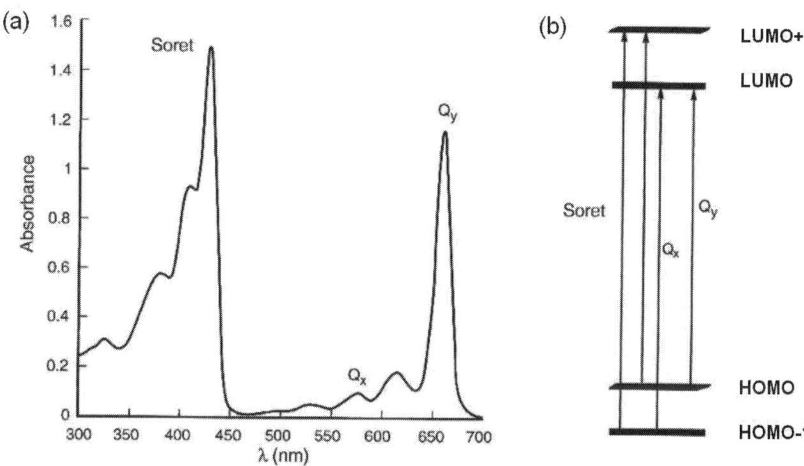

Figure 10.2 Absorption spectra of chlorophyll a. (a) The absorption peaks of chlorophyll *a* are in the red, yellow, and blue through the near-ultraviolet ranges. It is transparent to green light, which gives its characteristic green color. (b) Energy-level diagram of chlorophyll. The absorption peak in the red near 660 nm corresponds to the transition from the HOMO to the LUMO. The absorption peak in the yellow, around 570 nm, corresponds to the transition from one level below the HOMO to the LUMO. The peak in the violet, near 430 nm, corresponds to the transition from energy levels below the HOMO to energy levels above the LUMO. In a very short period of time, all those excitations are relaxed to the molecular state with one LUMO, which is about 1.88 eV above the ground state.

and C_2H_5 are bonded. Many other types of chlorophylls share the basic structure but with different groups at the external positions.

The absorption spectrum of chlorophyll *a* and its interpretation are shown in Fig. 10.2. It has three major groups of absorption peaks, centered at 662 nm (red), 578 nm (yellow), and 430 nm (blue). The green region is transparent, which gives rise to its characteristic green color. The energy diagram of chlorophyll *a* is shown in Fig. 10.2(b). In addition to the highest occupied molecular orbital (HOMO) and the lowest unoccupied molecular orbital (LUMO), two additional energy levels, one below the HOMO and one above the LUMO, are required to explain the absorption spectrum in the visible region. The absorption peak in the red, near 660 nm, corresponds to the transition from the HOMO to the LUMO. The absorption peak in yellow, around 570 nm, corresponds to the transition from one level below the HOMO to the LUMO. The peak in the violet, near 430 nm, corresponds to the transition from energy levels below the HOMO to energy levels above the LUMO. In a very short period of time, all those excitations are relaxed to the molecular state with one LUMO, which is about 1.88 eV above the ground state.

The energy stored in the excited chlorophyll molecule is transferred to an energy storage molecule, adenosine triphosphate (ATP). Then, ATP drives the process to synthesize sugar from carbon dioxide and water.

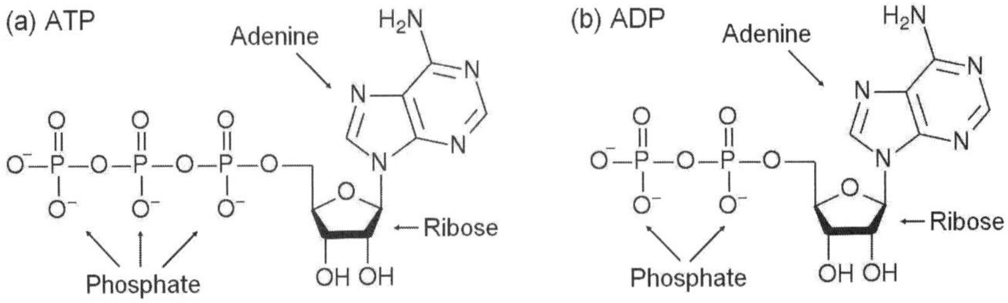

Figure 10.3 ATP and ADP. Both molecules contain an adenine, a ribose, and two or three phosphate groups. It takes energy to attach a phosphate group to ADP to form ATP. By detaching a phosphate group from ATP to recover ADP, energy is released. ATP is the universal "rechargeable battery" in biological systems.

10.1.2 ATP: Universal Energy Currency of Life

The structures of ATP and a related molecule adenosine diphosphate (ADP) are shown in Fig. 10.3. Both contain a nitrogenous base called *adenine*, a five-carbon sugar called *ribose*, and two or three *phosphate* groups. (Adenine is also one of the four nitrogenous bases which are the building blocks of DNA, the genetic code material.) These molecules are the universal rechargeable batteries in biological systems. ATP is the charged battery and ADP is the discharged battery. It takes some energy to attach a phosphate group to ADP to form ATP. By detaching a phosphate group from ATP to recover ADP, energy is released:

$$\mathrm{ATP}^{4-} + \mathrm{H_2O} \longrightarrow \mathrm{ADP}^{3-} + \mathrm{HPO_4^{2-}} + \mathrm{H}^+ + 0.539\,\mathrm{eV}. \tag{10.3}$$

The molecules ADP and ATP were isolated in 1929 from muscle tissues. In 1940, Fritz Lipmann (who won the Nobel Prize for Physiology and Medicine, 1953) proposed that ATP is the universal energy currency in cells. For example, when a human being is doing an aerobic exercise, glucose is oxidized by the oxygen in the blood into carbon dioxide and water, at the same time releasing energy. The energy is temporarily stored as ATP. Then the energized ATP drives the contraction of muscles.

The central role of ATP in photosynthesis was proposed by Daniel Arnon in the 1950s. His idea was met with a lot of skepticism. Then, a colleague in the same university, Berkeley, Melvin Calvin, did a series of experiments and discovered the process of photosynthesis and verified Arnon's hypothesis.

The molecule ATP plays such an important role in the energetics of life that further elucidation of its synthesis mechanism resulted in the 1997 Nobel Prize in Chemistry for Paul D. Boyer and John E. Walker and was named "The Molecule of the Year" by *Science* magazine in 1998.

10.1.3 NADPH and NADP$^+$

Besides transferring energy, reduction of CO_2 into carbon hydride is also required in the synthesis of glucose. The reduction agent in this process is NADPH, which can

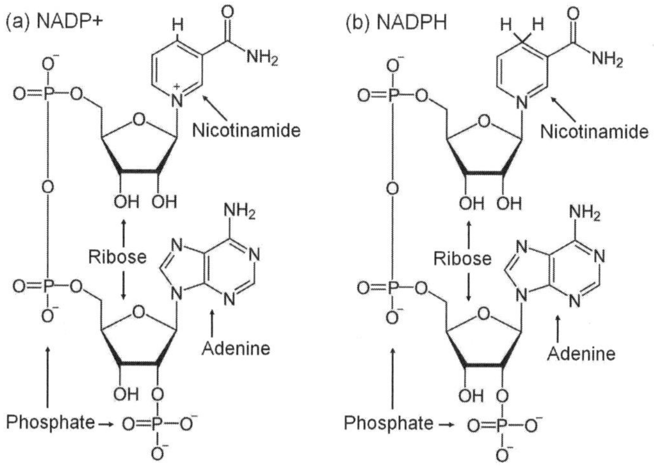

Figure 10.4 NADPH and NADP$^+$. The reduction agent in living systems. Both molecules contain a nicotinamide group, an adenine group, two ribose groups, and three phosphate groups. NADPH has two hydrogen atoms at a site on the nicotinamide. During reduction, it releases a hydrogen atom with an electron and becomes NADP$^+$.

release a hydrogen atom with an electron and become NADP$^+$. Both molecules contain a nicotinamide group, an adenine group, two ribose groups, and three phosphate groups (Fig. 10.4).

10.1.4 Calvin Cycle

During 1940s and 1950s, Melvin Calvin and his colleagues performed a series of experiments using a carbon isotope and worked out the pathway of photosynthesis. By adding $^{14}CO_2$ to a liquid containing the green alga *Chlorella pyrenoidosa*, using two-dimensional paper chromatograms, the subsequent molecules of the photosynthesis process were traced by the radioactivity of the ^{14}C atoms. The details of the process, the *Calvin cycle*, is rather complicated. Interested readers are recommended to check the books of Blankenship [11] or Chapter 24 of Voet [86]. Here we highlight some key points of the Calvin cycle with respect to energy transfer processes.

Figure 10.5 shows key steps in the Calvin cycle. The most significant discovery of Calvin's experiments is that the assimilation of carbon from CO_2 results in the generation of two identical three-carbon molecules (3-phosphoglycerate) by inserting a carbon atom and then cracking a five-carbon molecule (ribulose-1,5-biphosphate). The first step of the Calvin cycle is called *carbon fixation*. The next step is to reduce the carboxyl group to form triose phosphate by NADPH. The entire process of generating glucose requires that carbon fixation be repeated six times. In each loop, the five-carbon molecule (ribulose-1,5-biphosphate) must be regenerated to prepare for the next carbon assimilation process. It takes 9 – 10 photons to complete the process of fixing each carbon atom into the final product, for example, glucose.

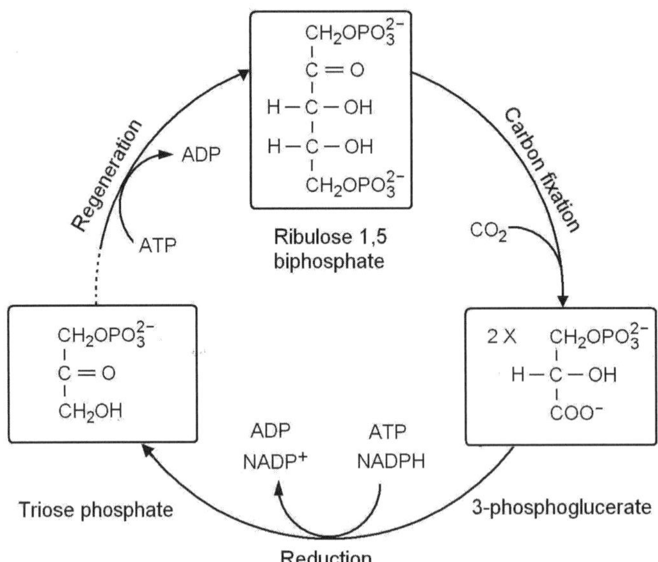

Figure 10.5 Key steps in the Calvin cycle. The Calvin cycle has three major steps. The first step, *carbon fixation*, is the insertion of a CO_2 molecule into a molecule with five carbon atoms and two phosphate groups. The resulting molecule is split, generating two identical molecules with three carbon atoms each. In a subsequent step, the molecule is reduced by the action of NADPH. The process repeats six times to generate a glucose molecule. In each loop, the five-carbon molecule (ribulose-1,5-biphosphate) must be regenerated to prepare for the next carbon fixation step.

10.1.5 C4 Plants versus C3 Plants

In the conventional Calvin cycle, there is an alternative reaction in the carbon fixation step: An oxygen molecule can take the role of the CO_2 and generate two different products, one with three carbon atoms and another with two carbon atoms. The process using oxygen, *photoperspiration*, reduces the efficiency.

In some plants, such as maize (corn) and sugarcane, a better process takes place in the carbon fixation step to circumvent photorespiration by a CO_2 pumping mechanism which generates two identical molecular fragments of four carbon atoms each. Because the initial products of carbon fixation is a 4-carbon-atom molecule instead of a three-carbon-atom molecule, such variation of the Calvin cycle is called a C4 cycle and plants such as maize and sugarcane are called C4 plants. Especially under high temperature and high insolation environments, the photosynthesis efficiency of C4 plants is significantly higher than the great majority of plants which have a C3 cycle [11].

10.1.6 Chloroplast

In plants, photosynthesis actions and ingredients are contained in disklike units called *chloroplasts*, see Fig. 10.6(a). The chloroplast, with a typical size of 5 μm, has a well-

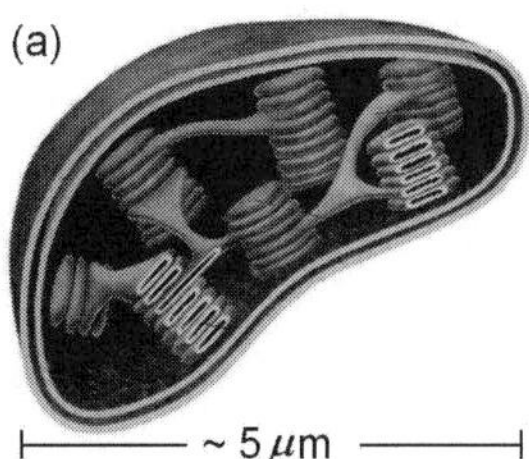

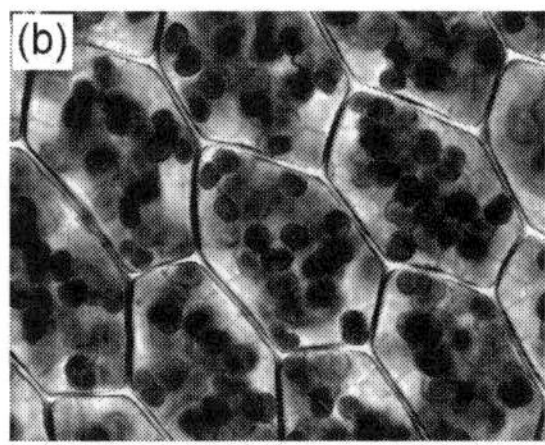

Figure 10.6 Chloroplast. (a) Chloroplast is the site of photosynthesis in plants. With a typical size of 5 μm, it has a well-defined structure to facilitate the flow of water, CO_2, and products. (b) A typical leaf cell contains 20 – 60 chloroplasts.

defined structure to facilitate the flow of water, CO_2, and products. Typically, each leaf cell contains 20 – 60 chloroplasts; see Fig. 10.6(b). A 1-mm^2 section in a typical corn leaf can have as many as a half million chloroplasts.

10.1.7 Efficiency of Photosynthesis

From an engineering point of view, the efficiency of photosynthesis is a critical parameter. By efficiency we mean the ratio of the chemical energy of the products of photosynthesis and the solar energy falling on the leaves.

Figure 10.7 shows the results of a study by Bolton and Hall [12]. The first loss is due to the wavelength range. The chlorophyll only absorbs less than one-half of the solar radiation, which is red, orange, and blue. The rest has no effect. The second loss is relaxation. As shown in Fig. 10.2, the excited molecule quickly relaxes to the state with only one LUMO, which is about 1.8 eV. The energy in the excited state of chlorophyll must be temporarily stored as usable chemical energy, the energy in ATP and NADPH, which is about 0.54 eV for each molecule. Sixty-eight percent of the energy is lost. The Calvin cycle is also not 100% efficient:. 35 – 45% of energy is lost. As a result, the net efficiency is about 5%.

Although the numerical value of the efficiency looks miserable, because of the enormous area of ground covered by plants, the total production of chemical energy by photosynthesis over the world each year is 3×10^{21} J, which is 6 times the total global consumption of energy in 2008.

To compare photosynthesis with other solar energy utilization processes, one often-used measure is *power density* in watts per square meter, defined as the chemical energy produced each year on a square meter of land divided by the number of seconds in a year. Note that the final useful product, for example, sugar or biodiesel, is only a small part of the total chemical product of photosynthesis. A large part of the products of photosynthesis, such as roots, branches, and leaves, is not useful. Table 10.1 is based on the data provided by a report published by United Nations Development Programme [72]. For comparison, data on total solar radiation and the power density of typical

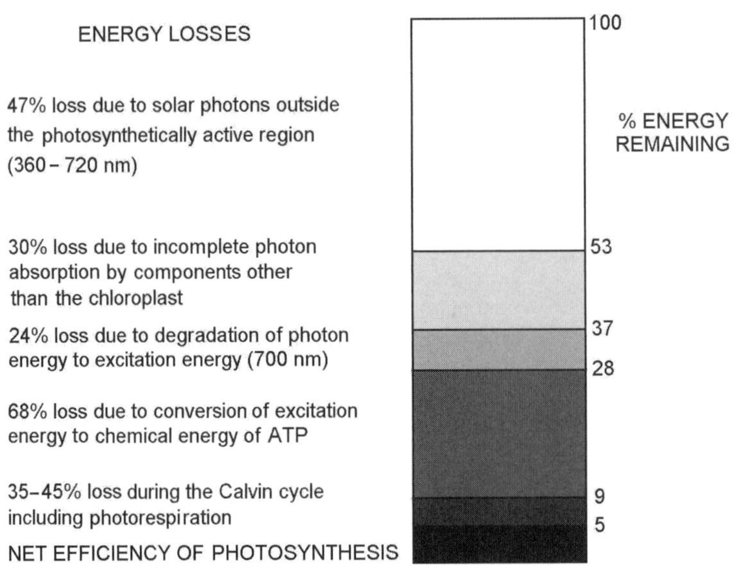

Figure 10.7 Efficiency of photosynthesis. The net efficiency of photosynthesis is about 5%. Notice that the largest percentage loss is due to the conversion of excitation energy in chlorophyll into ATP, that is, charging the biological rechargeable battery; and the limited wavelength range of the absorption spectrum of chlorophyll [12].

solar cells are also included. The average insolation is 1500 h per year. The average efficiency of crystalline silicon solar cell is 15%. The energy density of biomass is still much lower than that of typical solar cells. However, the cost of plants is much lower than solar cells.

Table 10.1: Power Density of Photosynthesis.

Item	Energy Density	Power Density
Parameter	(MJ/year/m^2)	(W/m^2)
Average solar radiation	5400	171
Average silicon solar cell	810	25.6
Wood (commercial forestry)	3 – 8	0.095 – 0.25
Rapeseed (northwest Europe)	5 – 9	0.16 – 0.29
Sugarcane (Brazil, Zambia)	40 – 50	1.27 – 1.58

Source: *World Energy Assessment: Energy and the Challenge of Sustainability*, UNDP 2000 [72].

10.2 Artificial Photosynthesis

For many decades, scientists have been trying to mimic the elegant process of photosynthesis to convert sunlight into fuel which can be stored and applied to, for example, transportation. The most studied approach is to use sunlight to split water into hydrogen and oxygen,

$$H_2O + 2.46\,\text{eV} \longrightarrow H_2 + \frac{1}{2}O_2. \tag{10.4}$$

Hydrogen can be used directly as a clean fuel. Once hydrogen is generated, by combining with carbon dioxide, liquid fuel could be generated. Therefore, if this process can be demonstrated, it should be a true revolution.

The current status and difficulties of this approach have been summarized in a review paper [6]. Direct cleavage of water into hydrogen and oxygen by sunlight is still a lofty dream. One experimentally verified approach to generate hydrogen and oxygen in significant quantity is to generate electricity using solar cells and then split water through electrolysis. Because of the high cost and low efficiency, it is not competitive with other means of energy storage, such as rechargeable batteries (see Chapter 12).

10.3 Genetically Engineered Algae

Although artificial photosynthesis is progressing slowly, an alternative approach using biotechnology seems extremely promising. The focus is on algae. As a source of biological fuel, algae have several advantages. First, they live in water and thus do not occupy arable land or require irrigation. Second, algae can have very high oil content, up to 50%. Third, the waste disposal problem could be minimal. The yield of oil per unit area per year for algae could be many times greater than even the most efficient land-based oil-producing plant, the oil palm. Recently, the use of genetically altered algae to produce liquid fuel has enjoyed much attention. Through gene modification combined with directed selection, new species or variations of algae could be created which will grow faster, contain more fuel, and be easy to harvest. For details, see a report by U.S. Department of Energy [25].

10.4 Dye-Sensitized Solar Cells

The principles of photosynthesis have inspired the invention of a novel type of solar cell, the *dye-sensitized solar cell* [32, 33, 64]. It has several advantages over the common crystalline silicon solar cell. The cost of materials and processing is greatly reduced because most of the process is by liquid-phase deposition instead of in a vacuum. In addition, it can be made on lightweight flexible substrates. To date, nearly 10% overall conversion efficiency from AM1.5 solar radiation to electrical power is achieved.

The sensitization of semiconductors to light of wavelength longer than that corresponding to the band gap has been used in photography and photo-electrochemistry.

The silver halides used in photography have band gaps on the order of 2.7 – 3.2 eV, and are not sensitive to most of the visible spectrum. Panchromatic films were made by adding dyes to sensitize silver halides, making them responsive to visible light.

Tne typical structure of a dye-sensitized solar cell is shown in Fig. 10.8. The most used semiconductor titanium dioxide (TiO_2), has many advantages for sensitized photochemistry and photoelectrochemistry: It is a low-cost, widely available, nontoxic and biocompatible material. As such it is also used in health care products as well as domestic applications such as paint pigmentation. The band gap, 3.05 eV, corresponding to a wavelength of 400 nm, lies in the near-ultraviolet region, which is too high for the solar spectrum. A dye is needed to mitigate this problem.

The ideal sensitizer for a single junction photovoltaic cell converting standard global AM1.5 sunlight to electricity should absorb all light below a threshold wavelength of about 920 nm. In addition, it must also carry attachment groups such as carboxylate or phosphonate to firmly graft it to the semiconductor oxide surface. Upon excitation it should inject electrons into the solid with a quantum yield of nearly unity. The energy level of the excited state should be well matched to the lower bound of the conduction band of the oxide to minimize energetic losses during the electron transfer reaction. Its redox potential should be sufficiently high that it can be regenerated via electron donation from the redox electrolyte or the hole conductor. Finally, it should be stable enough to sustain about 20 years of exposure to natural light. Much of the research in dye chemistry is devoted to the identification and synthesis of dyes matching these requirements while retaining stability in the photoelectrochemical environment. The

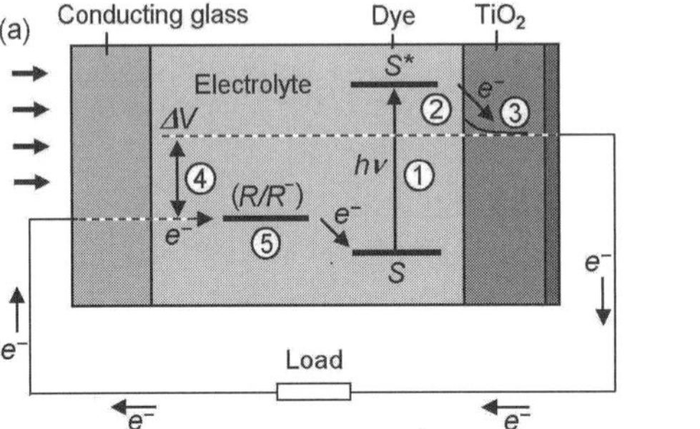

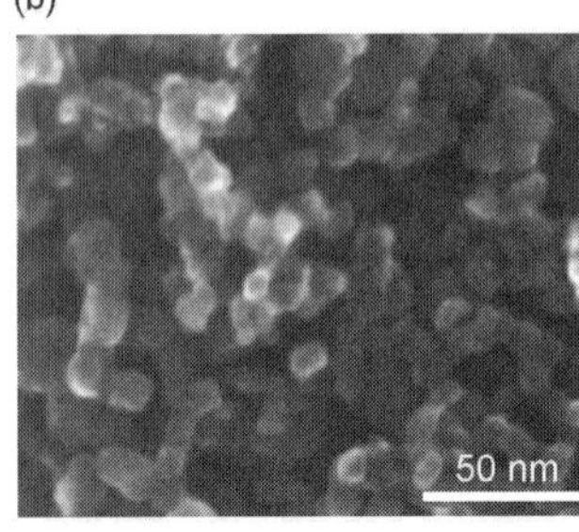

Figure 10.8 Structure of dye-sensitized solar cell. (a) The cell is built on top of a glass substrate with a conducting film. A nanostructured TiO_2 film of grain size about 15 nm and thickness about 10 μm is deposited on top of that conducting film. Dye molecules with a strong absorption band in the visible region are deposited on the surface of TiO_2 nanoparticles. The counter electrode is a film of transparent conducting oxide. The area between the cathode and the anode is filled with an electrode, typically a solution of lithium iodide. (b) a microscopic image of the TiO_2 film. The process of generating an electrical power is as follows: (1) Absorption of a photon by the dye to elevate an electron to the excited state, typically a LUMO. (2) Transfer of the electron to the TiO_2 film. (3) The electron relaxes to be at the bottom of the conduction band of TiO_2. (4) A photovoltage is generated by the cell, corresponding to the difference between the Fermi level in the semiconductor and the Nernst potential of the redox couple in the electrolyte. Adapted from Refs. [33] and [64].

attachment group of the dye ensures that it spontaneously assembles as a molecular layer upon exposing the oxide film to a dye solution.

One of the most studied and used dyes is the N3 ruthenium complex shown in Fig. 10.9. The strong absorption in the visible region makes the dye a deep brown-black color, thus the name "black dye." The dyes have an excellent chance of converting a photon into an electron, originally around 80% but improving to almost perfect conversion in more recent dyes. The overall efficiency is about 90%, with the "lost" 10% being largely accounted for by the optical losses in the top electrode. The spectral response for a dye-sensitized solar cell using an N3 ruthenium dye is shown in Fig. 10.9(b). The photocurrent response of a bare TiO_2 film is also shown for comparison.

The four-step process of generating electrical power is as follows (see Fig. 10.8(a)).

1. A photon is absorbed by the dye to elevate an electron to the excited state, typically a LUMO.

2. The electron is transferred to the TiO_2 film.

3. The electron relaxes to be at the bottom of the conduction band of TiO_2.

4. A photovoltage is generated by the cell, corresponding to the difference between the Fermi level in the semiconductor and the Nernst potential of the redox couple in the electrolyte.

Nevertheless, dye-sensitized solar cells have some disadvantages. First, the efficiency is about one-half of that of the crystalline silicon solar cells. Second, the necessity of a liquid-phase electrolyte made the solar cell mechanically weak. Third, the long-term stability of the organic materials needs to be improved.

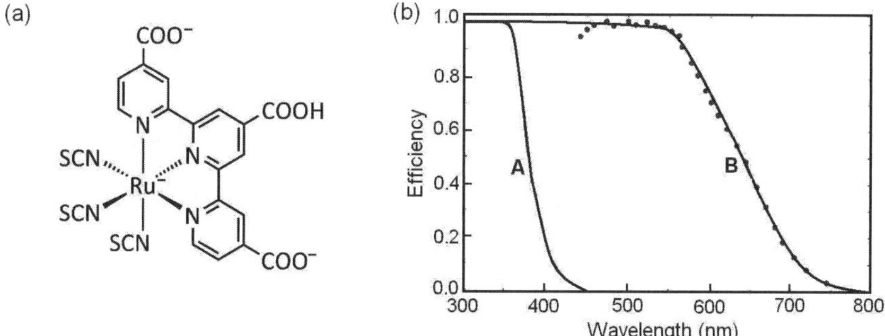

Figure 10.9 The N3 ruthenium dye and photocurrent spectrum. (a) Chemical structure of the N3 ruthenium complex used as a charge transfer sensitizer in dye-sensitized solar cells. (b) Photocurrent action spectra obtained with the dye as sensitizer, curve B. The photocurrent response of a bare TiO_2 films, A, is also shown for comparison. Adapted from Refs. [33] and [64].

10.5 Bilayer Organic Solar Cells

Another approach to mitigate the high cost of crystalline silicon solar cells is to use organic semiconductors, or semiconducting polymers, to replace the expensive purified silicon. Because of its high absorption coefficient in the visible region, a very thin film of organic material is sufficient. These polymers can be deposited by screen printing, inkjet printing, and spraying, as these materials are often soluble in a solvent. Furthermore, these deposition techniques can take place at low temperature, which allows devices to be fabricated on plastic substrates for flexible devices.

The basic structure of a bilayer organic solar cell is shown in Fig. 10.10(a). There are two layers of polymer films: a film of an absorbing polymer, the *electron donor*, and a film of *electron acceptor*. The double layer is sandwiched between the anode, a TCO film, and a metal back contact, the cathode. The process of generating a photocurrent has four steps, see Fig. 10.10(b). In the first step, a photon is absorbed by the polymer, the electron donor. An exciton, an electron–hole pair, is generated. In the second step, the exciton diffuses inside the absorbing polymer (the donor) toward the interface to the acceptor. In the third step, the electron transfers to the acceptor. Finally, the electron is collected by the cathode, or the back contact. Through the external electric circuit, the electron goes back to the anode (TCO) and eliminates the hole.

In the first successful bilayer organic solar cell, copper phthaocyanine (CuPc) is used for the absorbing polymer [80]. The chemical structure and absorption spectrum are shown in Fig. 10.11. It is a solid with dark blue color, as the red, yellow, green, and violet radiations are heavily absorbed. The absorption coefficient in some ranges is more than $10^6\,\text{cm}^{-1}$. Therefore, a very thin film of the absorbing polymer, typically around 100 nm, is used. A larger thickness is conversely a disadvantage because of the

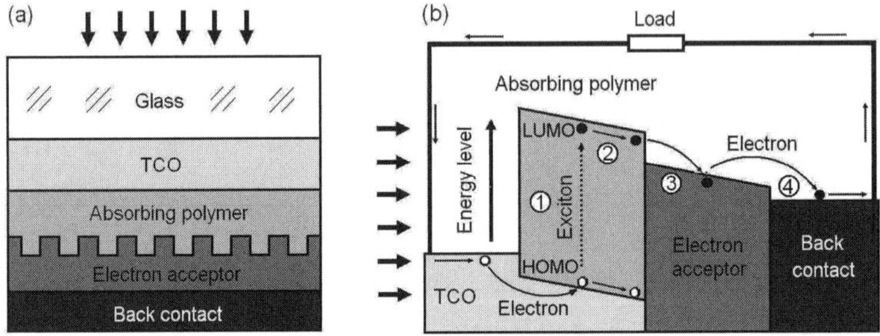

Figure 10.10 Bilayer organic solar cell. (a) A cross sectional view of the solar cell. Solar radiation comes from the top. Through a glass substrate and a transparent conducting oxide (TCO) film, light is absorbed by the absorbing polymer film, or the *electron donor*. The electron thus generated transfers to the *electron acceptor*, and then to a metal back contact or the cathode. (b) The working process. (1). A photon generates an exciton, typically a electron in LUMO and leaves a hole in a HOMO. (2) The exciton diffuses towards the acceptor. (3) The exciton dissociates into a free electron and a hole. (4) The electron moves to the cathode, then drives the external circuit. Adapted from Refs. [80, 38].

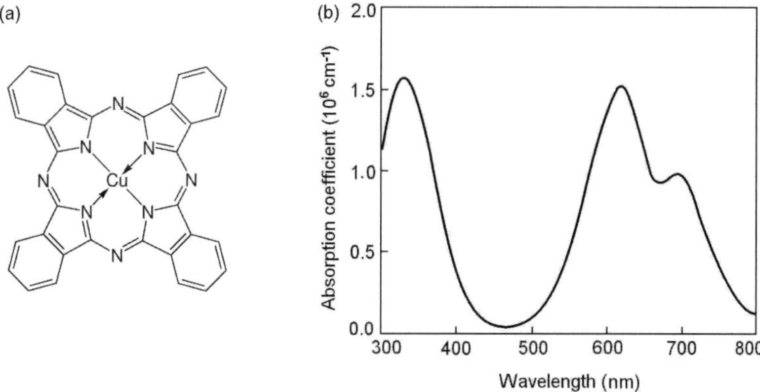

Figure 10.11 CuPc and its absorption spectrum. (a) Chemical structure of CuPc. (b) Absorption spectra of a solid thin film of CuPc. The red, yellow, green, and violet radiations are heavily absorbed. The material shows a dark blue color. Adapted from Ref. [38].

short diffusion length; see below.

The second process, exciton diffusion, deserves much attention. In contrast to semiconductor solar cells, the *diffusion lengths* of excitons in organic polymers are very short, typically 5 – 10 nm. Therefore, the lifetime of the excitons are very short. If the polymer is too thick, the excitons generated by photoexcitation might not reach the donor–acceptor interface and then disappear. In order to increase the probability for excitons to reach the acceptor, a nonplanar interface is often used; see Fig. 10.10(a).

In the third process, the exciton dissociates into a free electron and a free hole. The material for the acceptor should facilitate the dissociation and the final transfer of the carrier to the back contact, the cathode. In the first successful demonstration of a bilayer organic solar cell, a perylene tetracarboxylic derivative is used [80]. Later, C_{60} (buckminsterfullerene) and its derivatives are often used [15, 38].

Similar to dye-sensitized solar cells, the bilayer organic solar cells have disadvantages. Its efficiency is less than one-half that of the crystalline silicon solar cells. The long-term stability of the organic materials still needs to be improved.

Problems

10.1. Assume that (1) 50% of the leaf surface is stuffed with chlorophyll, (2) 30% of the solar photon hitting the chlorophyll generates one ATP, and (3) it takes 10 ATP to generate one sixth of glucose, a unit of CH_2O. What is the efficiency of that photosynthesis process?

Hint: Use the blackbody radiation formula to estimate the average photon energy of sunlight. Note that

$$\int_0^\infty \frac{x^2\,dx}{e^x - 1} = 2\,\zeta(3) = 2.404. \tag{10.5}$$

10.2. The crown of a typical sugar maple tree is roughly a sphere of 5 m radius. If the photosynthesis process of the leaves is devoted to making syrup, on a sunny summer day, how many kilograms of condensed maple syrup (60% sugar in weight) can this tree produce?

Hint: 1 eV equals 96.5 kJ/mol. Use the experimental value of photosynthesis efficiency (5%) to estimate the solar radiation required to produce 1 kg of syrup.

Chapter 11

Solar Thermal Energy

11.1 Early Solar Thermal Applications

One of the earliest documentations of solar thermal energy applications is in *Code of Zhōu Regulations* (Zhōu Lǐ), a government document on the organization and laws of the West Zhōu dynasty (eleventh century *B.C.* – 771 *B.C.*). An entry in that book says, "The fire-maker uses a solar igniter (yángsuì) to start a fire using sunlight." Mòzǐ, a philosopher and physicist living in the Zhōu dynasty (468 *B.C.* – 376 *B.C.*), expounded the imaging properties of concave mirrors. A good quantitative understanding had been achieved at that time. To date, six such solar igniters have been found from various West Zhōu dynasty tombs. Sixteen from the East Zhōu dynasty (770 *B.C.* to 221 *B.C.*) were also unearthed. Figure 11.1 shows an example, dated about 1000 *B.C.*, discovered in 1995 in Fúfōng county of Shǎnxī province. Its diameter is 90.5 mm. Its radius of curvature is 207.5 mm, and the focal length is 103.75 mm. The original mirror was rusty. By making a mold from the original and cast in bronze, after polishing, the replica can ignite a straw using sunlight in a few seconds. Similar solar igniters have been found in other Bronze Age cultures over the world.

According to a Greek legend, in 212 *B.C.*, Archimedes used mirrors to focus sunlight on ships of an invading Roman fleet at Syracuse and destroyed the fleet. The use of burning mirrors for military purposes was a favorite theme of Middle Age and Renaissance scholars. However, later experiments in the seventeenth through the nineteenth

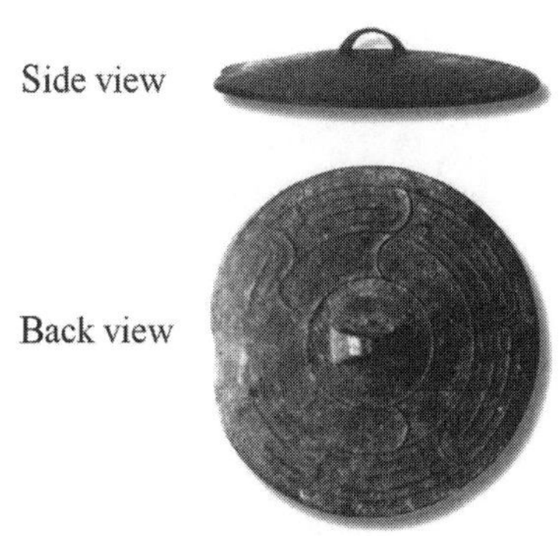

Figure 11.1 A 3000-years-old solar igniter. A bronze solar igniter (yángsuì) of the early Zhōu dynasty, dated 1000 B.C., discovered in 1995 from a Zhōu dynasty tomb in Fúfōng county of Shǎnxī province. With a radius of curvature 207.5 mm, its focal length is 103.75 mm. A replica of the original can ignite a piece of straw using sunlight in a few seconds.

centuries showed that even with modern technology and large mirrors the focused sunlight is not intense enough to burn ships at a reasonable distance. The story is probably a myth.

In 1767, French-Swiss scientist Horace Benedict de Saussure designed and built the first solar heat trapper that could be used for cooking [17]. Figure 11.2 shows the device. It is made of two wooden boxes, a small one inside a large one, with insulation (cork) in between. The inside of the small box is painted black. The top of the small box is covered by three separate sheets of glass, with air between adjacent glass sheets. By facing the top of the box toward the Sun, and moving the box to keep the glass perpendicular to sunlight, in a few hours, the temperature inside the small box reaches above 100°C. Therefore, it is a hot box heated by the Sun. To identify the source of the heat, de Saussure carried a hot box to the top of Mt. Cramont. He found that although the air temperature there is about 5 – 10°C lower than on the plain, the interior of the box could reach the boiling point of water as well. He attributed this effect to the clearness of the air on top of the mountain, where solar radiation is stronger.

Horace de Saussure's experiment is a demonstration of the *greenhouse effect*. It motivated Joseph Fourier to explain the equilibrium temperature of Earth by infrared absorption of Earth's atmosphere. Fourier explained his theory of the greenhouse effect by making an analogy to de Saussure's hot box where the glass sheets keep heat inside.

However, de Saussure's hot box was too slow to heat up, and the temperature could not become high enough for cooking, for example, 150°. This was probably the reason it did not become a popular product. The first mass-produced solar thermal device is the solar oven invented by W. Adams in the 1870s in Bombay, India [3]. He added a modest solar energy concentration device to de Saussure's hot box, as shown in Fig. 11.3. Eight glass mirrors (A) form an octagonlike reflector. Concentrated sunlight

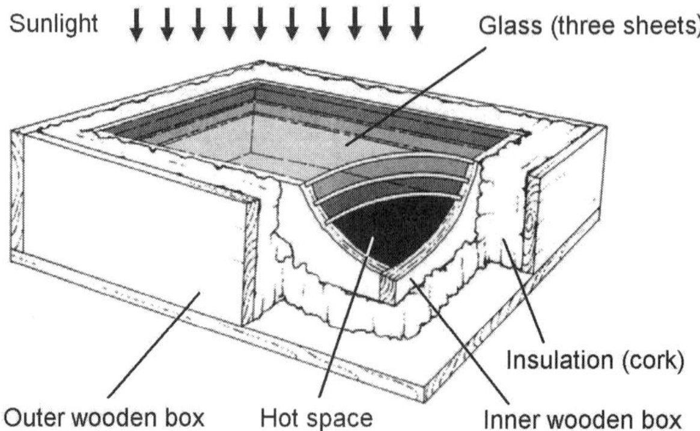

Figure 11.2 The hot box of Horace de Saussure. In 1767, Horace de Saussure designed and built the first solar cooking device, the hot box. Using three sheets of glass, the sunlight could come in but heat could not escape, making the temperature inside the box rise to the boiling point of water. Adapter from Butti and Perlin [17], courtesy of John Perlin.

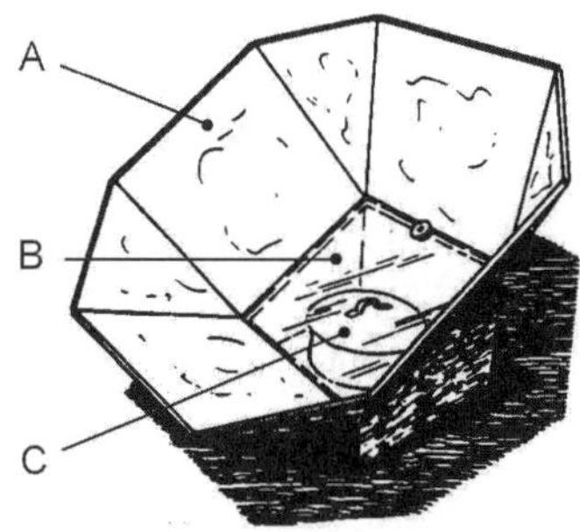

Figure 11.3 Adams solar oven. In 1878, W. Adams invented a solar oven in Bombay, India. Eight mirrors made of silvered glass (A) form an octagonal reflector. Sunlight is concentrated and floods into a wooden box covered with glass (B), which contains a pot (C). The box can be rotated by hand to align with sunlight. The temperature in the box could exceed 200°C [3].

floods into the wooden box covered with glass (B), which contains a pot (C). The mirrors and the glass are inclined so that the rays of the Sun fall perpendicular to the box. When the Sun moves, the box can be rotated by hand to align with sunlight. In a paper published in 1878 in *Scientific American* [3], Adams reported, "The rations of seven soldiers, consisting of meat and vegetables, are thoroughly cooked by it in two hours, in January, the coldest month of the year in Bombay, and the men declare the food to be cooked much better than in the ordinary manner." His solar oven was then mass produced in India and became quite popular. In the United States, the Adams solar oven has become a popular product for camping and an educational device for teenage students.

A solar oven currently in widespread use in third world countries is shown in Fig. 11.4. It is a parabolic reflector made of cast iron and plated with chromium. Two steel ribs support a steel holder for a pot. The pot is always positioned near the focal point of the parabolic dish. By manually aligning the axis of the parabolic dish with the sunlight, the pot can receive a maximum amount of solar heat. This type of solar oven is mass produced in Eastern China and more than 10 thousand units are sold annually in Tibet.

In the twentieth century, the solar water heater was invented and improved, and became increasingly popular [?]. A brief history of the solar water heater is presented in Chapter 1. On the other hand, concentrated solar thermal electricity is a bright

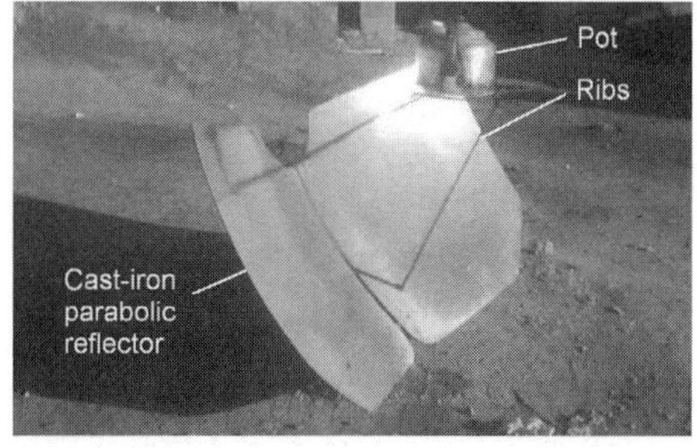

Figure 11.4 Cast-iron solar oven. Cast-iron parabolic reflector supported by a tripod with a joint which can be turned in two axes. Two steel ribs support a steel holder for a pot. By manually aligning the axis of the parabolic dish with the sunlight, the pot can always receive the maximum solar heat.

spot in solar power generation. In the following sections, we present the physics of solar thermal energy applications.

11.2 Solar Heat Collectors

For all solar thermal applications, the first step is to convert solar radiation energy into heat. From both materials and mechanical structure points of view, the key requirement is to absorb as much sunlight as possible and to lose as little heat energy as possible. Three methods are used: selective absorption surface, vacuum to block heat conduction and convection, and focused sunlight to change the ratio of the absorbing surface area and the emitting surface area.

11.2.1 Selective Absorption Surface

In early solar thermal applications, such as water heaters in the early twentieth century, the absorbing surface was painted a dull black. As a blackbody, it could absorb the maximum amount of solar radiation. When the body became hot, it lost energy by radiation.

The blackbody radiation spectra of the Sun and a hot body, for example, a pot of boiling water or a solar thermal absorber at 400°, could be separable on the energy scale; see Table 2.3 and Fig. 2.4. It becomes more intuitive to make a plot of the relative spectral power density on a logarithmic scale of wavelength (see Problem 2.3),

$$\begin{aligned} u(\lambda, T) &= \frac{2\pi hc^2}{\lambda^5 \left[e^{hc/\lambda k_B T} - 1\right]} \\ &= \frac{3.75 \times 10^8}{\lambda^5 \left[e^{\lambda_T/\lambda} - 1\right]} \left(\frac{\mathrm{W}}{\mathrm{m}^2}\right) \left(\frac{1}{\mu\mathrm{m}}\right), \end{aligned} \tag{11.1}$$

where the power density is expressed in watts per square meter and the spectrum is expressed in terms of wavelength in micrometers. Figure 11.5 shows the relative spectral power density of AM1.5 solar radiation and from hot bodies. As shown, the peak of solar radiation is around 0.5 μm. On the other hand, the radiation from a hot body at 400°C is 4 μm and that of boiling water is 8 μm. The blackbody radiation spectrum from the hot liquid is well separated from that of solar radiation. Therefore, by designing a material having a high absorptivity for wavelength shorter than 2 μm and a low emissivity for longer wavelength, the absorption of solar radiation can be maximized and radiation loss can be minimized.

In the solar energy literature, the property of the surface is usually denoted by its *reflectance* at different wavelengths, $R(\lambda)$. The absorption surface for solar thermal applications is opaque. Energy conservation requires that the sum of reflectivity and absorptivity be

$$R(\lambda) + A(\lambda) = 1; \tag{11.2}$$

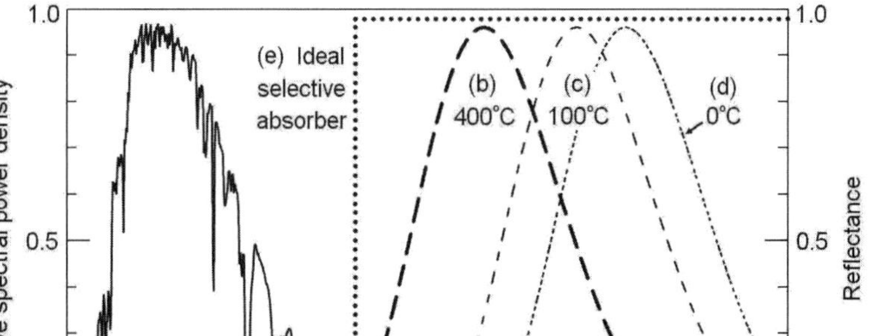

Figure 11.5 Spectral power density of solar radiation and hot bodies. (a) The spectral density of solar radiation concentrates to wavelengths shorter than 2 μm. (b) – (d) The spectral density of bodies on Earth. Even at 400°, it is concentrated in wavelengths greater than 2 μm. (e) The ideal selective absorber is a perfect blackbody for $\lambda < 2\,\mu$m and a perfect miror for $\lambda > 2\,\mu$m.

see Eq. 5.2. According to Kirchhoff's law, the emissivity of a surface equals its absorptivity (Eq. 5.3),

$$E(\lambda) = A(\lambda). \tag{11.3}$$

Therefore, the requirement of an ideal selective absorbing surface is low reflectivity for wavelengths shorter than 2 μm and a high reflectivity for longer wavelengths.

Two dimensionless quantities are defined to characterize the global performance of theselective absorption surface as follows. The solar absorptivity α is defined as

$$\alpha = \frac{\displaystyle\int_0^\infty u_\odot(\lambda)\,[1 - R(\lambda)]\,d\lambda}{\displaystyle\int_0^\infty u_\odot(\lambda)\,d\lambda}, \tag{11.4}$$

where the solar radiation power density $u_\odot(\lambda)$ is the AM1.5 direct normal spectral power density. The thermal emissivity ε is defined as

$$\varepsilon = \frac{\displaystyle\int_0^\infty u(\lambda, T)\,[1 - R(\lambda)]\,d\lambda}{\displaystyle\int_0^\infty u(\lambda, T) d\lambda}, \tag{11.5}$$

where $u(\lambda, T)$ is the blackbody radiation of the solar device at temperature T defined in

Eq. 11.1. The photothermal conversion efficiency can be calculated using the formula,

$$\eta = \alpha - \varepsilon \frac{\sigma T^4}{C I_\odot}, \tag{11.6}$$

where σ is the Stefan–Boltzmann constant, T is the temperature of the solar heat collector, C is the concentration factor, and $I_\odot$ is the solar radiation power density, with a typical value of 1 kW/m^2 on a sunny day under normal incidence conditions.

Since the concept of *selective absorption surface* was proposed in the 1950s, it became an intensive international research project, especially in the United States, Israel, and Australia. For a review with an extensive list of references, see Ref. [41]. For a systematic treatment of theory and processes, see Ref. [47]. The ideal selective absorption curve is the dotted line in Fig. 11.5, with $\alpha = 1$ and $\varepsilon = 0$. Nevertheless, very good approximations have been achieved. There is still much research interest in this, especially for solar thermal electricity applications, where the temperature of the fluid should be as high as possible. According to basic thermodynamics, the maximum efficiency is the Carnot efficiency (Eq. 6.19),

$$\eta_{\rm c} = 1 - \frac{T_{\rm L}}{T_{\rm H}}. \tag{11.7}$$

For example, if the fluid heated by solar radiation is 40°C higher than ambiant temperature at 20°C (293 K), the maximum efficiency of a heat engine is 12%. The temperature of the fluid should be 400 – 500°C in the solar thermal collector. However,

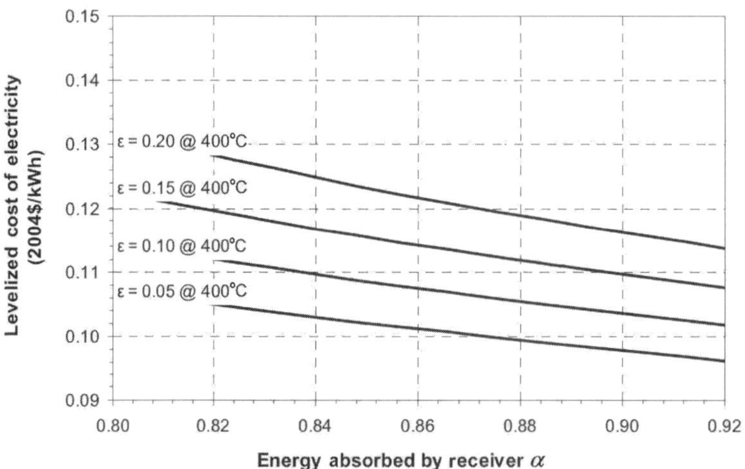

Figure 11.6 Effect of selective absorption on solar thermal devices. Estimated effect of solar absorptivity α and thermal emissivity ε on the levelized cost of electricity generated by a solar thermal system. As shown, improvement of the selective absorption coating is a critical factor in achieving grid parity. After Ref. [41].

at that temperature, the radiation loss is substantially higher. Figure 11.6 shows an estimate of the effect of solar absorptivity α and thermal emissivity ε on the levelized cost of electricity generated by a solar thermal system. As shown, the improvement of the selective absorption coating is a critical factor in achieving grid parity.

The requirements for a selective absorption coating are as follows:

1. High solar absorptivity, ideally $\alpha = 90\%$ to 97%
2. Low thermal emissivity, ideally $\varepsilon = 3\%$ to 10%
3. Durable at working temperature, for example, 30 years at 400°
4. Stable in air
5. Low-cost large-scale manufacturing

There are several types of selective absorption surfaces:

1. Intrinsic selective absorbers, such as semiconductors
2. Metal mirror coated with a thin absorption layer
3. Thin film with a window of transmittivity on a transparent cover.
4. Multilayer interference film

Because of the high stability, high contrast, and relative ease of mass production, a metal mirror coated with a thin absorption layer is the most popular type. The mirror can be made of any metal of high reflectivity for infrared radiation, such as copper, aluminum, nickel, and stainless steel. The absorption film is made of metal oxides, and named "dark mirrors."

Table 11.1 lists several well-studied and commercialized dark mirrors. the first three cases, black nickel, black chrome, and black copper, are produced with a liquid-phase process and are stable in air up to a few hundred degrees Celsius. The Al–AlN_x system must be processed in a vacuum chamber, and is stable in a vacuum up to 500°C. It is particularly convenient for vacuum tube thermal absorbing systems because the

Table 11.1: Selective Absorbing Surfaces

System	Mirror	Absorbant	α	ϵ
Black nickel	Ni or steel	NiS–ZnS	0.88 – 0.96	0.03 – 0.10
Black chrome	Cr	Cr_2O_3	0.97	0.09
Black copper	Cu	Cu_2O–CuO	0.97 – 0.98	0.02
Aluminum nitride	Al	Al_3N_4	0.97	0.10

Source: Ref. [41].

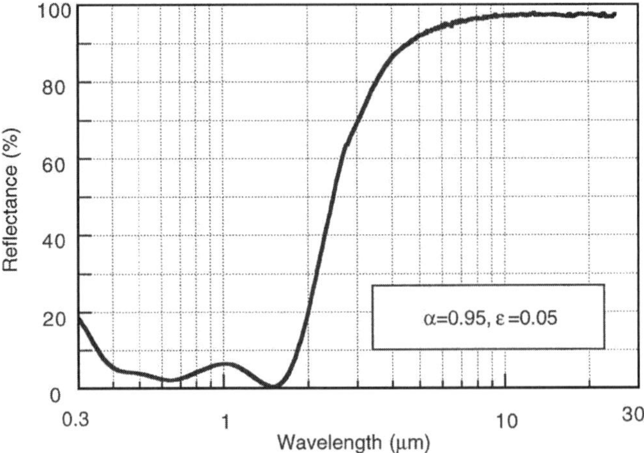

Figure 11.7 Reflectance curve for cermet selective-absorbing surface. The measured reflectance spectrum of a surface with AlN cermet composite on stainless steel base [89, 90]. It approaches the ideal behavior shown in Fig. 11.5.

deposition of aluminum and AlN can be made in the same processing chamber. The first step, aluminum deposition, is performed in a vacuum. Then, by controlled bleeding of nitrogen, good-quality AlN_x film can be formed. Because the selective coating is working under a vacuum, it does not deteriorate.

For high-temperature systems, for example, solar thermal electricity system working at 400 – 500°C, especially with a requirement of stability in air, a cermet coating is preferred. A cermet is a composite material composed of ceramic (cer) and metallic (met) materials which could have the properties of both a ceramic, such as high-temperature resistance and hardness, and a metal, such as the ability to undergo plastic deformation. Therefore, properly designed cermet coatings can withstand high temperature and are stable in air [42, 89, 90]. Figure 11.7 shows the reflectivity spectrum of a surface with an AlN cermet composite on a stainless steel base. As shown, it approaches the ideal behavior illustrated in Fig. 11.5.

11.2.2 Flat-Plate Collectors

As discussed in Section 1.5.1, the first successful solar water heater, the Day-and-Night, uses flat-panel solar heat collectors. It is relatively easy to build and rugged enough to withstand the elements. In fact, thousands of such flat-panel solar heat collectors built in 1920s, are still working properly in Florida after more than 80 years of exposure to the harsh weather. In the whole solar hot water system, only the water tank needs to be updated using foam polyurethane as insulating material. Figure 11.8 shows a schematic. It is essentially a de Saussure hot box (Fig. 11.2), hosting a copper plate with copper pipes soldered on, painted black. Usually one or two sheets of glass are installed for top-side heat insulation. Water can be heated up to 60° or 80°. It is a rugged device which can last for many decades.

The standard treatment of the flat-panel solar heat collectors is based on the Hottel–Whillier model [39]. Details can be found in Duffie and Beckman [23, 24], Lunde [54], and other publications [44, 46, 70].

Most solar thermal collectors are covered with glass. The normal-incidence transmittance of the glass cover is (see Chapter 9),

$$\tau = \frac{(4n)^{2N}}{(1+n)^{4N}}, \tag{11.8}$$

where N is the number of sheets and n is the refractive index of glass, typically $n = 1.5$ (see Section 9.4).

Consider a solar collector plate with total area A and absorptivity α. If the portion of effective absorption area is F and the solar power density is P_0, the input power to the plate is

$$Q_\text{I} = FA\,\tau\alpha\,P_0. \tag{11.9}$$

Because of the solar power, the temperature of the plate is elevated from ambient temperature T_a to T_p. If the temperature difference is not too great, the heat loss is proportional to the temperature difference. In all practical cases, the area of the panel, FA, is much greater than the area of the edge. For clarity and brevity, the edge effect is neglected. The heat loss is also proportional to the area of the hot plate,

$$Q_\text{L} = U_\text{L} FA\,(T_p - T_a). \tag{11.10}$$

Here, U_L is the *combined heat loss coefficient*. The efficiency of the solar thermal energy collector is

$$\eta = \frac{Q_\text{I} - Q_\text{L}}{P_0 A} = F\left(\tau\alpha - \frac{1}{P_0} U_\text{L}\,(T_p - T_a)\right)^+. \tag{11.11}$$

The plus sign indicates that only a positive value of the expression is taken. In other words, if the expression is negative, the value is taken as zero.

Therefore, the problem of the efficiency of the solar heat collector is reduced to the evaluation of the heat loss coefficient U_L. For flat-panel collectors, the problem is

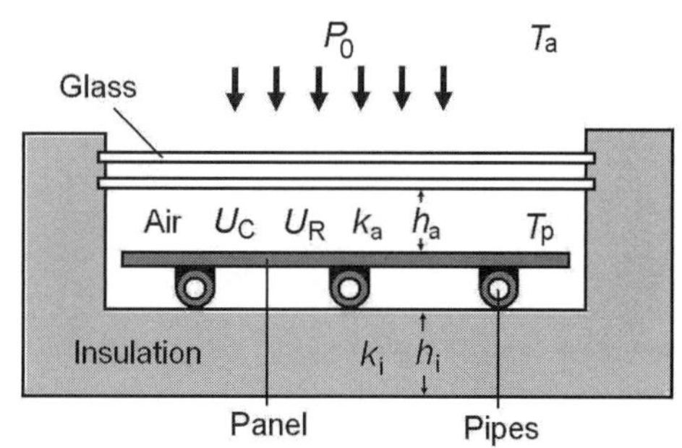

Figure 11.8 Flat-plate solar heat collector. It is essentially a de Saussure hot box, Fig. 11.2, hosting a copper plate with copper pipes soldered on, painted black. Usually one or two sheets of glass are installed for top-side heat insulation. Water can be heated up to 60° or 80° by sunlight. It is a quite rugged device which could last for many decades.

nontrivial. Many factors contribute to the loss through the front cover, and some of them are quite difficult to quantify:

1. *Conduction through the back-side insulation* is easy to quantify, as the corresponding loss factor equals k_i/h_i, the thermal conductivity and thickness of the insulation material.

2. *Conduction through the air space* is also easy to quantify, as the corresponding loss factor equals k_a/h_a, the thermal conductivity and thickness of air.

3. *Convection in the air space* is difficult to quantify. It has a complicated dependence on spacing and the tilt angle.

4. *Convection outside the glass cover* depends not only the temperature but also the wind speed of ambient air.

5. *Radiation.* The thermal radiation of the panel first reaches the glass panel. It essentially absorbs most of the radiation. The glass panel, being heated, radiates again to the ambient.

The transmittivity spectrum of common window glass is shown in Fig. 11.9. It is transparent for visible and near-infrared radiation, but opaque for ultraviolet and far-infrared radiation. The blackbody radiation from the hot plate is almost completely absorbed by the glass.

The *combined heat loss coefficient* can be estimated as

$$U_L = \frac{k_a}{h_a} + \frac{k_i}{h_i} + U_C + U_R. \tag{11.12}$$

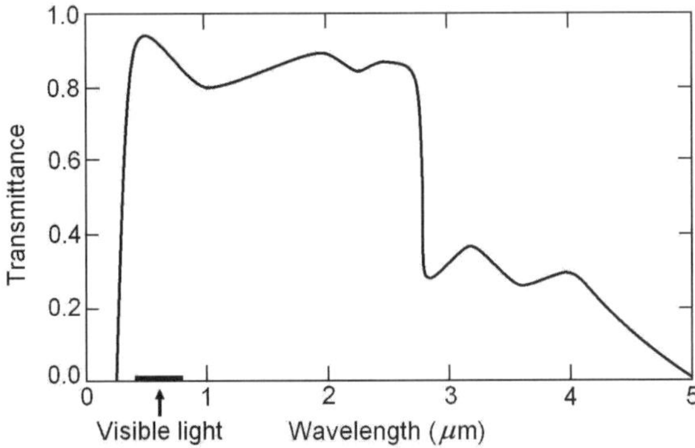

Figure 11.9 Transmittance of window glass. The window glass is transparent for the visible and near-infrared radiation, but opaque for ultraviolet and far-infrared radiation. The blackbody radiation from the hot plate is almost completely absorbed by the glass. Plotted using the data from American Institute of Physics Handbook.

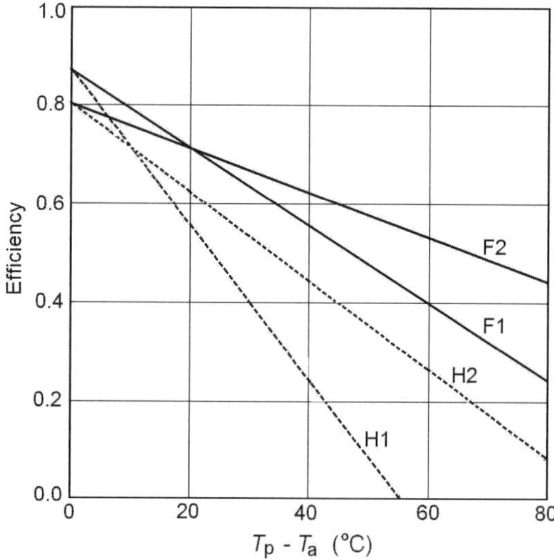

Figure 11.10 Efficiency of flat-plate collectors. Dependence of efficiency on solar radiation power and temperature rise. Curve F1 shows the efficiency under full sunlight, 1 kW/m^2, with one glass cover; F2 for two glass covers. Curve H1 shows the efficiency under one-half of full sunlight, 0.5 kW/m^2, with one glass cover; H2 for two glass covers. As the temperature of the plate increases, the efficiency deteriorates rapidly due to the heat loss via the top glass cover. At some point, the heat loss exceeds the solar energy received by the panel, and the efficiency becomes zero.

Where k_a and k_i are the thermal conductivities of the air and insulation material, respectively, h_a and h_i are the thicknesses of the air and insulation wall, respectively (see Fig. 11.8), U_C is the convection heat loss coefficient, and U_R is the radiation heat loss coefficient. The first two terms can be estimated using the typical parameters listed in Table 11.2,

$$\frac{k_a}{h_a} + \frac{k_i}{h_i} \approx 1.36\,(\mathrm{W/m^2 \cdot K}). \tag{11.13}$$

However, the last two terms are usually much larger than the first two. An effective

Table 11.2: Typical Parameters of Flat-Plate Solar Heat Collectors

Parameter	Description	Symbol	Unit	Value
Cover glass	Refractive index	n	—	1.50
Panel	Absorptance	α	—	0.95
Insulation	Thermal conductivity	k_i	W/m^2·K	0.02
Insulation	Thickness	d_i	meter	0.05
Air space	Thermal conductivity	k_a	W/m^2·K	0.024
Air space	Thickness	d_a	m	0.025

way to reduce convection and radiation loss is to increase the number of sheets of glass. According to the calculations of Duffie and Beckman [23, 24], under normal conditions (wind speed 5.0 m/s, average plate temperature 60°C, slope 45°, ambient temperature minus 20 – 40°C), the top loss coefficient is 6.9 W/m^2·K for one cover, 3.5 W/m^2·K for two covers, and 2.4 W/m^2·K for three covers. However, more covers result in more loss of transmittance. From Eq. 11.8, the transmittance of one sheet of glass is 0.92. It is reduced to 0.85 for two sheets, and 0.782 for three sheets. Therefore, using three covers does not represent an advantage.

The efficiency also depends on the power density of solar radiation. Because the heat loss is independent of the solar radiation, the weaker the solar radiation, the lower the efficiency. Figure 11.10 shows a typical dependence of efficiency on solar radiation power and temperature rise. Curve F1 shows the efficiency under full sunlight, 1 kW/m^2, with one glass cover; F2 shows it for two glass covers. Curve H1 shows the efficiency under one-half of full sunlight, 0.5 kW/m^2, with one glass cover; H2 for two glass covers. For simplicity, the area ratio F is assumed to be 1, which is approximately true in most practical cases. As the temperature of the plate increases, the efficiency deteriorates rapidly due to the heat loss via the top glass cover. At some point, the heat loss exceeds the solar energy received by the panel. And the efficiency becomes negative: The temperature of the plate falls instead of rises.

11.2.3 All-Glass Vacuum-Tube Collectors

As shown in the previous section, the single most important factor affecting the efficiency of solar heat collectors is the heat loss through the top cover. As early as 1911, William L. R. Emmet invented vacuum tube heat collectors (U.S. Patent 980,505) which could in principle completely resolve the problem of top-cover heat loss. It took 80 years to make them suitable for mass production.

Figure 11.11 shows a modern evacuated-tube solar thermal collector. It is made of two concentric glass tubes sealed at one end. The space in between is evacuated

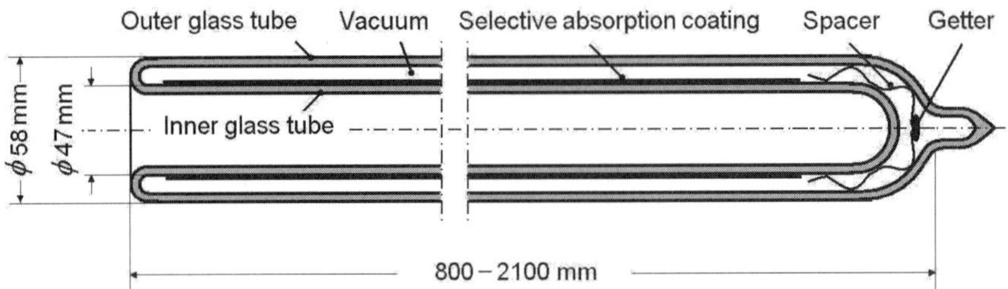

Figure 11.11 Evacuated-tube solar thermal collector. It is made of two concentric glass tubes sealed at one end. The space in between is evacuated to better than 10^{-4} Pa, or 10^{-6} Torr. A metal spacer is placed as a source for the getter, typically a mixture of barium and titanium. After it is sealed, the getter is evaporated onto the inner surface of the glass tubes. A high vacuum thus can be maintained. A selective absorption coating is applied on the outer surface of the inner glass tube.

to a medium high vacuum. A metal spacer is placed between the tubes to support the tubes and as a source for getter, typically a mixture of barium and titanium. After being sealed, the getter support is heated from outside using microwave power to evaporate the getter onto the inner surface of the glass tubes. A high vacuum thus can be maintained. Good-quality evacuated tubes should have a vacuum better than 10^{-4} Pa, or 10^{-6} Torr. A selective absorption coating is applied on the *outer surface* of the *inner glass tube*. Therefore, the selective coating is always under high vacuum. It has a significant advantage over the flat-panel heat collector as a selective coating stable in air is not required and the coating could stay intact virtually forever. Furthermore, an antireflection film can be applied on top of the selective absorption coating even if the film is not stable in air.

Another advantage over the flat-panel heat collector is that the materials are inexpensive and abundant and can be mass produced at very low cost. To date, 200 million evacuated tubes are produced annually in China.

An important consideration is that the area ratio F is much less than 1 for vacuum tube collectors, because the diameter of the inner tube determines the absorption area, and when the tubes are installed on a system, there should be spacing between adjacent outer tubes. Typically the space is 20 mm. The typical area ratio $F = 47/(58 + 20) \approx 0.6$. The smaller factor F actually is not a serious disadvantage. First, in residential applications, there is always more roof space than needed. Second, An important consideration is the *cost* of the solar thermal energy collector. Because there is empty

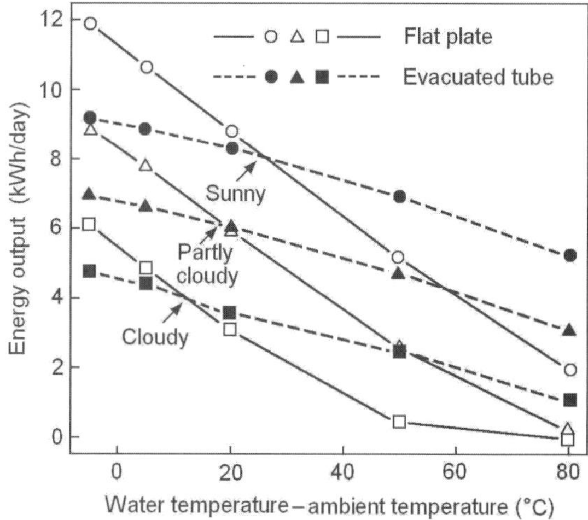

Figure 11.12 Performance comparison of flat-panel and evacuated-tube collectors. For sunny weather and low $T_p - T_a$ requirements, the flat-panel system clearly has an advantage because of the high area ratio F. For cloudy weather and high $T_p - T_a$ requirements, the evacuated-tube system is better because convection and conduction losses are eliminated. After a report by William Ferguson on Wikipedia.

space between the tubes, there is no additional cost. Third, because of the empty space, there is no additional weight on the roof. Finally, if the sunlight is not perpendicular to the plane of the tubes, up to an angle of incidence $\theta = \arccos(F)$, maximum power can be maintained.

Figure 11.12 compares the performance of solar water heaters using flat-panel collectors and evacuated-tube collectors. The flat-panel solar water heater is the Thermo-Dynamics S42-P and the evacuated-tube system is the SunMaxx 20EVT. The result was reported by William Ferguson on Wikipedia. For sunny weather and low $T_p - T_a$ requirements, the flat-panel system clearly has an advantage because of the high value of area ratio F. For cloudy weather and high $T_p - T_a$ requirements, the evacuated-tube system is superior because convection and conduction losses are eliminated. Furthermore, high-performance selective absorption coating can be used because there is no requirement of air stability, and the radiation loss becomes negligible. The heat loss observed here for vacuum-tube systems is due to the heat loss of the water tank. The general trend is consistent with the analysis in this section.

11.2.4 Thermosiphon Solar Heat Collectors

Evacuated-tube solar energy collectors are used primarily for direct-flow solar water heaters, where the usable water goes directly into the tubes. It has very high efficiency. However, the hot water could be contaminated by the system, and the pressure

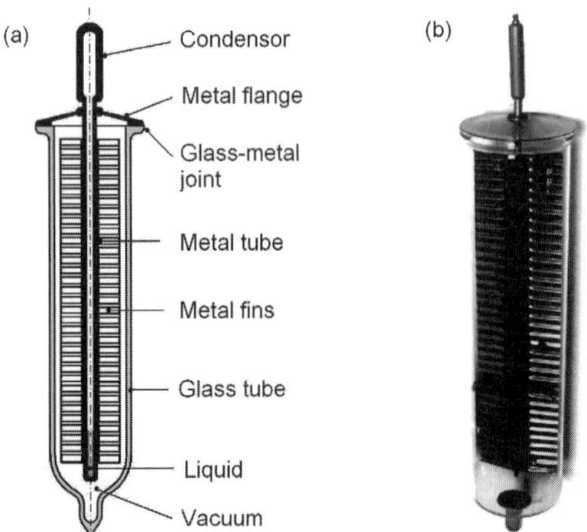

Figure 11.13 Thermosiphon solar heat collector. (a) Structure of a thermosiphon solar heat collector. At the center is a sealed metal tube, typically made of copper. A small amount of volatile liquid is filled in the metal tube, typically water. The metal tube is connected with metal fins, covered with selective absorption coatings. The metal tube is mounted on a metal flange, typically stainless steel. A glass-metal joint is formed between the flange and the glass tube. A vacuum is drawn in the glass tube. (b) Photograph of the collector. With sunlight falling on the metal fins, the liquid in the metal tube is evaporated, than condensed at the top, which is thermally connected to a heat load.

comes directly from gravitation. For systems requiring pressurized hot water and more stringent sanitation, thermosiphon solar heat collectors are used.

Figure 11.13(a) shows a cross section of a thermosiphon solar heat collector. At the center is a sealed metal tube typically made of copper. A small amount of volatile liquid is in the metal tube, typically water. The metal tube is connected with metal fins covered with selective absorption coatings. The metal tube is mounted on a metal flange, typically stainless steel. A glass–metal joint is formed between the flange and the glass tube. A vacuum is drawn in the glass tube. Figure 11.13(b) is a photo of the device. The tube must be installed at a tilted position with the evaporator at the top. With sunlight falling on the metal fins, the liquid in the metal tube is evaporated, then condensed at the top, which is thermally connected to a heat load.

A key technical problem is formation of the glass–metal joint. A widely used technology is a metal gasket of relatively low melting point, such as tin, lead, or aluminum. By heating the joint under pressure at a temperature lower but close to the melting point of the metal gasket, a good joint can be formed. The high vacuum in the tube constantly exerts pressure on the glass–metal joint; therefore, the probability of a leakage is small. Compared with all-glass vacuum tubes, thermosiphon solar heat collectors have several advantages. First, because there is no running water in the tubes, it can withstand bitter cold without breaking the glass. Second, because the thermal mass of the tube is much smatter than the water in the all-glass tubes, the start-up time is much shorter. Third, even if one of the glass tubes is broken, for example, by hail impact, there is no water leakage. Fourth, because hot water does not flow in the tubes, the tank can be pressurized and run high-standard clean water. Lastly, because the liquid returns to the bottom of the siphon tube by gravity, there is a *thermal diode effect* — the heat only flows from the collector to the tank and cannot be reversed. However, because of the metal structure and the glass–metal joint, the cost is much higher than the all-glass tubes. Therefore, it is used in high-end solar water heaters.

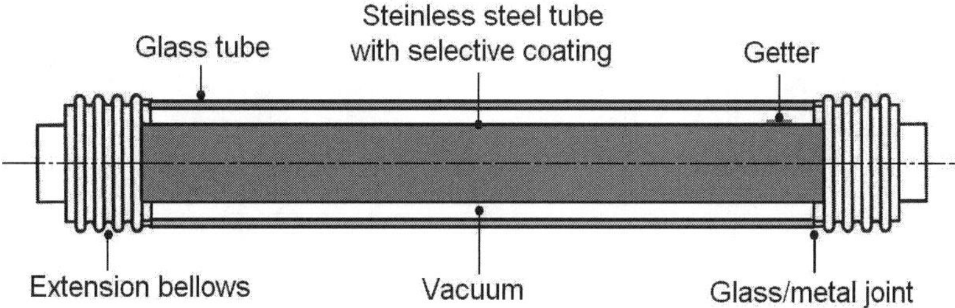

Figure 11.14 High-pressure vacuum tube collector. At the core is a stainless steel tube coated with a selective absorption film. Both ends are fitted with extension bellows. Through a glass–metal joint, each side is attached to an end of the glass tube. A vacuum is drawn between the stainless tube and the glass tube. A getter is included to maintain a good vacuum.

11.2.5 High-Pressure Vacuum Tube Collectors

For solar thermal applications, the working fluid inside the tube is not hot water. It is either oil at high temperature (300°C or more) or superheated steam at high pressure (10 – 100 atm). The glass tube would not withstand such temperature and pressure. The inner tube must be made of a strong metal, typically stainless steel. The outer tube must be transparent, made of glass. Therefore, there is a problem of mismatch of thermal expansion coefficients, and a metal–glass joint is required. The typical structure of such a solar heat collector is shown in Fig. 11.14. At the core is a stainless steel tube coated with a selective absorption film. Both ends are fitted with extension bellows. Through a glass-metal joint, each side is attached to an end of the glass tube. A vacuum is drawn between the stainless steel tube and the glass tube. A getter is included to maintain a good vacuum.

11.3 Solar Water Heaters

The most popular solar water heater is the direct-flow system using all-glass evacuated-tube solar heat collectors; see Section 1.5.3, especially Fig. 1.32. Usually it is installed on the roof of a single-family house or an apartment building. The hot water flows simply by gravitation. Because of low manufacturing cost as a result of large-scale mass production, whenever such a system can could be used, the investment can be recouped in a few years without government subsidy. After the initial cost is paid off, the system can work properly for 20 – 30 years with no maintenance. In China alone,

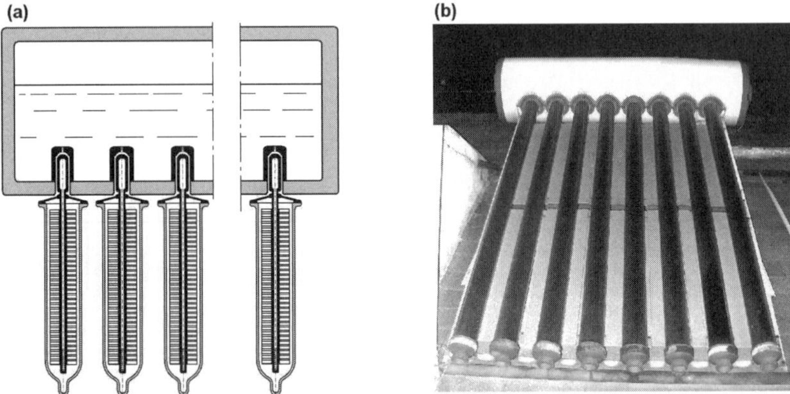

Figure 11.15 Solar water heater with thermosiphon collectors. (a) Design of the solar water heater. The evaporators of the thermosiphon solar heat collector are in thermal contact with the water in the tank through copper blocks. (b) A system on a roof after more than 10 years of operation. No degradation is observed. Photo taken by the author.

more than 10 million such systems have been installed.

This simple and elegant system does have some disadvantages. The hot water cannot be pressurized. Any contamination in the system would appear at the outlet. If a single tube breaks, all the water will flow out immediately. Many improved systems are developed and used.

11.3.1 System with Thermosiphon Solar Heat Collectors

By using the thermosiphon solar heat collectors instead of the all-glass evacuated tubes, the running hot water makes no contact with the heat collectors. Heat transfer is through the evaporators of the collector and the (usually copper) blocks inside the water tank, see Fig. 11.15(a). The water inside the tank can be cleaned thoroughly. If one of the collector tube breaks, the rest of the collector tubes will keep the system running, although the power is slightly reduced, until a replacement collector tube is reinstalled. Figure 11.15(b) shows a photo of such system on a roof after more than 10 years of operation. No degradation is observed.

11.3.2 System with Pressurized Heat-Exchange Coils

A much cleaner system can be formed by installing a heat exchange coil inside the insulated water tank. The water in the evacuated tubes and the insulated water tank is then used for heat exchange only; see Fig. 11.16. The usable hot water runs only in the heat exchange coil and thus can be pressurized and will not be contaminated by the heat collecting system. The insert in Fig. 11.16 shows some detail of the heat exchange coil.

Figure 11.16 Solar water heater with pressurized heat exchange coils. The hot water only runs in the heat exchange coil inside the insulated tank. Therefore, the hot water can be pressurized and will not be contaminated by the heat-collecting system. Insert shows some detail of the heat exchange coil.

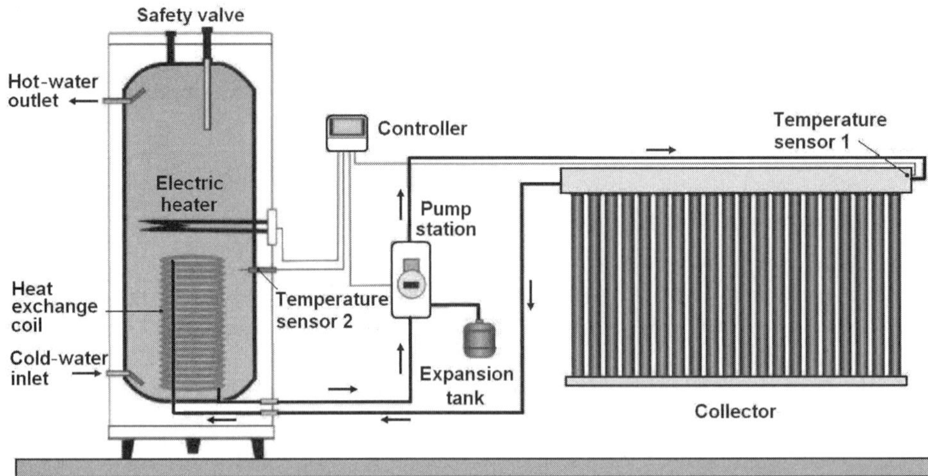

Figure 11.17 System with separate heat exchange tank. The solar heat collector is filled with a heat exchange fluid, circulated by a pump. The temperature of the water tank is sensed by thermometer 2. The temperature of the outlet water can be preset using a controller, which controls the pump station.

11.3.3 System with a Separate Heat-Exchange Tank

Using a separate water tank and a heat exchange coil inside the tank, the usable hot water is completely isolated from the heat collectors. The typical structure of such a system is shown in Fig. 11.17.

The solar heat collector is filled with a heat exchange fluid which can be water or an antifreeze fluid, for example, a mixture of water and glycerol. In a flat-panel solar collector, in regions with freezing temperature, antifreeze fluid is a necessity. In evacuated-tube collectors, because of the superb insulation, ordinary water can be used. The heat exchange fluid is circulated with a pump. The temperature of the water tank is sensed by a thermometer (2). The temperature of the outlet water can be preset using a controller, which controls the pump station.

11.4 Solar Thermal Power Systems

A more important area of solar thermal application is electric power generation. According to the second law of thermodynamics, the upper limit of the efficiency of converting heat to mechanical power is the Carnot efficiency (Eq.6.19),

$$\eta_{\mathrm{c}} = 1 - \frac{T_{\mathrm{L}}}{T_{\mathrm{H}}}, \tag{11.14}$$

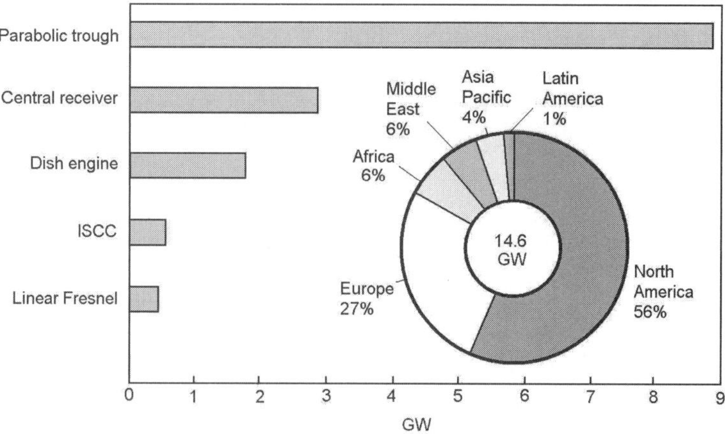

Figure 11.18 Accumulated installation of concentration solar power systems. Installation of these systems up to 2009, including parabolic trough, central receiver with heliostats, paraboloidal dish, ISCC, and CLFR systems [88].

where T_L is the temperature of the cold reservoir and T_H is the temperature of the hot source. In order to improve efficiency, the temperature of the hot source should be as high as possible. Since T_L cannot be lower than the atmospheric temperature, on average 300 K, in order to achieve an efficiency of 50%, T_H should at least be 600 K, or 327°C. Without concentration, the temperature can hardly reach 150°C.

To achieve such a high temperature, concentration of solar radiation is a necessity. Three commonly used configurations for stand-alone solar thermal power systems were developed in the 1970s: the parabolic trough concentration system, a system of a central receiver with heliostats, and the paraboloidal dish concentration system. Recently, the Integrated Solar Combined Cycle (ISCC) system has attracted much attention. The Compact Linear Fresnel Reflector (CLFR) system, suitable for the ISCC system, has observed a rapid development. Figure 11.18 shows the accumulated installation of those systems up to 2009 [88].

11.4.1 Parabolic Trough Concentrator

As shown in Fig. 11.18, to date, about two-thirds of electricity generated by concentrated solar thermal power plants is from parabolic trough systems. The structure of the system is shown in Plate 11. Parabolic mirrors are mounted on an axis to track the Sun. At the focal line of the parabolic mirrors is a linear collector, typically a high-temperature evacuated tube. See Plate 11.

To date, the largest solar power plants are the Solar Energy Generating Systems (SEGS) in the California Mojave Desert. The nine SEGS power plants built between 1984 and 1990 have a total capacity of 354 MW. The total area of the 232,500 parabolic mirrors is 6.5 km^2, with a total length of 370 km. The sunlight bounces off the mirrors

Figure 11.19 Aerial photo of SEGS system. The axes of the parabolic mirrors are north–south, which turn from east to west every day. About 3000 mirrors are replaced each year because of wind damage. Several damaged mirrors are shown in this photo. To avoid shadows from neighboring mirrors, the distance between adjacent mirrors is about twice the width of the mirrors.

and is directed to a central tube filled with synthetic oil which heats to over 400°C. The reflected light focused at the central tube is 71 – 80 times more intense than ordinary sunlight. The synthetic oil transfers its heat to water, which boils and drives the Rankine cycle steam turbine, thereby generating electricity. Synthetic oil is used to carry the heat (instead of water) to keep the pressure within manageable parameters. See Plate 12.

Figure 11.19 is an aerial photo of the SEGS. The axis of the parabolic mirrors is north–south, which turns from east to west every day. The most expensive parts are the curved mirrors. Wind damage occurs frequently, and about 3000 mirrors are replaced each year. The mirrors are periodically cleaned by a special machine. To avoid shadows from neighboring mirrors, the distance between mirrors is about twice the width of the mirrors. The cost of electricity from the SEGS is still not competitive with coal-burning power plants.

11.4.2 Central Receiver with Heliostats

To achieve high power at a centralized receiver, hundreds or thousands of *heliostats*, mirrors mounted on a two-dimensional shaft to track the position of the Sun by a centralized computer distributed on a typically circular or oval field; see Fig. 11.20. Because very high temperatures can be reached (e.g. 565°C), superheated steam or molten salt (e.g. 60% $NaNO_3$ + 40% KNO_3) is usually used as the working material. Finally, it drives a standard Rankine cycle steam turbine to generate electrical power [73].

The first pilot project, Solar One, in the Mojave Desert, was completed in 1981, and was operational from 1982 to 1986. It has 1818 mirrors, each 40 m^2. Oil or water was used as the working fluid. An efficiency of 6% was demonstrated. A photo of Solar

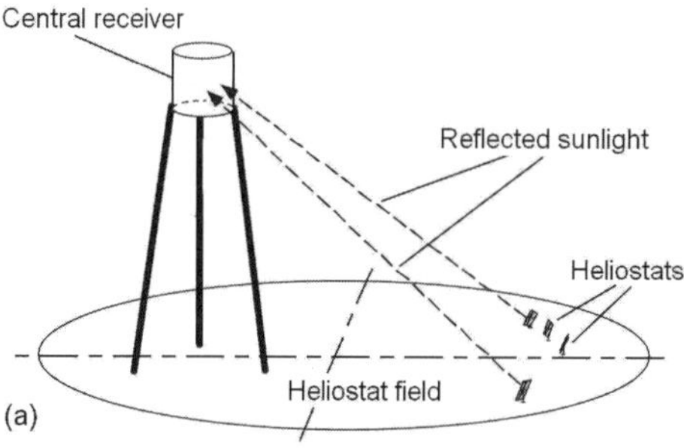

Figure 11.20 Solar power plant with central receiver. To achieve high power at a centralized receiver, hundreds of *heliostats* are used, that is, mirrors mounted on a two-dimensional shaft to follow the position of the Sun by a centralized computer. The temperature of the central receiver can reach 500°C or more.

One is shown in Fig. 11.19(b). In 1995, Solar One was upgraded to Solar Two. Molten salt, 60% sodium nitride and 40% potassium nitride were used as the working material. Ten megawatts of power is demonstrated, and its efficiency was improved to 16%. On November 25, 2009, the Solar Two tower was demolished.

To be used as a stand-alone power plant, an energy storage system is required. Latent heat of molten salt is used for energy storage. However, the cost per kilowatt-hour is increased from \$0.08 – \$0.15 to \$0.15 – \$0.20 by adding the energy storage unit [73].

11.4.3 Paraboloidal Dish Concentrator with Stirling Engine

The third type of concentration solar power plant is the paraboloidal dish concentrator with Stirling engine, as shown in Plate 10. According to a report released by Sandia National Laboratory in February 2008, the Stirling engine system has shown a solar-

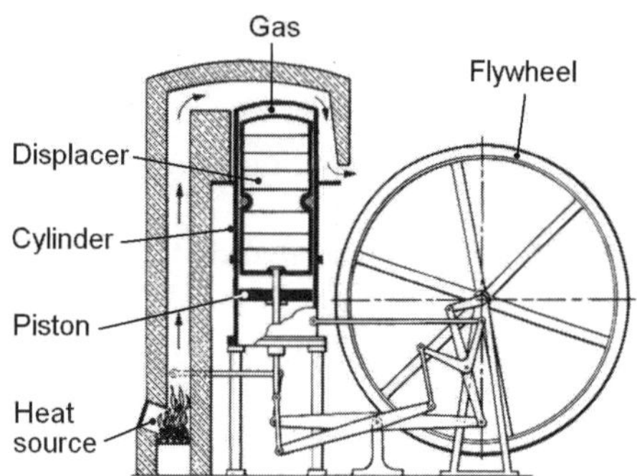

Figure 11.21 Stirling engine. Invented in 1816 by Robert Stirling, the engine without a valve is the simplest heat engine. A fixed amount of gas, typically hydrogen, is used as the working medium. It can be driven by a single heat source, with an efficiency close to the Carnot limit. It is the ideal heat engine driven by concentrated sunlight.

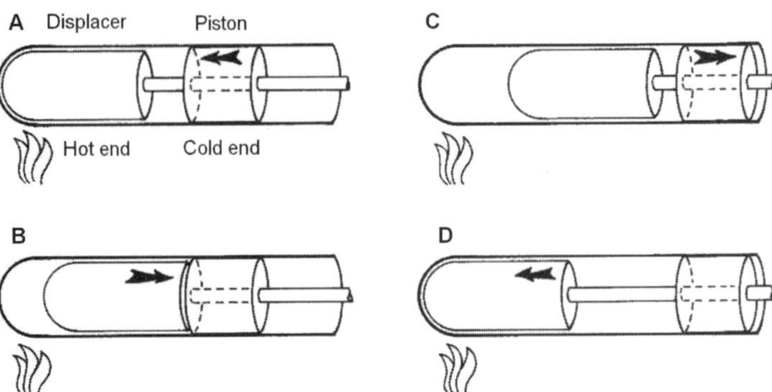

Figure 11.22 Working principle of Stirling engine. (A) The gas is cold, the displacer is in the innermost position, and the piston pushes into the cylinder. (B) The hot source heats up the gas. (C) The expanding hot gas pushes the piston outwards. (D) The displacer returns to the innermost position and the gas is cooled by the surrounding.

to-grid energy conversion efficiency of 31.25%, the highest of any solar-to-electricity conversions, recorded on a perfectly clear and cold New Mexico winter day [48].

The Stirling engine was invented in 1816 by Robert Stirling, a priest studying mechanical engineering as a hobby who built the first such engine in his home machine shop. A schematic diagram of the Stirling engine is shown in Fig. 11.21. This engine is different from the two popular heat engines, the steam engine and the internal combustion engine. Similar to the steam engine, it uses an *external heat source*. However, instead of constantly evaporating water into steam and then discarding it, the Stirling engine uses a fixed body of gas in a closed cylinder. It is the simplest heat engine: There are no valves. It can approach the Carnot efficiency. It can be operated by any type of single heat source; thus concentrated sunlight is perfect. In Fig. 11.21, the heat is provided by burning wood or coal. A piston is tightly fit in a cylinder which drives a flywheel through a crankshaft. A gas displacer, loosely fit in the cylinder, is driven by the crankshaft with a phase shift in the motion of the piston. The working media, the gas, is always contained in the cylinder. The details of its working cycles are shown in Fig. 11.22. During step (A), the gas is cold, the displacer is in the innermost position, and the piston pushes into the cylinder. Then the hot source heats up the gas in step (B). During step (C), the expanding hot gas pushes the piston outward. Finally, step (D), the displacer returns to the innermost position, and the gas is cooled by the surrounding.

In order to achieve effective operation, the gas must have high thermal conductivity. The most frequently used gas is hydrogen. However, the diffusion coefficient of hydrogen in steel is very high. Therefore, either a special material with low diffusion coefficient for hydrogen is used to construct the cylinder or the hydrogen is periodically supplemented.

The Stirling engine is not suitable for vehicle applications because of its large volume and the requirement of an effective cooling mechanism.

11.4.4 Integrated Solar Combined Cycle

In previous sections, solar thermal power systems designed for stand-alone power stations were described. Because of the intermittent nature of solar energy, extensive energy storage facilities are required, such as using sensible heat and latent heat during phase transition as well as batteries. However, large-scale energy storage is very expensive. Moreover, pure solar power plants often require special equipment, such as the Stirling engine, DC–AC converters, or molten-salt heat transfer facilities.

A recent trend in the exploration of solar energy is not to *compete* with traditional power stations, such as fossil-fuel-based boilers or nuclear-powered boilers, but to *supplement* them. It is called *Integrated Solar Combined Cycle* ISCC. In this model, the solar field is an addition to a traditional power plant, as shown in Fig. 11.23. The solar field can take water as the input, heat it to generate superheated steam, then supply the steam either at the maximum-temperature point (high-temperature operation), or a point with less than maximum temperature (medium- or low- temperature operation); see Fig. 11.23.

A significant advantage of the ISCC is that the solar energy component is built using a traditional, well-developed power generation technology with limited additional investment while taking full advantage of the inclusion of solar energy. It is even possible to retrofit an existing fossil fuel power plant by adding a solar component *without interrupting the operation of the power plant.* Thus, the ISCC is a winning

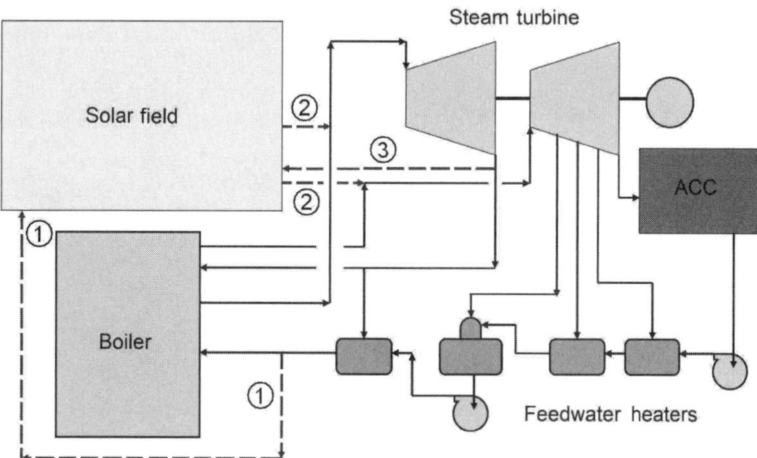

Figure 11.23 Schematics of integrated solar combined cycle. By adding a solar field to a traditional steam-turbine-based power station, a low-cost utilization of solar energy is created. In a high-temperature ISCC implementation, the sunlight is utilized to generate superheated steam, indicated by dashed lines in the Figure. (1) By feeding water to the solar field, superheated steam is supplied to the turbines either at a high temperature, through path (2), or at a lower than maximum tamperature, through path (3). After a report from Bechtel [84].

combination for both traditional and solar plants in terms of reduced capital cost and continuous power supply. Another advantage of the ISCC is that it produces energy when most needed during peak times of the day and the year where air conditioners are run at full capacity. Therefore, by adding a solar field, the nominal capacity of the power plant can be substantially reduced for the same service in a given area. Since 2008, eight such power stations have been constructed, notably in northern Africa. The percentage of solar power ranges from 5 to 20%.

11.4.5 Linear Fresnel Reflector (LFR)

The parabolic trough concentrator uses large, individually designed mirrors to form a huge mechanical system. The cost of the mechanical system is high, and it is vulnerable to wind damage. Cleaning and repair are also costly. A solution to this problem is the *linear Fresnel concentrator* (LFR), see Fig. 11.24. Instead of using large, custom-designed mirrors, long and narrow flat mirrors are used. These mirrors are mounted on a one-dimensional axis, and turn individually to reflect the sunlight onto the small linear concave mirror and then concentrated it onto the evacuated-tube absorber. The heat transfer fluid is usually water, to generate superheated steam up to about 365°C directly to run the turbine. In such an arrangement, the mirrors can be placed very close to the ground. There are several advantages. First, there are many linear receivers in the system. If they are close enough, then individual reflectors have the option of directing reflected solar radiation to at least two receivers. This additional variable in reflector orientation allows much more densely packed arrays and lower absorber tower heights, because patterns of alternating reflector orientation can be set up such that

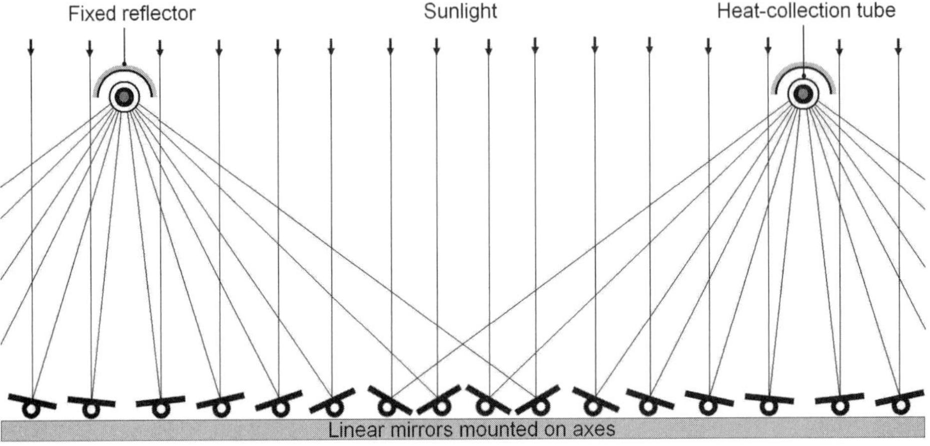

Figure 11.24 Linear Fresnel reflector system. An array of linear (narrow and long) mirrors mounted on an axis which can be rotated to align to the sunlight individually. The sunlight is further concentrated by a small linear concave mirror to the evacuated-tube absorber.

Figure 11.25 Linear Fresnel reflector system: a model. A model at the Himin Solar Energy Museum, Dézhōu, China. Photo taken by the author. Courtesy of Himin Solar Energy Group.

closely packed reflectors can be positioned without mutual blocking. The shadow effect can be reduced, as shown in Fig. 11.24, and the efficiency of land usage is improved. The mechanical structures are close to the ground, and the cost is substantially reduced. Furthermore, cleaning and repair become much simpler than for the parabolic trough system. Figure 11.25 is a model of a LFR system in Himin Solar Energy Museum, Dézhōu, China.

Problems

11.1. Based on the algebraic definition of a parabola, $y = x^2/4f$, prove that a parabola is the locus of equal distance from a focus to a directrix. See Fig. 11.26.

11.2. Using the geometric definition in Problem 11.1, prove that all light rays parallel to the y-axis would be reflected by a parabolic surface to the focus. See Fig. 11.26.

11.3. A vacuum solar heat collector has an internal diameter of 45 mm and a length of 1800 mm and is filled with water of 20°C. The axis of the tube is perpendicular to the sunlight. Assuming the efficiency is 90%, with full sunlight, how long does it take to boil the water inside the tube?

11.4. A solar hot water system consists of 24 vacuum tubes of outer diameter 58 mm and inner diameter of 47 mm, and length 1800 mm. It is connected to an insulated tank containing 200 liters of water. In a sunny day, with the sunlight perpendicular to the plane of vacuum tubes, and the efficiency is 90%, how long it takes to heat up the water by 20°C? (See Figure 1.32, Figure 11.11, and Figure 11.17).

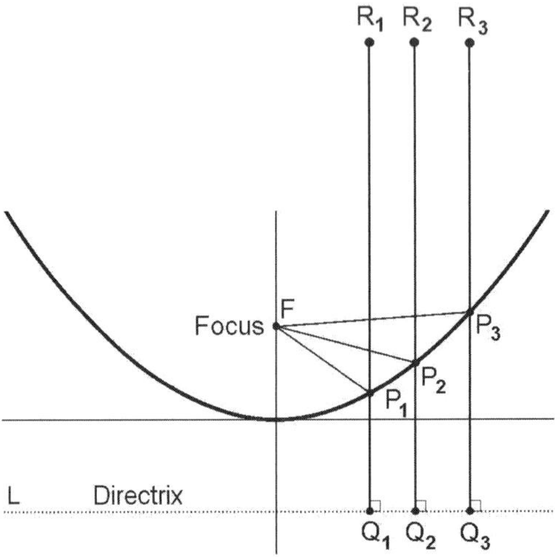

Figure 11.26 Focusing property of a parabola.

11.5. Kyoto box is a simple solar oven used extensively in Africa; see Fig. 11.27. Assuming a square box with four reflectors of $L = 75$ cm, on a sunny day with direct sunlight from the zenith, what is the optimum angle θ? What is the total solar power received by the box? If the efficiency is 70%, how long it will take to heat 1 gallon (3.785 liters) of water from 25°C to the boiling point?

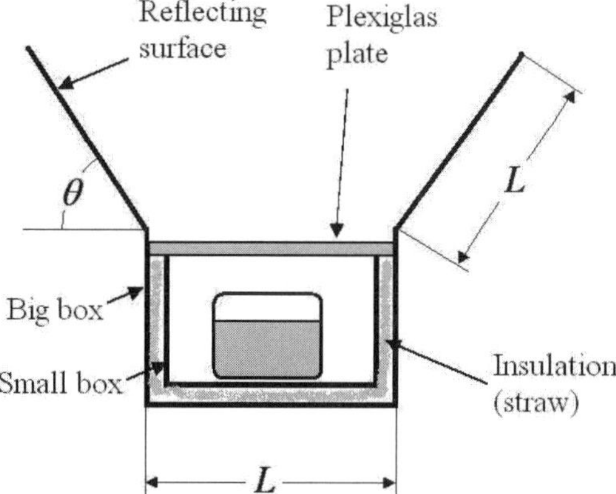

Figure 11.27 Kyoto box. A simple solar oven used extensively in Africa.

Chapter 12

Energy Storage

Because solar energy is intermittent, if it is the main supply, energy storage is a necessity. In general, there are two types of energy storage: utility-scale massive energy storage and the application-related, distributed energy storage. For utility-scale energy storage, the most effective method is using a reversible hydroelectric plant, which stores mechanical energy as potential energy of water in a high-level reservoir. We have discussed this in Section 1.3.1. The two most promising methods of distributed energy storage are thermal energy storage and rechargeable batteries. Especially for transportation (automobiles and small vessels) the rechargeable batteries will become the dominating energy storage device. Compressed air and flywheels are also important but will not be as widespread as thermal and rechargeable batteries. In this chapter, we will focus on thermal energy storage and rechargeable batteries.

12.1 Sensible Heat Energy Storage

Storage of energy as heat content of matter is inexpensive and easy to implement. It can be applied to space heating and cooling as well as for power generation. Two types of thermal energy are used: sensible thermal energy, essentially proportional to temperature difference, and phase transition thermal energy, such as the latent heat during freezing and melting, which could maintain a fixed temperature with energy content much greater than sensible thermal energy. Phase-change materials (PCM) are well suited for the storage of solar energy.

Sensible heat energy storage utilizes the heat capacity and the change in temperature of the material during the process of charging or discharging — the temperature of the storage material rises when energy is absorbed and drops when energy is withdrawn. One of the most attractive features of sensible heat storage systems is that charging and discharging operations can be expected to be completely reversible for an unlimited number of cycles, that is, over the lifespan of the storage.

In sensible heat energy storage, the thermodynamic process of the material is almost always *isobaric*, or under constant pressure, typically atmospheric pressure. A solid or liquid is usually used. The specific heat of gas is too low and not practical for thermal energy storage. The heat Q delivered by the material from initial temperature T_1 to final temperature T_2 is

$$Q = M \int_{T_1}^{T_2} c_p \, dT, \tag{12.1}$$

where M is the mass, c_p is the isobaric specific heat. In most applications, the density and specific heat can be treated as a constant. Equation 12.1 can be simplified to

$$Q = M c_p \left(T_2 - T_1 \right). \tag{12.2}$$

The quantity of material required for the storage tank and the heat losses are approximately proportional to the surface area of the tank. The storage capacity is proportional to the volume of the tank. Larger tanks have a smaller surface area–volume ratio and therefore are less expensive and have less heat losses per unit energy stored.

An important issue in thermal energy storage is thermal conduction, or temperature equalization in the medium. In liquids, heat conduction has two major paths: conduction and convection. Temperature in a liquid medium can become equalized much faster than in a solid. Therefore, liquid is preferred whenever applicable. Table 12.1 shows some thermal properties of commonly used liquids for sensible heat thermal energy storage.

12.1.1 Water

As shown in Table 12.1, water has the largest heat capacity both per unit volume and per unit weight. And it is free. Therefore, it is logical to use water as the material for sensible heat storage. A typical case is the hot-water tank used in most homes. The tank is typically insulated by foam polyurethane, which has a thermal conductivity κ = 0.02 W/mK and density ρ = 30 kg/m^3.

Table 12.1: Thermal Properties of Some Commonly Used Materials

Materials	Density ρ	Heat capacity c_p	Product ρc_p	Temperature range ΔT
	10^3kg/m^3	10^3J/kg·K	10^6J/m^3·K	°C
Water	1.00	4.19	4.19	0 to 100
Ethonal	0.78	2.46	1.92	-117 to 79
Glycerine	1.26	2.42	3.05	17 to 290
Canola Oil	0.91	1.80	1.64	-10 to 204
Synthetic Oil	0.91	1.80	1.64	-10 to 400

Source: *American Institute of Physics Handbook*, 3rd Ed., American Institute of Physics, New York, 1972.

If the entire tank is filled with water, the volume is

$$V = \frac{1}{4}\pi D^2 L \tag{12.3}$$

and the heat capacity is

$$C_p = \rho c_p V. \tag{12.4}$$

The total surface area of the tank is

$$A = \frac{1}{2}\pi D^2 + \pi D\, L. \tag{12.5}$$

The rate of heat loss is

$$\frac{dQ}{dt} = \frac{\kappa A}{\tau}\left(T_w - T_a\right), \tag{12.6}$$

where $T_w - T_a$ is the difference of water temperature T_w and ambient temperature T_a. The rate of temperature loss is

$$\frac{dT}{dt} = \frac{2\kappa(D + 2L)}{\tau\rho c_p D L}\left(T_w - T_a\right). \tag{12.7}$$

The rate of temperature drop through the tank skin is proportional to the total surface area and inversely proportional to the volume. If the tank is too thin or too flat, then the heat loss is high. Therefore, for a tank of fixed volume, there should be an optimal ratio L/D to minimize the heat loss. Intuitively, the condition should be $L \approx D$. In Problem 12.1, one can show that the intuition is correct: The optimum condition is $L = D$, and Eq. 12.7 becomes

$$\frac{dT}{dt} = \frac{6\kappa}{\tau\rho c_p D}\left(T_w - T_a\right). \tag{12.8}$$

It verifies another qualitative argument: The larger the dimension, the better the tank could preserve temperature.

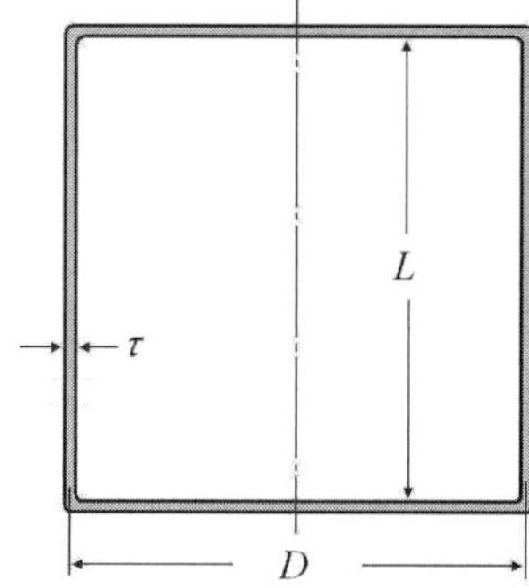

Figure 12.1 Water in an insulating tank. Calculation of energy storage behavior of an insulated water tank. The energy loss is proportional to the total surface area, and the energy content is proportional to the volume. The rate of temperature drop is minimized when the diameter D equals the length L. With a tank of linear dimension about 1 m with a 5-cm-thick foam polyurethane insulation, it take 8 h for the water temperature to drop by 1°C.

Here is a numerical example. If $D = L = 1$ m, $\tau = 5$ cm $= 0.05$ m, $T_w = 80°C$, and $T_a = 20°C$, the rate of temperature drop is

$$\frac{dT}{dt} = \frac{6 \times 0.02}{0.05 \times 4.19 \times 10^6 \times 1} \times 60 = 3.4 \times 10^{-5}\ °\text{C/s} = 0.124\ °\text{C/h}. \tag{12.9}$$

The temperature takes 8 h to drop by 1°C. Such an energy storage unit is extensively used in hot-water systems.

12.1.2 Solid Sensible Heat Storage Materials

In contrast to water and other liquids, solid materials can provide a larger temperature range and can be installed without a container. However, thermal conductivity becomes a significant parameter. Table 12.2 shows the thermal properties of typical solid materials. For many items, such as soil and rock, the values are only approximate or an average, because those materials vary widely. For example, the thermal parameters of soil could vary by one order of magnitude depending on the water content.

As shown in Table 12.2, materials with high thermal conductivities usually have low heat capacity. To use solid materials with high heat capacities, a long temperature equalizing time is expected.

Table 12.2: Thermal Properties of Solid Materials

Material	Density ρ (10^3kg/m^3)	Heat Capacity c_p (10^3J/kg·K)	Product ρc_p (10^6J/m^3·K)	Thermal Conductivity k (W/m·K)
Aluminum	2.7	0.89	2.42	204
Cast iIron	7.90	0.837	3.54	29.3
Copper	8.95	0.38	3.45	385
Earth (wet)	1.7	2.1	3.57	2.5
Earth (dry)	1.26	0.795	1.00	0.25
Limestone	2.5	0.91	2.27	1.3
Marble	2.6	0.80	2.08	2.07 - 2.94
Granite	3.0	0.79	2.37	3.5
Bricks	1.7	0.84	1.47	0.69
Concrete	2.24	1.13	1.41	0.9 - 1.3
Wood (oak)	0.48	2.0	0.96	0.16

Source: *American Institute of Physics Handbook* , 3rd Ed., American Institute of Physics, New York, 1972; and Ref. [31].

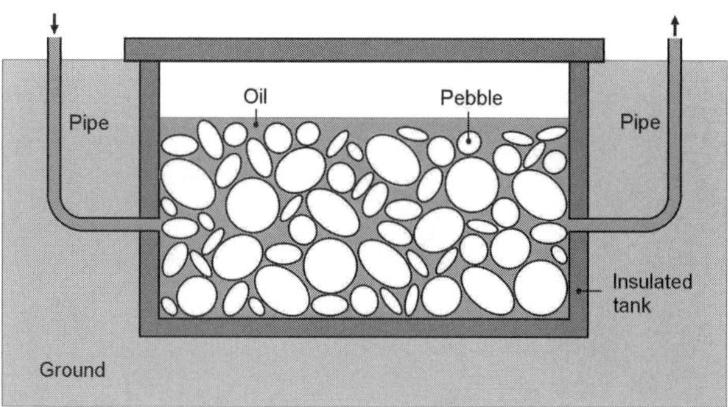

Figure 12.2 Rock-bed thermal energy storage system. Using a mixture of synthetic oil and pebbles, a thermal energy storage system at high temperature (e.g. 400°C) can be built with reasonable cost. The heat conduction is mainly through convection of oil, and the pebbles provide heat capacity.

12.1.3 Synthetic Oil in Packed Beds

Because the temperature range of water is limited, in order to store sensible heat at higher temperature, for example, in solar power generation systems, synthetic oil should be used. However, synthetic oil is expensive. A compromised solution is to use a mixture of synthesized oil and inexpensive solid materials, such as pebbles. Figure 12.2 shows such a thermal energy storage system schematically. A thermal energy storage system at high temperature (e.g. 400°C) can be built with limited cost. The heat conduction is mainly through convection of oil, and the pebbles provide heat capacity.

12.2 Phase Transition Thermal Storage

In sensible heat thermal energy storage systems, the process of charging or discharging of energy is related to a change of temperature, and the temperature is related to the amount of heat energy content. The storage density is limited by the heat capacity of the material. Using phase-change materials (PCMs), a considerably higher thermal energy storage density can be achieved that is able to absorb or release large quantities of energy ("latent heat") at a constant temperature by undergoing a change of phase. Theoretically, three types of phase changes can be applied: solid–gas, liquid–gas and solid–liquid. The first two phase changes are generally not employed for energy storage in spite of their high latent heats, since gases occupy large volumes. Large changes in volume make the system large, complex, and impractical. Solid–liquid transformations involve only a small change in volume (often only a few percent) and are therefore appropriate for phase-change energy storage.

Table 12.3: Commonly Used Phase-Change Materials.

Materials	Transition Temperature °C	Density ρ 10^3kg/m^3	Latent Heat h 10^6J/m^3
Water-ice	0	1.00	335
Paraffin wax	58 – 60	0.90	180 – 200
Animal fat	20 – 50	0.90	120 – 210
$CaCl_2$ (6→2)H_2O	29	1.71	190.8
Na_2SO_4 (10→0)H_2O	32.4	1.46	251
$Ba(OH)_2$ (8→0)H_2O	72	2.18	301
$MgCl_2$ (6→4)H_2O	117	1.57	172

Source: Refs. [28] and [31].

In general, the heat absorbed or released during the phase transition is

$$\Delta Q = H_2 - H_1, \tag{12.10}$$

where H is the enthalpy before and after the transition. The latent heat h, or specific enthalpy, is defined by

$$H = hV, \tag{12.11}$$

where V is the volume of the material.

Table 12.3 shows the thermal properties of several commonly used PCMs. A good PCM should provide energy storage at the desired temperature, have a large latent heat and a small volume change, and be non-flammable, noncorrosive, nontoxic, and inexpensive.

In general, PCMs are more expensive than sensible heat systems. They undergo solidification and therefore cannot generally be used as heat transfer media in a solar collector or the load. A separate heat transport medium must be employed with a heat exchanger in between. Many PCMs have poor thermal conductivity and therefore require large amount of heat exchange liquid. Others are corrosive and require special containers. These increase the system cost.

12.2.1 Water–Ice Systems

As shown in Table 12.3, the latent heat of the freezing of water or melting of ice is one of the highest. The water-ice system has already used in industry to save energy in air-conditioning systems. Figure 12.3 is a photo of a water–ice energy storage system, named Ice Bear, designed and manufactured by Ice Energy, Inc. The system has a large insulated tank filled with water and a lot of copper heat exchange coils. During the night, the refrigerator uses inexpensive electricity and cool air to make ice from the water. As we shown in Chapter 6, the lower the ambient temperature, the higher the

Figure 12.3 Ice Bear energy storage system. An insulated tank is filled with water and many copper heat exchange coils. During the night, the refrigerator uses inexpensive electricity and cool air to make ice from the water. As shown in Chapter 6, the lower the ambient temperature, the higher COP. Therefore, to make a well-defined mass of ice during the night, the electricity cost is much lower than in the hot daytime. Courtesy of Ice Energy, Inc.

coefficient of performance (COP). Therefore, to make a well-defined mass of ice during the night, the electricity cost is much lower than in the hot daytime. During the hot daytime, the system uses the ice to cool the building. With this system, the efficiency of energy storage can be better than 90%. The overall energy savings can be as high as 30%.

Below is a numerical example of a water–ice system. Using the insulated container in Fig. 12.1, the total latent heat is

$$Q = \Delta H = H_2 - H_1, \tag{12.12}$$

Using Eq. 12.3,

$$\Delta H = \frac{1}{4}\pi h D^2 L. \tag{12.13}$$

For a tank of $D = L = 1$ m, the total enthalpy of phase transition is

$$\Delta H = 335 \times \pi \times 4 \times 10^6 \,\mathrm{J} = 2.63 \times 10^8 \,\mathrm{J}. \tag{12.14}$$

If the ambient temperature is 20°C, for such a tank, according to Eq. 12.6, the rate of heat loss is

$$\frac{dQ}{dt} = \frac{3 \times \pi \times 0.02}{2 \times 0.05} \times 20 = 1.89\,\mathrm{W}. \tag{12.15}$$

If at the beginning the tank is full of ice, then it can keep the temperature at 0°C for $2.63 \times 10^8/1.89 = 1.39 \times 10^8$ s, or 4.4 years. Therefore, using a moderate means, the energy storage is efficient.

The freezing temperature of pure water is 0°C. If the working temperature is other than 0°C, a mixture of water and other materials can do the job. An example is solar-operated refrigerator (U.S. Patent 7,543,455). The freezer should work at around -10°C. Because the refrigerator is used for food and medicine, the additive must be

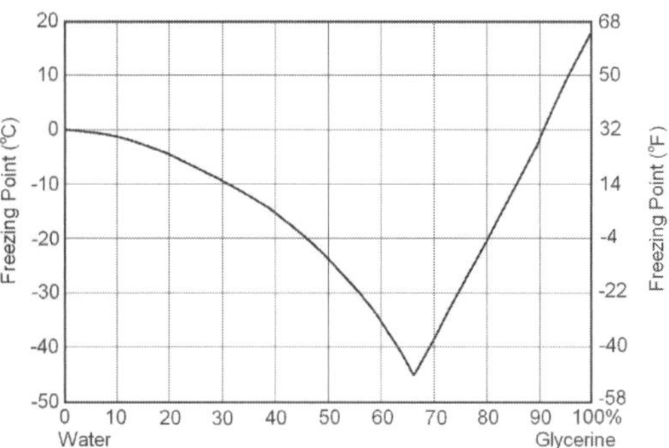

Figure 12.4 Freezing point of water–glycerin system. By mixing water and glycerin, the freezing point can be lowered from 0 to -40°C depending on the ratio. By mixing with a few percent of alcohol, the freeing point ca be lowered further.

nontoxic. Glycerin and ethanol are good additives because both are ingredients of food and medical substances. Figure 12.4 shows the freezing points of a mixture of water and glycerin. By mixing with alcohol, the freezing temperature can be lowered further.

Glycerin is a byproduct of biodiesel production, and there is a surplus of raw glycerin. The impurities in raw glycerin as byproducts of biodiesel production are water, salt, fatty acids and alcohol, and there is no need to use high-purity glycerin. Therefore,

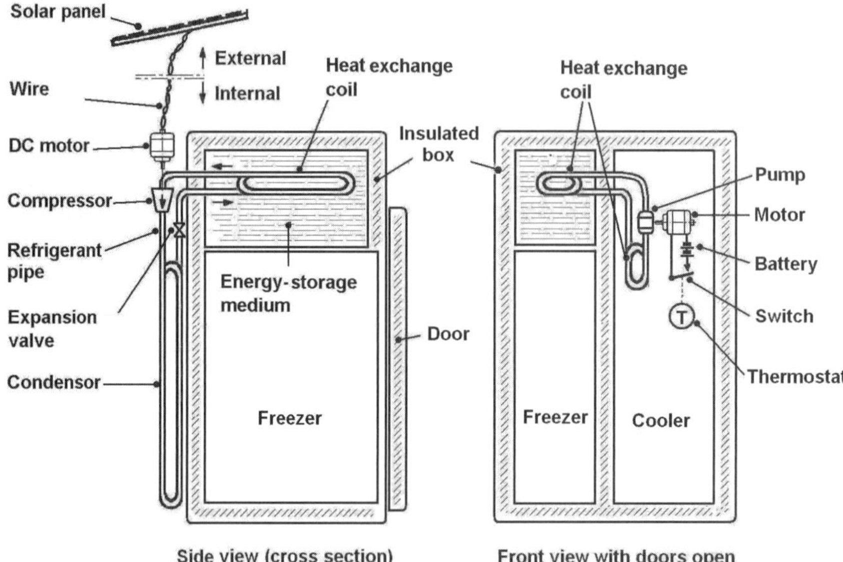

Figure 12.5 Solar-powered refrigerator. The design of a solar-powered refrigerator. The composition of the water–glycerin–alcohol system determines the working temperature of the freezer. The temperature of the cooler is controlled by a thermostat to the brine circulation, which can be set by the user. See U.S. Patent 7,543,455.

economically it is advantageous. Figure 12.5 shows the design of a solar-powered refrigerator. The composition of the water–glycerin–alcohol system determines the working temperature of the freezer. The temperature of the cooler is controlled by a thermostat to the brine circulation, which can be set by the user.

12.2.2 Paraffin Wax and Other Organic Materials

Paraffin wax is a byproduct of petroleum refining. The melting point of paraffin wax ranges from 50 to 90°C. Currently, paraffin wax only has a few commercially valuable applications, such are candles and floor wax. For such applications, only these with melting temperature between 58 and 60° are usable. But the supply is abundant. The melting temperature of paraffin matches the range needed for space heating and domestic hot water. It is also nontoxic and noncorrosive. One problem is its low thermal conductivity. This can be mitigated with encapsulation; see Section 12.2.4.

Other organic materials have similar properties as paraffin wax. An example is animal fat. Lard and chicken fat are considered harmful to human health because they can increase blood triglyceride and cause obesity. In some sense they are wastes of the food-processing industry. Animal fat is nontoxic and noncorrosive, thus it can be safely utilized for energy storage in residential environments.

12.2.3 Salt Hydrates

Many inorganic salts crystallize with a well-defined number of water molecules to become salt hydrates. Heating a salt hydrate can change its hydrate state. For example, hydrated sodium sulfate (Glauber's salt) undergoes the transition at 32.4°C

$$\mathrm{Na_2SO_4 10H_2O} + \Delta Q \longrightarrow \mathrm{Na_2SO_4} + 10\mathrm{H_2O}. \tag{12.16}$$

In general, the transition is

$$\mathrm{Salt}\, m\mathrm{H_2O} + \Delta Q \longrightarrow \mathrm{Salt}\, n\mathrm{H_2O} + (m-n)\mathrm{H_2O}. \tag{12.17}$$

Thus, at the melting point the hydrate crystals break up into anhydrous salt and water or into a lower hydrate and water. The latent heat could be quite large, thus the storage density could be very high. If the water released is sufficient, a water solution of the (partially) dehydrated salt is formed.

These salt hydrates can be used in solar-operated space-heating or hot-water systems to provide uniform temperature over a longer period of time.

12.2.4 Encapsulation of PCM

To mitigate the problem of low thermal conductivity of PCMs, the material is often encapsulated in various forms. Figure 12.6 shows an example of a PCM encapsulated in flat or tubular parcels. A heat transfer fluid is required to make it operational.

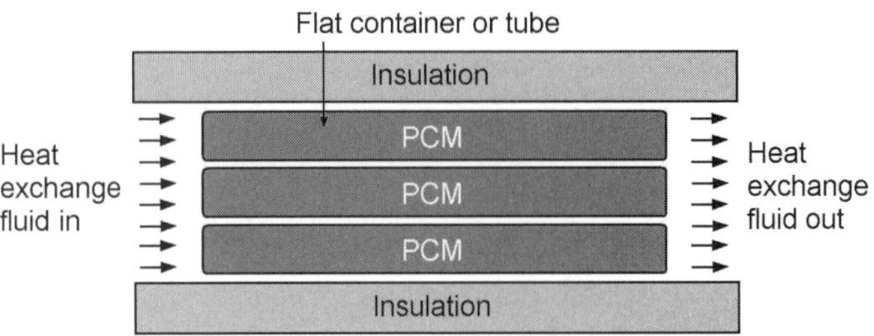

Figure 12.6 Encapsulation of PCM. An example of PCM encapsulated in flat or tubular parcels, to mitigate the low thermal conductivity of PCM.

12.3 Rechargeable Batteries

In the last decades, the technology of rechargeable batteries has observed a phenomenal expansion. To date, the century-old lead–acid rechargeable battery has been constantly improved and is still in widespread use. New types of batteries, especially lithium ion rechargeable batteries, are experiencing an explosive growth, and will soon become the dominating distributed storage device for electricity.

Table 12.4 gives the specifications of several types of rechargeable batteries. One of the basic parameters is nominal voltage. Cells with higher nominal voltage are certainly advantageous, because fewer cells are needed to construct the desired system. Energy density and specific energy are also significant parameters. For static applications such as street lights, a smaller specific energy is not a serious problem. For automobile applications, energy density and specific energy are critical parameters. Lifetime is also a critical parameter for automobile applications. Not shown here is the unit cost. Currently, lead–acid batteries are still the least expensive and thus widely used.

Table 12.4: Comparison of Rechargeable Batteries

Type	Voltage (V)	Energy Density (Wh/liter)	Specific Energy (Wh/kg)	Lifetime (cycles)
Lead–acid	2.1	70	30	300
NiMH	1.4	240	75	800
$LiCoO_2$	3.7	400	150	1000
$LiMn_2O_4$	4.0	265	120	1000
$LiFePO_4$	3.3	220	100	3000

(a) Discharge

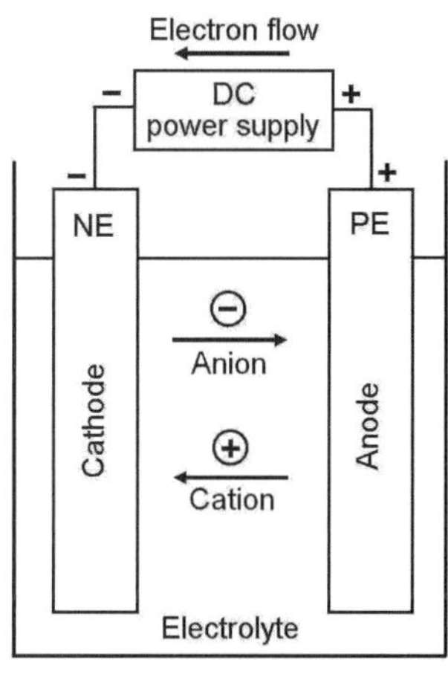

(b) Charge

Figure 12.7 Electrochemistry of rechargeable batteries. (a) Discharging process. By connecting the cell to an external load, electrons flow from NE, the anode, to PE. The electric circuit is completed in the electrolyte by the flow of anions and cations to the anode and cathode, respectively. (b) Charging process. An external DC power supply forces electrons to flow into NE, then the circuit is completed in the electrolyte by the flow of anions and cations.

12.3.1 Electrochemistry of Rechargeable Batteries

The basic structure and the charging–discharging processes of rechargeable batteries are shown in Fig. 12.7. For reference, definitions are provided as below. For more details, see, for example, *Handbook of Batteries* [51].

Cell The basic electrochemical unit converting electrochemical energy to electrical energy.

Battery One or more electrochemical cells connected in series or parallel to provide electrical power.

Primary cells or batteries One-time source of electricity, cannot be recharged after usage. Are discarded after usage.

Secondary (rechargeable) cells or batteries Can be recharged electrically after usage to their original condition.

Oxidation Loss of electron(s).

Reduction Gain of electron(s).

Redox Reduction and oxidation.

Anion Negative ion — after gaining electron(s).

Cation Positive ion — after losing electron(s).

Anode Oxidation takes place. During charge, it is the positive electrode (PE). During discharge, it is the negative electrode (NE).

Cathode Reduction takes place. During charge, it is NE. During discharge, it is PE.

Electrolyte Typically an ionic conducting liquid.

Figure 12.7(a) shows the discharging process. By connecting the cell to an external load, electrons flow from the negative electrode (NE), the anode, which is oxidized, through the external load to the positive electrode (PE), the cathode, where the electrons are accepted and the cathode material is reduced. The electric circuit is completed in the electrolyte by the flow of anions (negative ions) and cations (positive ions) to the anode and cathode, respectively.

Figure 12.7(b) shows the charging process. An external DC power supply is connected to the battery. The external electrical field forces electrons to flow into the negative electrode (NE), where reduction takes place. On the other hand, oxidation takes place at the positive electrode (PE), where the electrons flow out from. As the anode is, by definition, the electrode at which oxidation occurs and the cathode the one where reduction takes place, the positive electrode is the anode and the negative electrode is the cathode.

12.3.2 Lead–Acid Batteries

To date, the most widely used rechargeable battery is the lead–acid battery. Every automobile should have one with six cells. For a fully charged lead–acid battery, the positive electrode is made of PbO_2 and the negative electrode is made of pure lead. The electrolyte is diluted sulfuric acid. After discharging, both the positive electrode and the negative electrode become $PbSO_4$. Sulfuric acid is thus consumed. By measuring the specific gravity of the electrolyte, and thus the concentration of sulfuric acid, the state of discharging, thus the energy remaining, can be determined.

The electrochemistry is as follows. During discharging, at the positive electrode, PbO_2 is reduced,

$$PbO_2 + H_2SO_4 + 2e^- \longrightarrow PbSO_4 + 2OH^-. \tag{12.18}$$

At the negative electrode, lead is oxidized,

$$Pb + H_2SO_4 \longrightarrow PbSO_4 + 2H^+ + 2e^-. \tag{12.19}$$

During charging, at the positive electrode, $PbSO_4$ is oxidized,

$$PbSO_4 + 2OH^- \longrightarrow PbO_2 + H_2SO_4 + 2e^-. \tag{12.20}$$

At the negative electrode, $PbSO_4$ is reduced,

$$PbSO_4 + 2H^+ + 2e^- \longrightarrow Pb + H_2SO_4. \tag{12.21}$$

Lead is a heavy and toxic metal. Sulfuric acid is a dangerous liquid. The lifetime is short (300 cycles). Therefore, it is not suitable for portable devices and automobiles. However, since lead is intrinsically an inexpensive metal and can be recycled, the overall cost is low. It is expected that such batteries will still be extensively used in the foreseeable future, for example, as energy storage devices in remote areas or in the basements of residential buildings to store solar electricity.

12.3.3 Nickel Metal Hydride Batteries

In recent decades, nickel metal hydride rechargeable batteries have been widely used in automobiles and relatively large portable electronic devices. The positive electrode is nickel hydroxide, and the negative electrode is an intermetallic compound. The most common metal has the general form AB_5, where A is a mixture of rare earth elements, lanthanum, cerium, neodymium, praseodymium and B is nickel, cobalt, manganese, and aluminum.

The electrochemistry is as follows. During discharging, at the positive electrode, NiOOH is reduced,

$$NiOOH + H_2O + e^- \longrightarrow Ni(OH)_2 + OH^-. \tag{12.22}$$

At the negative electrode, metal hyride is oxidized,

$$MH + OH^- \longrightarrow M + H_2O + e^-. \tag{12.23}$$

During charging, at the positive electrode, $Ni(OH)_2$ is oxidized,

$$Ni(OH)_2 + OH^- \longrightarrow NiOOH^+ H_2O + e^-. \tag{12.24}$$

At the negative electrode, metal is reduced,

$$M + H_2O + e^- \longrightarrow MH + OH^-. \tag{12.25}$$

When overcharged at low rates, the oxygen produced at the positive electrode passes through the separator and recombines at the surface of the negative. Hydrogen evolution is suppressed and the charging energy is converted to heat. This process allows NiMH cells to remain sealed in normal operation and to be maintenance free.

NiMH batteries have been applied extensively in electric automobiles, such as the General Motors EV1, Honda EV Plus, Ford Ranger EV, and Vectrix Scooter. Hybrid vehicles such as the Toyota Prius, Honda Insight, Ford Escape Hybrid, Chevrolet Malibu Hybrid, and Honda Civic Hybrid also use them. NiMH technology is used extensively in rechargeable batteries for consumer electronics.

12.3.4 Lithium-Ion Batteries

Currently the lithium ion battery is the most rapidly developing energy storage device. Soon after its invention in 1991, the CoO_2–Li ion battery became the predominant power source for small portable electronics such as mobile phones, digital cameras, and laptop computers. It is widely believed to be the best candidate for powering automobiles because it has the highest specific energy and the longest lifetime; see Table 12.4.

In some sense, the electrochemistry of Li ion batteries is the simplest. The only ion involved is the lithium cation, Li^+. It has the smallest radius and the highest standard potential, -3.01 eV. The negative electrode is made of graphite, where the small Li ion intercalates into the space between adjacent sheets of grapheme. The positive electrode is a transition metal oxide, where the base metal can have different valence states to allow the lithium atom to join in or leave out.

Figure 12.8 shows the electrochemical processes in a Li ion cell. When fully charged, most of the lithium ions are buried in the planes of graphite. During the discharging process, as shown in Fig. 12.8(a), the lithium ions leave the negative electrode, drift through the electrolyte, pass the microporous separation film, and combine with the metal oxide in the positive electrode. At the end of the discharging process, most of the lithium ions are combined with the metal oxide in the positive electrode. During the charging process, as shown in Fig. 12.8(b), the lithium ions are forced by the external voltage to leave the metal oxide, drift through the electrolyte, pass the microporous separation film, and are intercalated into graphite.

Most Li ion batteries for small electronic devices, such as cell phones and digital cameras, use CoO_2 as the basis of the positive electrode. During discharging, at the positive electrode, the lithium ion is combined with CoO_2,

$$\mathrm{Li}_{1-x}\mathrm{CoO}_2 + x\mathrm{Li}^+ + x\mathrm{e}^- \longrightarrow \mathrm{LiCoO}_2. \tag{12.26}$$

At the negative electrode, lithium ions are *extracted*,

$$\mathrm{Li}_x\mathrm{C}_6 \longrightarrow \mathrm{C}_6 + x\mathrm{Li}^+ + x\mathrm{e}^-. \tag{12.27}$$

During charging, at the positive electrode, lithium ions are extracted,

$$\mathrm{LiCoO}_2 \longrightarrow \mathrm{Li}_{1-x}\mathrm{CoO}_2 + x\mathrm{Li}^+ + x\mathrm{e}^-. \tag{12.28}$$

At the negative electrode, lithium ions are intercalated into graphite

$$\mathrm{C}_6 + x\mathrm{Li}^+ + x\mathrm{e}^- \longrightarrow \mathrm{Li}_x\mathrm{C}_6. \tag{12.29}$$

In the above reactions, $0 \leq x < 1$ is the fraction of lithium ion reacted.

The CoO_2-based Li ion battery has very high specific energy. For applications where weight is an important factor, it is preferred. However, cobalt is expensive. In addition, it has been recorded that for large-size CoO_2-based Li ion batteries, explosion has occurred. For power applications, Li ion batteries based on manganese oxide and iron phosphate are more preferred.

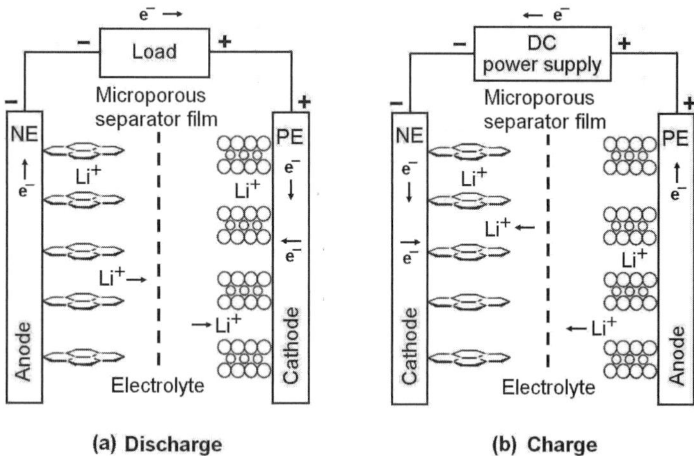

Figure 12.8 Electrochemical processes in a Li ion cell. (a) during discharging, lithium ions leave the negative electrode, drift through the electrolyte, pass the microporous separation film, and combine with the metal oxide in the positive electrode. (b) during charging, lithium ions are forced by the external voltage to leave the metal oxide and intercalated into graphite.

$LiFePO_4$ was discovered by John Goodenough's research group at the University of Texas in 1996 as a material for the positive electrode of Li ion batteries. Because of its low cost, nontoxicity, high abundance of iron, excellent thermal stability, safety characteristics, good electrochemical performance, and high specific capacity (170 mA·h/g, or 610 C/g), it gained acceptance in the marketplace.

While $LiFePO_4$ cells have lower voltage and energy density than $LiCoO_2$ Li ion cells, this disadvantage is offset over time by the slower rate of capacity loss (aka greater calendar life) of $LiFePO_4$ when compared with other lithium ion battery chemistries. For example, after one year on the shelf, a $LiFePO_4$ cell typically has approximately

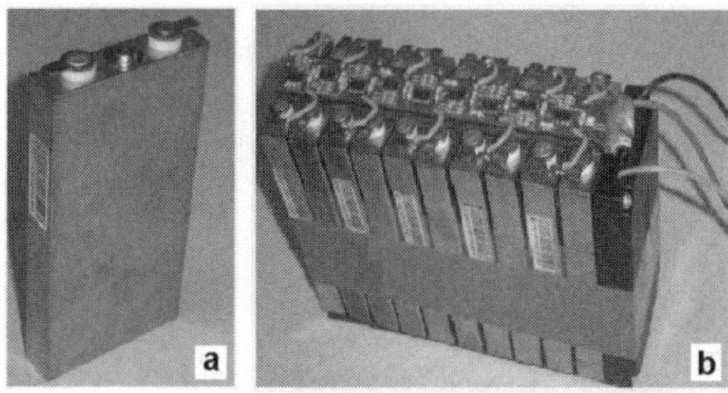

Figure 12.9 Power Li ion batteries. (a) Single Li ion battery with nominal voltage of 3.7 V. (b) With 10 Li ion batteries connected in series, the nominal voltage is 37 V. Photo taken by the author. Courtesy of Phylion Battery Co., Sūzhōu, China.

the same energy density as a $LiCoO_2$ Li ion cell. Beyond one year on the shelf, a $LiFePO_4$ cell is likely to have higher energy density than a $LiCoO_2$ Li ion cell due to the differences in their respective calendar lives.

The basic electrochemistry of the $LiFePO_4$ battery is as follows. Iron has two oxidation states, Fe(II) and Fe(III). The Fe(III) compounds are often strong oxidizers. For example, the standard method of etching copper to make printed circuit boards is to use $FeCl_3$,

$$2FeCl_3 + Cu \longrightarrow 2FeCl_2 + CuCl_2. \tag{12.30}$$

Therefore, both $FePO_4$ and $LiFePO_4$ are stable compounds. Because the lithium ion is very small, these two compounds has negligible differences in volume per mole and share the same crystallographic structure.

The charging and discharging processes are as follows. In a fully charged $LiFePO_4$ battery, the positive electrode is mostly $FePO_4$, and graphite in the negative electrode is filled with lithium atoms. During discharging, at the positive electrode, the lithium ions are squeezed into $FePO_4$,

$$FePO_4 + Li^+ + e^- \longrightarrow LiFePO_4. \tag{12.31}$$

At the negative electrode, lithium ions are extracted from graphite,

$$LiC_6 \longrightarrow C_6 + Li^+ + e^-. \tag{12.32}$$

During charging, at the positive electrode, lithium ions are extracted from iron phosphate,

$$LiFePO_4 \longrightarrow FePO_4 + Li^+ + e^-. \tag{12.33}$$

At the negative electrode, lithium ions are intercalated into graphite,

$$C_6 + Li^+ + e^- \longrightarrow LiC_6. \tag{12.34}$$

12.3.5 Mineral Resource of Lithium

As Li ion batteries become a major component of automobiles in the future, the problem of the mineral resource of lithium is currently of interest. First, let us estimate how much lithium is needed to equip all the automobiles in the world, and then compare it with the known mineral resources.

The atomic weight of lithium is 6.94 g/mol. Each mol of lithium has 96,490 coulombs of electrical charge. The working voltage of the Li ion battery is 3.5 V. Therefore, the specific capacity of lithium is

$$P = \frac{96490 \times 3.5}{3600 \times 6.94} = 13.5\,\mathrm{kWh/kg}. \tag{12.35}$$

If each car needs a battery of 30 kWh capacity, then 2.2 kg of lithium is sufficient. Currently, there are 600 million cars in the world. The total amount of lithium required is then 1.32 million tons.

According to the 2009 U.S. Geological Survey [2], the world's total verified lithium reserve base is 11 million tons, which could equip all the cars in the world many times. Recently, according to New York Times [71], a huge amount of lithium deposition was found in Afghanistan. According to an internal Pentagon report, Afghanistan could become the "Saudi Arabia of lithium." Because currently most of the lithium resource is from the concentrated brine in high-altitude saline lakes, the discovery in Afghanistan is not surprising.

12.4 Solar Energy and Electric Vehicles

According to the Energy Information Administration, in the United States, transportation uses 26.5% of the total energy, or 67.6% of the petroleum. To reduce the use of fossil energy, the transition to electric cars using rechargeable batteries, especially using Li ion batteries as storage medium and solar energy as the source, is the best approach. Electric cars have many desirable features:

- The intrinsic efficiency of electric motors is very high, typically 90%.
- The round-trip efficiency of energy storage in rechargeable batteries, especially Li ion batteries, is very high, typically around 90%.
- The mechanical structure of electric cars is much simpler than either the Otto engine cars or the diesel engine cars.
- The regenerative brake can be implemented naturally. Actually, the efficiency enhancement of hybrid cars, such as the Toyota Prius, is mainly due to the regenerative brake.
- As battery technology is progressing rapidly, the manufacturing cost of electric cars will decrease rapidly.
- They can make a natural connection to solar electricity.

Table 12.5: Efficiency of Several Automobiles.

Automobile	Miles per Gallon	km/kWh	kWh/km
BMW hydrogen car	10	0.45	2.24
Rangerover	20	0.89	1.19
Toyota Camry	32	1.43	0.70
Toyota Prius	55	2.46	0.41
Chevy Volt	150	6.67	0.15

Source: *Sustainable Energy – without hot air*, David J. C. MacKay [56].

- They are virtually noise free.

Let us first examine some technical data. In the United States, the efficiency of cars is measured by miles per gallon of gasoline, or mpg. Ine SI units, the most convenient measure is either kilometers per kilowatt-hours of energy, or the energy in kilowatt-hours required to drive 1 km. Because the energy content of gasoline is approximately 1.3×10^8 J/gallon and one mile is 1.609 km, a simple calculation gives 1 km/kWh = 22.37 mpg. Table 12.5 gives the measured data for several popular cars.

For several decades, hydrogen and fuel cell has been considered as an alternative to the Otto engine and the diesel engine for automobiles. Grandiose expressions such as the hydrogen age, hydrogen economy, and hydrogen era have been used. However, according to an analysis by Joseph J. Romm [74], a deputy energy secretary during the Clinton administration in charge of hydrogen projects, based on his hands-on experience, in the foreseeable future the use of hydrogen will not become a commercially viable method of energy storage. It is prohibitively expensive and notoriously dangerous. It is especially unsuitable for automobiles, because the density of storage of the highly compressed hydrogen is only one tenth that of gasoline. Besides, fuel cells have low efficiency (compared with rechargeable batteries) and low lifetime and use expensive precious metals.

In the future, solar photovoltaics will become the main source of electricity, a technology that works well with electric cars. In particular, solar photovoltaics can charge the batteries in electric cars without going through the grid. This approach has in-

Figure 12.10 Solar-powered electric car charging station in Kyoto. The charging station, supplied by Nissin Electric Co., has a battery pack, an inverter, and a rapid charging setup. Courtesy of Kyoto University.

Figure 12.11 Solar-powered electric car charging station in Tennessee. In the United States, taking another step in building its electric vehicle charging infrastructure, Tennessee is now home to the first of several solar-powered EV-charging stations [10].

herent advantages. It avoids the cost and energy loss due to the DC–AC inverter and the AC–DC inverter. Furthermore, the intermittency of solar energy is no longer a disadvantage because to charge batteries a very stable power source is not required. The advantage of solar-powered battery charger can be further improved by using the battery-swap procedure: Spare batteries are charged when there is sunlight. A car with a depleted battery can come to the charging station to swap for a fully charged one in a few minutes, probably even faster than filling the gas tank. Figure 12.10 shows a solar-powered electric car charging station in the city of Kyoto supplied by Nissin Electric Co. It has a battery pack, an inverter, and a rapid charging setup.

In the United States, both General Motor and Nissan will mass-produce electric cars, Chevy Volt and Nissan Leaf, in the state of Tennessee. Therefore, it makes sense for people in Tennessee to start laying a plug-in groundwork. That has already happened in Pulaski, Tennessee. In August 2010, the first solar parking lot opened — that is, parking spaces with an electric vehicle charger that is powered by solar energy; see Fig. 12.11. At the opening ceremony, after remarks highlighting American energy independence and job creation, Congressman Lincoln Davis switched on the array, sending the sun's energy into the grid. It was stated that "All the components are American made. It is an example of how American small business and manufacturing are growing in the new green economy [10]."

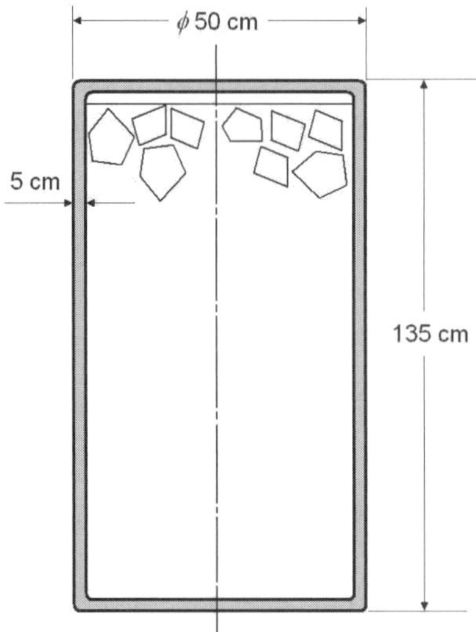

Figure 12.12 A typical domestic hot water tank.

Problems

12.1. A typical domestic hot water tank has a dimension of diameter $D = 50$ cm and height $H = 135$ cm, with a $\tau = 5$ cm thick insulation made of rigid foam polyimide, see Figure 1. If the difference of the external and internal temperatures is 45°C, what is the energy loss of this tank in watts? (The thermal conductivity of rigid foam polyimide is $k = 0.026$ W/(m°C).)

12.2. By storing hot water in that tank at 65°C and the environment temperature is 20°C, how long it takes to cool the water temperature down by 1°C?

12.3. By storing ice at 0°C in that insulated tank, with external temperature of 25°C, how long it takes for all the ice to melt? (The latent heat of ice is 335 kJ/liter.)

Chapter 13

Building with Sunshine

In an interview with *U.S. News and World Report*(March 2009) Steven Chu, the U.S. Energy Secretary, said this about the importance of improving building design to conserve energy [30]:

> People sometimes say energy efficiency and conservation are not sexy; they're low tech. That's actually not true. They can be very sexy and very high tech. ... Say you're building a new home. McKinsey has done a study that says for $1,000 extra in material and labor that investment could pay for itself in one and a half years and you could save a tremendous amount of energy. In terms of new homes and buildings, part of it is regulatory. But you've also got to convince people — they've got to believe in their heart and soul — that a small, minor upfront cost will actually decrease their monthly bill.

The details are in two reports by McKinsey published in 2009, entitled "*Unlocking Energy Efficiency in the U.S. Economy*" [60] and "*China's Green Revolution*" [59], which addressed the problems and suggestions for the two largest fossil fuel energy users in the world. The central conclusion is [60]:

> Energy efficiency offers a vast, low-cost energy resource for the U.S. economy — but only if the nation can craft a comprehensive and innovative approach to unlock it. Significant and persistent barriers will need to be addressed at multiple levels to stimulate demand for energy efficiency and manage its delivery across more than 100 million buildings and literally billions of devices. If executed at scale, a *holistic approach* would yield gross energy savings worth more than $1.2 trillion, well above the $520 billion needed for upfront investment in efficiency measures....

In fact, the most efficient approach to save energy is to apply passive design of the buildings into the architecture. In general, buildings consume about 40% of the world's energy. Using the holistic approach of house design, up to 50% of such energy can be saved. The principles include the following [59]:

1. Orient or position the building to absorb solar heat in cold regions and optimize solar heat in hot regions.

2. Build windows that open to reduce dependency on air conditioners, fans, and heaters.

3. Use smaller heaters or coolers, given that passive design reduces the need for them.

There are two challenges for such an approach [59]. The first is to embed the approach in the mindset of professionals in the building sector. Currently, active heating and cooling systems, often using fossil fuels, are still considered the panacea. The second is to overcome a lack of effective dialogue between architects and civil engineers. Energy efficiency is still seldom taught in the curriculum of architects. Civil engineers, who may be more aware of energy efficiency considerations, tend to focus on executing a design and rarely get involved in the decisions that led to it.

The *holistic approach* to building design utilizing solar energy is not new. Many ancient cultures practiced it, but it has almost been forgotten in the developed world. To build a better future in the twenty-first century, we need a renaissance of the art of holistic building design.

13.1 Early Solar Architecture

13.1.1 Ancient Solar Architecture

Archeological evidence has shown that many ancient cultures built their houses according to the principles of passive solar design. According to Socrates, the ideal house should be cool in the summer and warm in the winter [17]. He also suggested that this goal can be achieved partially by orienting the opening of the house toward the south and providing a portico to create shade in the summer. Almost all important ancient cultures were located in the Northern hemisphere, and the houses were designed with large windows to the south but full walls in the north.

Holistic design principles, which were well-documented in the ancient Chinese literature since the West Zhōu dynasty (eleventh century *B.C.* – 771 *B.C.*), known as Fēngshŭi, meaning "wind and water," but reference to the Sun is one of the cardinal principles. In recent decades, the discipline of Fēngshŭi has become popular in the Western world. Determining true south is of fundamental importance in Fēngshŭi. In *Code of Zhōu Regulations* (Zhōu Lĭ), written *circa* eleventh century BC, a method of determining true south was documented as follows: "by marking the point of sunrise and the point of sunset, then taking the middle point, the south point can be found." It is much more accurate than the magnetic compass.

13.1.2 Holistic Architecture in Rural China

Figure 13.1 shows a typical peasant house in rural northern China which embodies some principles of holistic design utilizing solar energy. It was probably designed over 1000

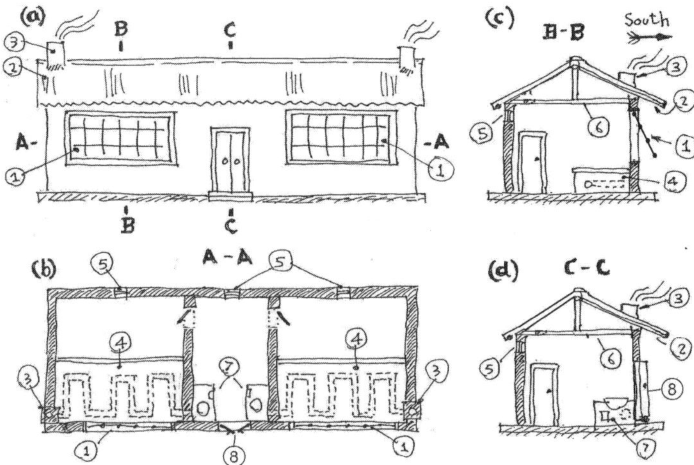

Figure 13.1 Peasant house in rural northern China. The long axis of the house is east–west. Windows (1) are on the south side. Due to the long eaves (2), it has full sunlight in the winter and is in the shade in summer. The heat from the oven (7) is stored in the adobe beds (4), which can stay warm for the entire night. The wind from the front window to the small north windows (5) keeps it cool in the summer.

years ago and is still used in less developed regions of northern China. I have personally lived in such houses for several years and have been impressed by its comfort.

The long axis of the house is east–west. Large glass windows (1) are always on the south side. The windows can be swung partially open from the lower side. The eaves (2) are designed such that in the winter the windows exposes to full sunlight and in the summer the windows are in shade. Near the top of the northern wall, there are small windows (5). In the summer, by opening the small northern windows and the south window partially, wind will automatically blow from the lower opening of the south window to the small north windows, which can cool the house by ventilation. In the winter, the smoke from the oven (7) flows to the chimney (3) through the zigzagged tunnels in the adobe bed (4), called *kàng*, which can stay warm for the entire night. In the summer, cooking is often done in the sheds outside the house; thus the adobe bed, directly connected to the ground, can remain cool. The space between the ceiling (6) and the roof also helps to keep a stable temperature. To keep the rooms warm, in the winter, the front door (8) is often equipped with a heavy curtain.

13.2 Building Materials

The design of a building involves an interplay of architectural and civil engineering. In this section, some basic concepts of civil engineering for building design are presented.

13.2.1 Thermal Resistance

The thermal resistance, or the R-value, is a measure of the insulation property of a panel or sheet of building material, such as a wall panel, a door, or a window. It is the ratio of temperature difference on the two surfaces and the heat flux (heat flow per unit area). In SI unit, the R-value (RSI) is the temperature difference (in Kelvin) required to leak heat energy in watts per unit area ($\mathrm{W/m^2}$). Therefore, the unit of RSI is $\mathrm{K{\cdot}m^2/W}$. In British units, the R-value is the temperature difference (in Fahrenheit) required to leak heat energy in Btu per hour per unit area. Therefore, the unit of R-value is $\mathrm{F{\cdot}h{\cdot}f^2/Btu}$). The conversion relation is 1 RSI = 5.678R, 1 R = 0.1761RSI.

For a wall consisting of several layers, the R-value is additive:

$$R = \sum_i R_i, \tag{13.1}$$

where R_i is the R-value of the i-th layer of the material. Following are some examples:

- Single-pane glass window: R-1 (RSI = 0.18)
- Double-pane glass window: R-2 (RSI = 0.35)
- Above window with low-emissivity coating: R-3 (RSI = 0.52)

13.2.2 Specific Thermal Resistance

The insulation property of a material, for example, concrete, fiberglass wool, or foam polyurethane, is measured by its *specific thermal resistance*, or the r-value. In SI units (rSI), it is defined as the temperature difference (in Kelvin) per unit thickness (in meter) required to leak heat energy in watts per unit area ($\mathrm{W/m^2}$). In British unit (r), it is defined as the temperature difference (in Fahrenheit) per unit thickness (in inches) required to leak heat energy in Btu per hour per unit area ($\mathrm{Btu/h{\cdot}f^2}$). The conversion relation is 1 rSI = 0.125r, 1 r = 8rSI.

Table 13.1: Specific Thermal Resistance

Material	r-Value (Btu)	rSI (SI unit)
Wood panels	2.5	20
Fiberglass or rock wool	3.1 – 3.6	25 – 30
Icynene spray	3.6	30
Polyurethane rigid panel	7	56
Poured concrete	0.08	0.64

For a wall consisting of several layers, the R-value is the sum of the product of the thickness and the r-value of each layer:

$$R = \sum_i t_i r_i, \tag{13.2}$$

where t_i is the thickness of the ith layer and r_i is the specific thermal resistance of the material of the ith layer.

Table 13.1 lists the r-values of some common building materials. In the United States, the most commonly used insulation material is fiberglass wool. It has a reasonably high r-value and is lightweight, inexpensive, chemically inert, and nonflammable. Foam polyurethane has a much higher r-value. It is supplied as either rigid panels or as a spray material that fills the space of the wall, for example, between the plywood panel and the sheetrock panel of a wall. The commercial name of the spray is Icynene. In terms of thermal insulation, concrete is much worse than those standard insulation materials.

The R-values of some typical wall insulation panels can be calculated using Eq. 13.2 and the data in Table 13.1. Some typical results are shown in Table 13.2.

13.2.3 Heat Transfer Coefficient: The U-Value

To calculate the total heat loss of a building or a room, the R-value is not convenient. The inverse of R-value, the heat transfer coefficient, or the U-value, is used. In the SI unit system, it is defined as the heat loss in watts per unit square meter per Kelvin temperature difference (W/K·m^2). In the British unit system, it is defined as the heat loss in Btu per hour per unit square foot per degree Fahrenheit temperature difference (Btu/h·f^2). The relation of conversion is 1U = 5.678 USI, 1 USI = 0.1761U. The relation with the R-value is

$$U = \frac{1}{R}. \tag{13.3}$$

For a room or building with walls, windows, doors, and so on, the heat loss Q can be computed as the sum of the product of the area and the U-value of each component:

$$Q = \Delta T \sum_i A_i U_i = \Delta T \sum_i \frac{A_i}{R_i}, \tag{13.4}$$

Table 13.2: Typical R-Values of Wall Insulation

Material	R-Value (Btu)	RSI (SI unit)
4-in. fiberglass butts	R-13	2.3
6-in. fiberglass butts	R-19	3.3
12-in. fiberglass butts	R-39	6.7
6-in. icynene	R-22	3.8
10-in. concrete wall	R-0.8	0.14

where A_i is the area and U_i is the U-value of the ith component and ΔT is the temperature difference.

13.2.4 Thermal Mass

The stability of room temperature depends on how much heat can be retained inside the room. It is the sum of of the thermal masses of the components,

$$M = \sum_j V_j \rho_j c_{pj}, \tag{13.5}$$

where V_j is the volume and $\rho_j c_{pj}$ is the specific heat capacity per volume of the jth component. The rate of temperature decline is

$$\frac{dT}{dt} = -\frac{Q}{M} = -\frac{\Delta T}{M}\sum_i A_i U_i = -\frac{\Delta T}{M}\sum_i \frac{A_i}{R_i}. \tag{13.6}$$

Therefore, the greater the thermal mass, the slower the temperature decline.

Tables 13.1 and 13.2 list thermal properties of commonly used building materials. From those tables we find that water is by far the best material for sensible energy storage. Concrete and bricks are the worst, but unfortunately, are the materials often used.

13.2.5 Glazing

Glass windows let sunlight into a room, and thus provide warmth in the winter. However, as seen in Table 13.2, even high-grade windows are poor insulators. Double-pane glass windows have an RSI-value of 0.35, but the mediocre 6-in. fiberglass butts provides RSI = 3.3. Therefore, there is a one order of magnitude difference.

It has become fashionable to use a lot of glass in modern buildings. However, this requires a lot of electricity in the summer to cool the building and a lot of natural gas

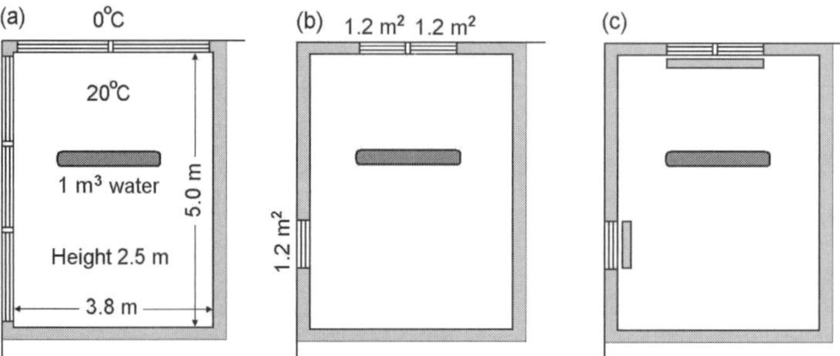

Figure 13.2 Effect of glazing on insulation. (a) All external walls are glazed with double-pane glass windows. (b) With three 1.2–m^2 double-pane glass windows. (c) After the windows are covered with polyurethane panels.

(or worse electricity) in winter to keep it warm. Too much glazing results in a waste of energy.

Figure 13.2 shows an example. A room at the corner of the house has a 1 m^3 water tank as the thermal storage medium. The size of the room is 3.8 m × 5.0 m × 2.5 m. Assuming that the room temperature is 20°C and the external temperature is 0°C. The thermal mass $M = 4.19 \times 10^6$ J/K. The heat loss through the internal walls can be neglected. As shown in Fig. 13.2(a), if all external walls are glazed with double-pane glass windows, then the total area of glass is 22 m^2. The rate of heat loss is

$$Q = \frac{20 \times 22}{0.35} = 1257 \text{ W}. \tag{13.7}$$

The rate of temperature drop is

$$\frac{dT}{dt} = -\frac{1257}{4.19 \times 10^6} = -3 \times 10^{-4}\,\text{K/s} = -1.08\,\text{K/h}. \tag{13.8}$$

Therefore, even with such a large thermal mass, without heating, the room will become very cold after several hours.

Figure 13.2(b) shows a house with three 1.2-m^2 double-pane windows. The rest of the wall is insulated by conventional 6-in. (150 mm) fiberglass butts. The rate of heat loss is

$$Q = 20 \times \left(\frac{22 - 3.6}{3.3} + \frac{3.6}{0.35} \right) = 317\,\text{W}. \tag{13.9}$$

The rate of temperature drop is

$$\frac{dT}{dt} = -\frac{317}{4.19 \times 10^6} = -7.6 \times 10^{-5}\,\text{K/s} = -0.27\,\text{K/h}. \tag{13.10}$$

After 8 hours, the temperature will drop by 2.2°C.

At night there is no sunlight, and the windows become solely a drain of energy. By covering the windows with 25-mm-thick polyurethane rigid panels, with R =6.5, the R-value of windows is increased to 8.5, or RSI = 1.5. The rate of heat loss is

$$Q = 20 \times \left(\frac{22 - 3.6}{3.3} + \frac{3.6}{1.5} \right) = 159 \text{ W}. \tag{13.11}$$

The rate of temperature drop is

$$\frac{dT}{dt} = -\frac{159}{4.19 \times 10^6} = -3.32 \times 10^{-5}\,\text{K/s} = -0.136\,\text{K/h}. \tag{13.12}$$

Then, the temperature drop is about 1°C overnight. If there are people living in that room, the temperature drop will be even less because the heat generated by each person is roughly equal to an incandescent lightbulb, that is, 40 – 60 w. It will partially compensate for the heat loss through the walls and windows and make the temperature virtually constant.

The windows in this room are already large enough to admit sufficient sunlight to warm the room in the winter. From Fig. 4.7, in locations near latitude 40°, from November to February, the daily solar radiation on a south-facing window is greater than 6.5 kWh/m^2. The two 1.2-m^2 windows could admit more than 10 kWh of thermal energy on a sunny day. The solar thermal power is already much greater than the thermal loss, so the room would be kept warm enough from just the sunlight through the windows. Excessive glazing would probably make the room too hot during the day.

13.3 Example of Holistic Design

According to Steven Chu [30] and McKinsey [60, 59], when one tries to build a new home, with a little extra material and labor, a large amount of energy — and consequently a large amount of money — could be saved. I tested this statement by building my own home. I found that even the extra upfront investment money is not necessary. The key is to make a design that maximizes the benefit of sunlight. Since I was the

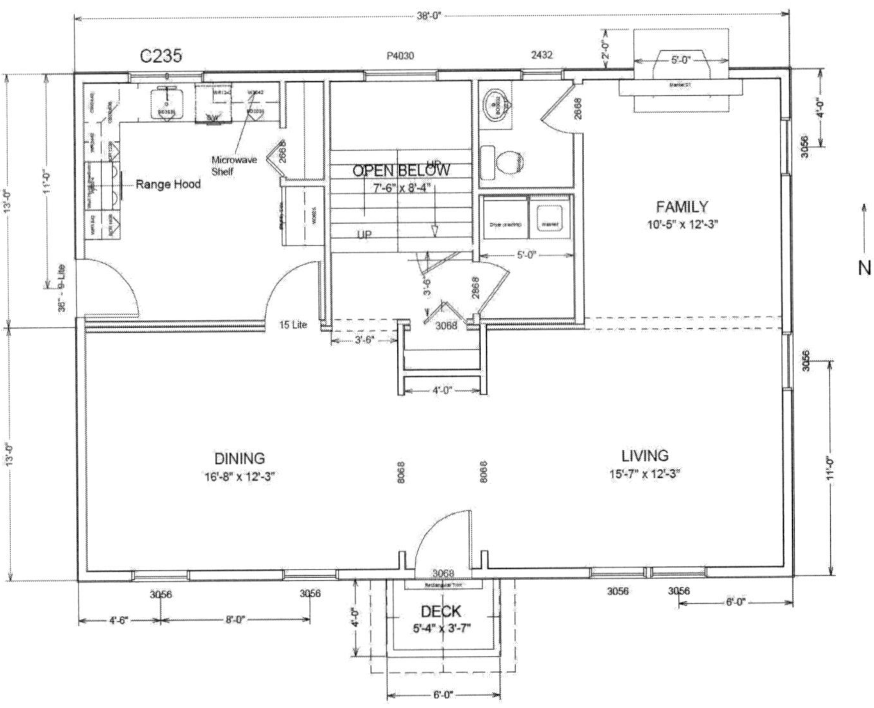

Figure 13.3 Design of a solar house: First floor. All large windows are facing south, and the entire grand hall is in the south. Less important rooms, such as the stairs and the bathroom, are in the north.

architect and the principal civil engineer of my house, the only extra cost was my time.

Before design, I read almost every book and paper I could find about solar-energy house. I was somewhat disappointed that many solar energy house designs were either too exotic, requiring unconventional materials and unconventional building technique, or emphasized energy saving to demonstrate a house that solely depends on solar energy while sacrificing comfort. In addition, the planning board and architecture review board of the town required that the house be in harmony with other houses around it, and the construction company insisted on using building materials that it could easily find.

In addition to building a comfortable home, I had some additional requirements. First, I have a Hamburg Steinway Model B, which requires a rather large room with good resonance and stable temperature and humidity. Second, I wanted to build a professional-grade home movie theater which could be operated any day of the year and any time of the day. Third, I want a playroom of more than 60 m^2 for my grandchildren.

Fortunately, I found an empty lot in North White Plains which fit the requirements: a 0.89 acre (3500 m^2) land sloping towards south. Because of the slope, in spite of several diligent attempts, the previous owners were not able to make a design that satisfied the requirements of the town planning board. By reviewing the previous

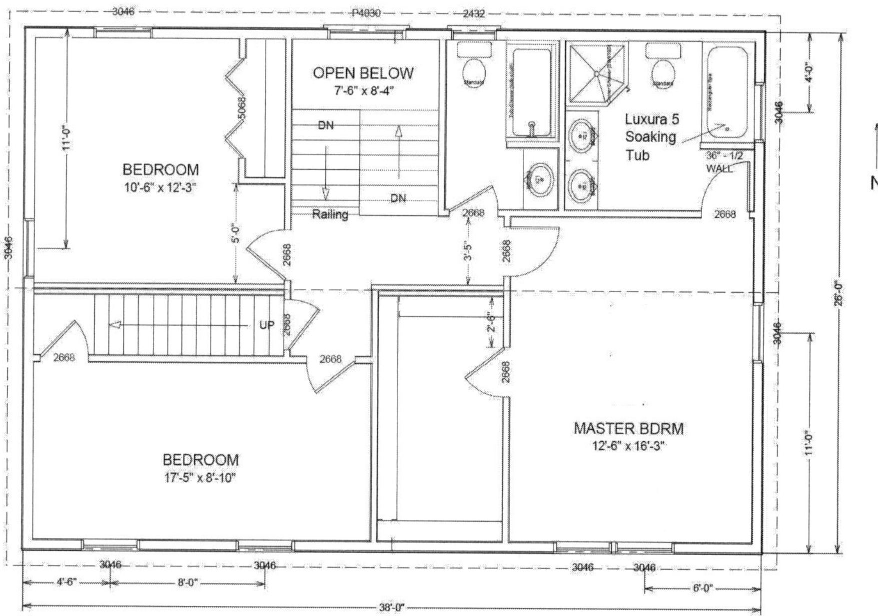

Figure 13.4 Design of a solar house: Second floor. All large windows are facing south, and the two most used rooms are in the south. Less important rooms, such as the stairs, the guest room and the bathrooms are in the north.

design documents and drawings, I found that the previous owners wanted to carve out a large flat space through the slope, which required extensive exploding and excavation plus an 8-m-high retaining wall, which would be very dangerous.

I decided to take advantage of the slope and bury part of the house beneath the surface. Therefore, part of the structure is in the earth, which can utilize the shallow geothermal energy — the solar energy stored in the ground. The front plane of the house faces true south, with nothing blocking the sunlight. All large windows face south to admit as much sunlight as possible. The north face of the house is close to the hill. Very few windows are needed. The idea was quickly approved by the planning board and architectural review board.

The design drawings are shown in Figs. 13.3 and 13.4. From outside to inside, it looks like a normal central-hall colonial. But solar energy is extensively utilized. The first floor, with a high ceiling, hosts a Steinway B as its focal point. All large windows face south. The less important rooms, such as the scissor stairs and the bathroom, are in the north. On the second floor, the two most used rooms, the master bedroom and the office (bedroom 2) are in the south, and all large windows are facing south. Less important rooms, such as the guest room (bedroom 3) and the bathrooms are in the north.

Westchester Module Home Construction Company worked out the details and built the house using some of the best conventional materials and building methods. All windows are double-pane airtight Anderson windows. The insulation for the ceiling is R39, and others are R19. The foundation is poured concrete. It was finished in August 2009. See Plate 18.

On cold and sunny winter days, sunlight from the large south windows often bring the temperature of the two most used rooms to above 22°C, with the thermostat almost always turned off. This is reasonable from a solar radiation point of view. As shown in Fig. 4.7, at locations near latitude 40°, on a sunny day in winter, sunlight can bring in more than 10 kWh of thermal energy. In the summer, because of the eaves, no sunlight

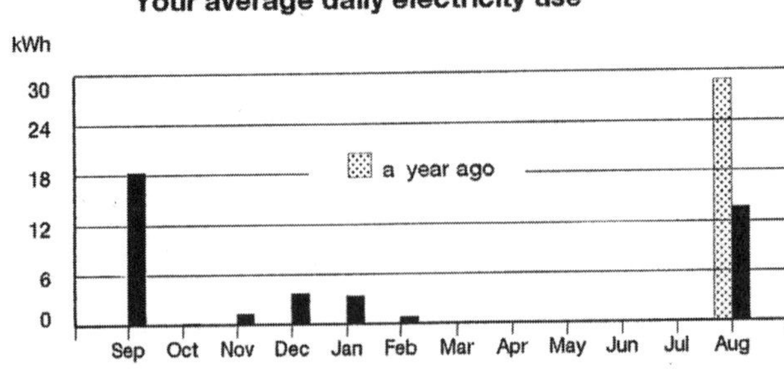

Figure 13.5 August 2010 Con Edison electricity bill. Since the installation of the 4.2-kW photovoltaic system in September 2009, it generated 6129 kWh of electricity. The net electricity consumption of the entire year is 1688 kWh.

gets into the rooms. In addition, there is a solar-powered attic fan on the roof, which helps to reduce the temperature in the attic and the entire house.

The movie theater was placed in the far corner of the foundation, which is deepest in the ground. Utilizing the solar energy stored in the ground, even without air conditioning and external heating, the temperature remains stable and comfortable throughout the year.

The 4.2-kW photovoltaics system also worked as planned. Because the orientation of the roof faces exactly south, the tilt angle equals the latitude, and sunlight is almost completely unobstructed, the efficiency is high. Figure 13.5 is from the August 2010 Con Edison electricity bill. The total electricity consumed for the year was 1600 kWh, about one-tenth that for a regular house of similar size.

For the period August 2009 – August 2010, the photovoltaic system generated 6129 kWh of electricity. In White Plains, the cost of electricity is \$0.226/kWh. For a full year, the value generated was \$1,385. The out-of-pocket cost of the photovoltaic system was \$9,800. Therefore, the system can be paid off in about seven years. The photovoltaic system can then generate free electricity for about 25 years.

Although the immediate goals of this solar house are already accomplished, several further solar energy experiments are in progress. The 4.2 kW photovoltaic system only occupies about one half of the roof. The other half is reserved for a DC photovoltaic system for charging electric cars and powering DC appliances. A solar air conditioning system is in planning stage. A solar thermal project is in progress in order to reduce the use of natural gas to a minimal.

13.4 Land Usage of Solar Communities

Solar architecture works if there is no object blocking the sunlight to the building. For isolated buildings, such a condition can be easily satisfied. However, in cities and towns, land is precious, sunlight may be blocked by nearby buildings. In order to fully utilize sunlight, there is a minimal spacing between adjacent buildings, which can be determined using the theory presented in Chapter 4.

Figure 13.6 shows a typical layout of a solar community in the Northern Hemisphere, which consists of an array of buildings with the long axis in the east-west direction. Most of the large windows are on the south face of the buildings. Let the vertical distance from the lowest edge of the ground-floor windows to the ridge of the building be H and the horizontal distance from the window to the ridge of the adjacent building be L. If the height of the Sun is h and the azimuth is A, using trigonometry, it can be shown that the *shading angle* θ from the adjacent building is determined by

$$\tan\theta = \frac{\cos A}{\tan h}. \tag{13.13}$$

Using the coordinate transformation formulas 4.15 and 4.19, we obtain

$$\tan\theta = \frac{\sin\phi\,\cos\delta\,\cos\omega - \cos\phi\,\sin\delta}{\sin\delta\,\sin\phi + \cos\delta\,\cos\phi\,\cos\omega}. \tag{13.14}$$

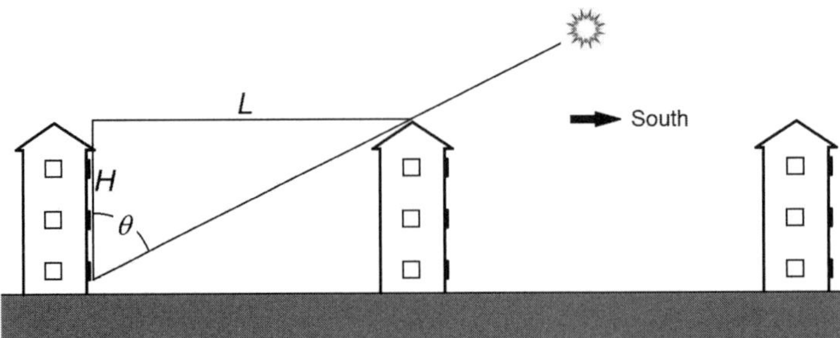

Figure 13.6 Minimum spacing between adjacent buildings. Typical layout of a solar community in the Northern Hemisphere, which consists of an array of buildings with the long axis in the east-west direction. Most of the large windows are on the south face of the buildings. The vertical distance from the lowest edge of the ground-floor window to the ridge of the building is H, and the horizontal distance from the window to the ridge of the adjacent building is L.

By introducing an angle ξ,

$$\tan\xi = \frac{\tan\delta}{\cos\omega}, \tag{13.15}$$

Eq. 13.14 can be reduced to

$$\tan\theta = \frac{\tan\phi\ - \tan\xi}{1 + \tan\phi\ \tan\xi}. \tag{13.16}$$

Finally, we obtain

$$\theta = \phi - \xi = \phi - \arctan\left(\frac{\tan\delta}{\cos\omega}\right). \tag{13.17}$$

In order to avoid sunlight being blocked by the adjacent building, the spacing L must be large enough such that

$$L > H\ \tan\theta. \tag{13.18}$$

Figure 13.7 shows the dependence of shading angle on date and time for an observer at a geographical latitude $\phi = 35°$. The definition of shading angle is such that when $\theta = 0°$ there is no shade and $\theta = 90°$ means full shade. At an equinox, the shading angle equals the latitude of the location. In winter, the shading angle is greater, which indicates more shade. In summer, the shading angle is smaller, which indicates less shade. The number on a curve indicates the order of solar term, which determines the date in a year, and thus the declination δ of the Sun, see Section 4.3.5.

As shown, in winter, for the early morning and late afternoon, the shading angle θ can reach 90°. Therefore, no matter how far the spacing is, full sunlight over the entire day is impossible. A compromise should be applied, for example, requiring that at the winter solstice, at least 6 h of full sunlight should be available. As shown in Fig. 13.7, this requires that $\theta = 66°$. For a three-story building with $H = 10$ m,

$$L > 10 \times \tan 66° \approx 22 \text{ m}. \tag{13.19}$$

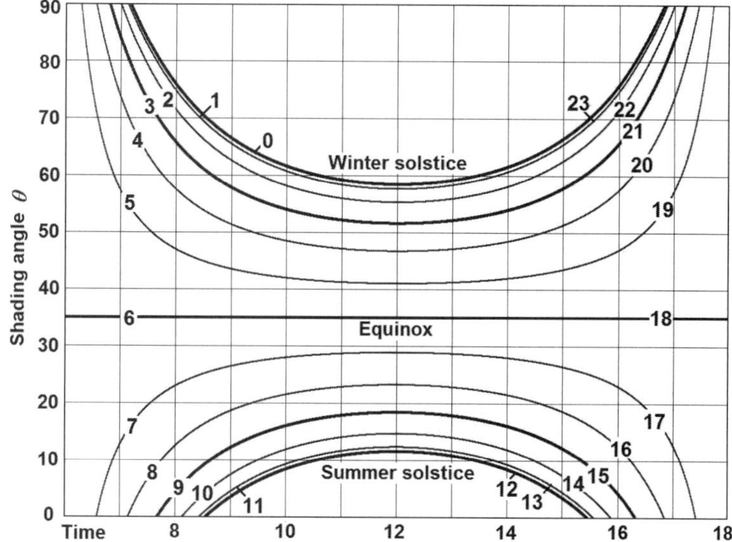

Figure 13.7 Dependence of shading angle on date and time. Dependence of shading angle on date and time for an observer at geographic latitude $\phi = 35°$. In winter, the shading angle is greater, indicating more shade. In summer, the shading angle is smaller, indicating less shade. The number on a curve indicates the order of the solar term, a convenient and accurate way to specify a date in a year with regard to the motion of the Sun, see Section 4.3.5.

Because of Eq. 13.17, for locations of a different latitude, the results can be obtained by simply adding the difference of latitude. For example, at a location of $\phi = 40°$, the spacing should be

$$L > 10 \times \tan 71° \approx 29 \text{ m}. \tag{13.20}$$

From the above discussions we can draw some general conclusions. In order to have full sunlight on the South face of a row of buildings, the spacing required between the buildings should be proportional to the height. This intuitively understandable result implies that tall buildings are not advantageous if passive solar heating is preferred. Besides, tall buildings usually require a concrete and steel structure, which further deteriorates thermal properties. For buildings of two to five stories, light building materials can be used, especially materials with superior insulation properties, which could be designed to be heated by passive solar energy.

Problems

13.1. Consider a house of dimension 12 m long, 8 m wide, and 5 m high, with typical insulation of 6-in. fiberglass with R-19 on peripherals and 12-in. fiberglass with R-39 on the ceiling and floor. In the summer, the average external temperature is 80°F (26.67°C), but the comfortable internal temperature is 68°F (20°C). By using the latent heat of the tank of ice to cool the house (equipped with a thermostat and circulating mechanism), for how long can the tank of ice keep the room temperature constant?

13.2. In the New York area (latitude 40°47′), on the south side of a house the is a window of height $H = 1.67$ m, see Fig. 13.8. In order to fully utilize the sunlight at the winter solstice and fully block the sunlight at the summer solstice, what should be the length of overhang L and the distance D from the top of the window to the base of the overhang?

13.3. A typical room in a well-insulated house has a dimension of 20 ft in length, 16 ft in width, and 8 ft in height. All walls, including ceiling and floor, are insulated by R39 material. In the winter, the outside temperature is 0°C and the room temperature is 20°C. What is the thermal loss in watts? If the room contains one ton of water as thermal mass, what is the rate of temperature loss in °C per hour?

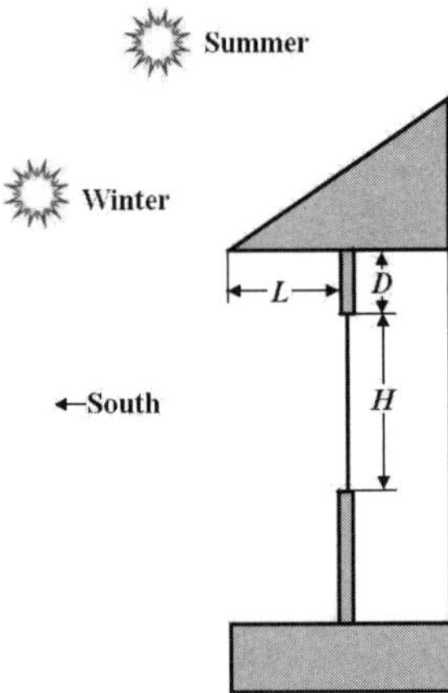

Figure 13.8 Calculation of eaves.

Appendix A

Energy Unit Conversion

Because energy is one of the most important quantities, there are many energy units which often creates confusion. Throughout this book, we used the SI units for all physical quantities. The SI unit of energy is joule, defined as the energy capable of pushing an object with one newton of force by one meter:

$$\mathrm{J} = \mathrm{N} \cdot \mathrm{m}. \tag{A.1}$$

The basic SI unit of power, the watt, equals one joule per second,

$$\mathrm{W} = \mathrm{J/s}. \tag{A.2}$$

The most frequently used units of energy and power are listed in Table A.1.

Electrical power is the product of the current in amperes and voltage in volts:

$$\mathrm{W} = \mathrm{A} \cdot \mathrm{V}. \tag{A.3}$$

Utility companies often use an energy unit derived from electrical power, the kilowatt-hour, or kWh:

$$1\ \mathrm{kWh} = 3600\ \mathrm{kJ} = 3.6\ \mathrm{MJ}. \tag{A.4}$$

Table A.1: Energy and Power Units

Name	Symbol	Equals
Kilojoule	kJ	10^3 J
Megajoule	MJ	10^6 J
Gigajoule	GJ	10^9 J
Etajoule	EJ	10^{18} J
Kilowatt	kW	10^3 W
Megawatt	MW	10^6 W
Gigawatt	GW	10^9 W
Terawatt	TW	10^{12} W

In microscopic physics, the electron-volt (eV) is often used as a unit of energy. Because the charge of an electron is $e = 1.60 \times 10^{-19}$ C, 1 eV $= 1.60 \times 10^{-19}$ J. Another unit frequently found in the literature is the calorie, abbreviated as cal. It is defined as the energy required to raise the temperature of one gram of water by one degree Celsius. Because at different temperatures the amount of energy required is different, the definition of the unit is ambiguous. A well-defined and widely accepted definition of the calorie is

$$1\ \text{cal} = 4.184\ \text{J}. \tag{A.5}$$

One kilocalorie (kcal) is defined exactly as 4184 J. Therefore, unless necessary, SI units are much preferred.

In the United States, the British thermal unit (Btu) is often used as the unit of energy. It is defined as the energy to raise the temperature of one pound water by one degree Fahrenheit. Similar to calorie, the exact value of the unit is ambiguous. The International Table defines the Btu as 1055.06 J, which is very close to 1 kJ. In dealing with renewable energy problems, we can take the approximation

$$1\ \text{Btu} \approx 1\ \text{kJ}. \tag{A.6}$$

For large amounts of energy, a frequently used unit is the exajoule (EJ), which is defined as 10^{18} J. In the United States, the corresponding unit for large amounts of energy is quadrillion Btu, abbreviated as quad, which is defined as 10^{15} Btu. Because 1 Btu is very close to 1 kJ, for practical purposes, we can consider the two units to be almost equivalent.

$$1\ \text{quad} \approx 1\ \text{EJ}. \tag{A.7}$$

Another unit for large amounts of energy is terawatt-hour (TWh). It equals one billion kilowatt-hours, or 0.0036 EJ. One terawatt-hour equals 3.6×10^{15} J. One exajoule equals 277.8 TWh.

In the utility industry, the gigawatt (GW) and terawatt (TW) are often used. One gigawatt-year equals 3.156×10^{16} J, roughly equals one 1/32 EJ. One terawatt-year equals 3.156×10^{19} J, roughly equals 32 EJ. In 2007, the world's energy consumption of energy is approximately 500 EJ, or an average power of 15 TW; the energy consumption of the United States is approximately 100 EJ, or an average power of 3 TW. In 2008, the energy consumption of the United States is slightly less than 100 EJ.

Several important approximate relations are worth noting:

- One barrier of crude oil is defined as 5.8×10^6 Btu.
- One cubic foot of natural gas is approximately 1000 Btu, or 1 MJ.
- One therm of natural gas is defined as 100,000 Btu, approximately 100 MJ.
- One kilojoule per mole is 0.01036 eV, approximately 10 meV.
- One kilocalorie per mole is 0.0434 eV.

Appendix B

Spherical Trigonometry

When we look into the sky, it seems that the Sun and all the stars are located on a *sphere* of a large but unknown radius. In other words, the location of the Sun is defined by a point on the *celestial sphere*. On the other hand, the surface of Earth is, to a good approximation, a sphere. Any location on Earth can be defined by a point on the *terrestrial sphere*; namely by the *latitude* and the *longitude*. In both cases, we are dealing with the geometry of spheres.

To study the location of the Sun with respect to a specific location on Earth, we will correlate the coordinates of the location on the terrestrial sphere of Earth with the location of the Sun on the celestial sphere. The mathematical tool of this study is *spherical trigonometry*. In this Appendix, we will give a brief introduction to spherical trigonometry, sufficient to deal with the problem of tracking the sunlight.

B.1 Spherical Triangle

A plane passing through the center of a sphere O cuts the surface in a circle, which is called a *great circle*. For any two points A and B on the sphere, if the line AB does not pass the center O, there is one and only one great circle which passes both points. The angle $\widehat{AOB}$, chosen as the one smaller than 180° or π in radians, is defined as the length of the arc AB. Given three points A, B, and C on the sphere, three great circles can be defined. The three arcs AB, BC, and CA, each less than 180° or π in radians, form a *spherical triangle*; see Fig. B.1.

Following standard notation, we denote the sides BC, CA, and AB by c, b, and a, respectively. The length of side a is defined as the angle $\widehat{BOC}$, the length of side b is defined as the angle $\widehat{COA}$, and the length of side c is defined as the angle $\widehat{AOB}$. The vertex angles of the triangle are defined in a similar manner: The vertex angle A is defined as the angle between a straight line AD tangential to AB and another straight line AE tangential to AC, and so on.

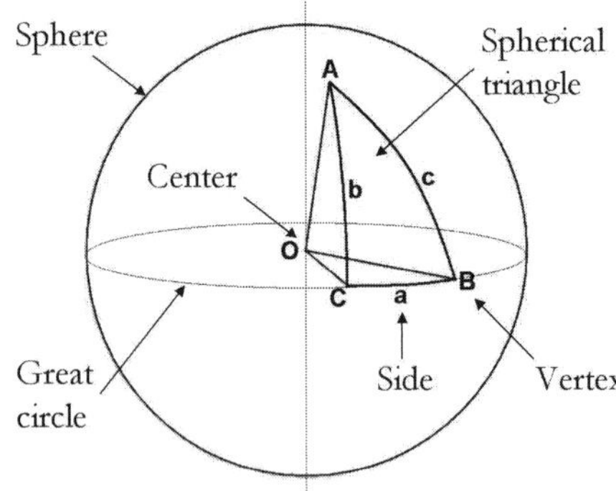

Figure B.1 The spherical triangle. A *great circle* is defined by the intersection of a plane that cuts the sphere in two equal halves with the sphere. Given three points A, B, and C on the sphere, three great circles can be defined. The three arcs AB, BC, and CA, each less than 180° or π in rad, form a *spherical triangle.*

B.2 Cosine Formula

In planar trigonometry, there is a cosine formula

$$a^2 = b^2 + c^2 - 2bc\cos A. \tag{B.1}$$

In spherical trigonometry, a similar formula exists:

$$\cos a = \cos b\cos c + \sin b\sin c\cos A. \tag{B.2}$$

When the arcs are short and the spherical triangle approaches a planar triangle, Eq. B.2 reduces to Eq. B.1. In fact, for small arcs,

$$\cos b \approx 1 - \frac{1}{2}b^2, \tag{B.3}$$

and

$$\sin b \approx b, \tag{B.4}$$

and so on. Substituting Eqs B.3 and B.4 into Eq. B.2 reduces it to Eq. B.1.

Here we give a simple proof of the cosine formula in spherical trigonometry by an analogy to that in planar trigonometry. To simplify notation, we set the radius of the sphere $OA = OB = OC = 1$. By extending line OB to intersect a line tangential to AB at a point D, we have

$$AD = \tan c; \quad OD = \sec c. \tag{B.5}$$

Similarly, by extending line OC to intersect a line tengential to AC at at a point E, we have

$$AE = \tan b; \quad OE = \sec b. \tag{B.6}$$

From the planar triangle DAE, using the planar cosine formula,

$$\begin{aligned} DE^2 &= AD^2 + AE^2 - 2\,AD\cdot AE\cos\widehat{DAE} \\ &= \tan^2 c + \tan^2 b - 2\tan b\tan c\cos A. \end{aligned} \tag{B.7}$$

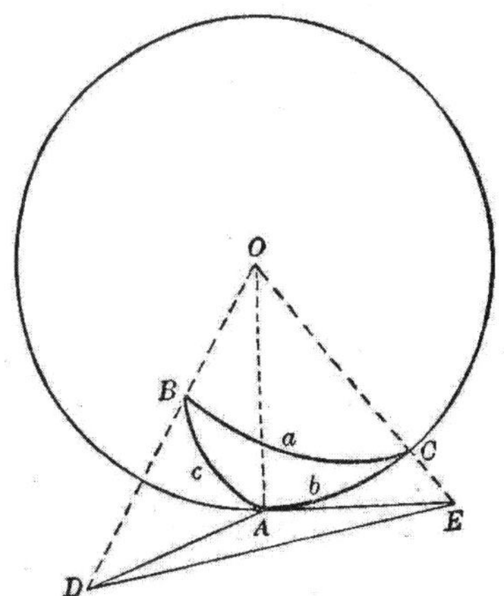

Figure B.2 Derivation of cosine formula. The derivation is based on the projection of a spherical triangle onto a plane and the cosine formula in planar trigonometry. A straight line tangential to arc AB intersects the extension of line OB at D. Another straight line tangential to arc AC intersects the extension of line OC at E. By applying the cosine formula in planar trigonometry on triangles ODE and ADE, after some brief algebra, the corresponding cosine formula is obtained.

Similarly, for the planar triangle DOE,

$$\begin{aligned} DE^2 &= OD^2 + OE^2 - 2\,OD \cdot OE \cos \widehat{DOE} \\ &= \sec^2 c + \sec^2 b - 2 \sec b \sec c \cos a. \end{aligned} \tag{B.8}$$

Subtrating Eq. B.7 from B.8, and notice that

$$\sec^2 b = 1 + \tan^2 b, \quad \sec^2 c = 1 + \tan^2 c, \tag{B.9}$$

we obtain

$$2 - 2 \sec b \sec c \cos a = -2 \tan b \tan c \cos A. \tag{B.10}$$

Multiplying both sides with $\cos b \cos c$ yields

$$\cos a = \cos b \cos c + \sin b \sin c \cos A. \tag{B.11}$$

Similarly, for vertices B and C,

$$\cos b = \cos c \cos a + \sin c \sin a \cos B; \tag{B.12}$$

$$\cos c = \cos a \cos b + \sin a \sin b \cos C. \tag{B.13}$$

B.3 Sine Formula

In planar trigonometry, there is the sine formula

$$\frac{a}{\sin A} = \frac{b}{\sin B} = \frac{c}{\sin C}. \tag{B.14}$$

In spherical trigonometry, a similar formula exists,

$$\frac{\sin a}{\sin A} = \frac{\sin b}{\sin B} = \frac{\sin c}{\sin C}. \tag{B.15}$$

Obviously, for small arcs, Eq. B.15 reduces to Eq. B.14.

To prove Eq. B.15, we rewrite Eq. B.2 as

$$\sin b \, \sin c \, \cos A = \cos a - \cos b \, \cos c. \tag{B.16}$$

Squaring, we obtain

$$\sin^2 b \, \sin^2 c \, \cos^2 A = \cos^2 a - 2 \cos a \, \cos b \, \cos c + \cos^2 b \, \cos^2 c. \tag{B.17}$$

The left-hand side can be written as

$$\sin^2 b \, \sin^2 c \, - \sin^2 b \, \sin^2 c \, \sin^2 A, \tag{B.18}$$

or

$$1 - \cos^2 b - \cos^2 c + \cos^2 b \, \cos^2 c \, - \sin^2 b \, \sin^2 c \, \sin^2 A. \tag{B.19}$$

Hence,

$$\sin^2 b \, \sin^2 c \, \sin^2 A = 1 - \cos^2 a - \cos^2 b - \cos^2 c + 2 \cos^2 a \, \cos^2 b \, \cos^2 c \, . \tag{B.20}$$

Define

$$Z = \frac{\sin^2 a \, \sin^2 b \, \sin^2 c}{1 - \cos^2 a - \cos^2 b - \cos^2 c + 2 \cos^2 a \, \cos^2 b \, \cos^2 c}. \tag{B.21}$$

Equation B.20 can be written as

$$\frac{\sin a}{\sin A} = \pm\sqrt{Z}. \tag{B.22}$$

By definition, in a spherical triangle, the sides and the vertical angles are always smaller than 180°. Therefore, in Eq. B.22, only a positive sign is admissible. Because Z is symmetric to A, B and C, we obtain

$$\frac{\sin a}{\sin A} = \frac{\sin b}{\sin B} = \frac{\sin c}{\sin C}. \tag{B.23}$$

B.4 Formula C

We rewrite the cosine formula B.11 in the following form and use Eq. B.13:

$$\begin{aligned} \sin b \, \sin c \, \cos A &= \cos a - \cos b \, \cos c \\ &= \cos a - \cos b \, (\cos a \, \cos b + \sin a \, \sin b \, \cos C) \\ &= \cos a \, \sin^2 b - \cos b \, \sin a \, \sin b \, \cos C. \end{aligned} \tag{B.24}$$

Dividing both sides by $\sin b$, one obtains *formula C*

$$\sin c \cos A = \cos a \sin b - \sin a \cos b \cos C. \tag{B.25}$$

Similarly,

$$\sin a \cos B = \cos b \sin c - \sin b \cos c \cos A, \tag{B.26}$$

$$\sin a \cos C = \cos c \sin b - \sin c \cos b \cos A, \tag{B.27}$$

and so on.

Problems

B.1. Show that if one of the arcs is 180°, then no spherical triangle can be constructed.

B.2. If one of the vertex angles of a spherical triangle, for example, C, is a right angle, show that for small arcs the cosine formula leads to the Pythagorean theorem.

B.3. Using the cosine and sine formulas, show that

$$\cot a \, \sin b = \cot A \, \sin C + \cos b \, \cos C, \tag{B.28}$$

$$\cot c \, \sin a = \cot C \, \sin B + \cos a \, \cos B, \tag{B.29}$$

and so on.

B.4. For a rectangular spherical triangle, where C=90°, show that

$$\sin a = \sin c \, \sin A, \tag{B.30}$$

$$\sin b = \sin c \, \sin B, \tag{B.31}$$

$$\tan a = \tan c \, \cos B, \tag{B.32}$$

$$\tan b = \tan c \, \cos A, \tag{B.33}$$

$$\tan a = \sin b \, \tan A, \tag{B.34}$$

$$\tan b = \sin a \, \tan B. \tag{B.35}$$

Appendix C

Quantum Mechanics Primer

Usually, introductory quantum mechanics starts with Schrödinger's equation and using partial differential equations as the mathematical tools. For example, the hydrogen atom problem is resolved with spherical harmonics and Laguerre polynomials. A possible shortcoming with this approach is that the readers become submerged in pages and pages of mathematical formulas and lose the conceptual understanding of the physics. Historically, before Erwin Schrödinger discovered the partial differential equation format, Heisenberg and Pauli developed the algebraic approach of quantum mechanics, and resolved several basic problems in quantum mechanics, including harmonic oscillator, angular momentum, and the hydrogen atom. From a pedagogic point of view, the succinct notation of the algebraic approach, especially the Dirac notation, could be conceptually more directly related to the underlying physics. From a practical point of view, to handle the problems with the utilization of solar energy, analytic approach for partial differential equations is not useful because numerical calculations and perturbation methods are the norm. Furthermore, the more advanced methods of quantum mechanics, such as quantum electrodynamics, rely on the algebraic method rather than the partial differential equation method.

This Appendix is a brief summary of the algebraic approach to quantum mechanics, exemplified by the problems of the harmonic oscillator, angular momentum, and hydrogen atom. For clarity, we use the Dirac notation, and adding a hat on an operator to distinguish it from a (in general complex) number.

C.1 Harmonic Oscillator

The Hamiltonian of a one-dimensional harmonic oscillator is

$$\hat{H} = \frac{1}{2m}\hat{p}^2 + \frac{m\omega^2}{2}\hat{q}^2, \tag{C.1}$$

where the momentum $\hat{p}$ and coordinate $\hat{q}$ satisfy the commutation relation

$$[\hat{p}, \hat{q}] \equiv \hat{p}\hat{q} - \hat{q}\hat{p} = i\hbar. \tag{C.2}$$

We introduce a pair of operators, the *annihilation operator*,

$$\hat{a} = \frac{1}{\sqrt{2m\hbar\omega}}\hat{p} + i\sqrt{\frac{m\omega}{2\hbar}}\,\hat{q}, \tag{C.3}$$

and the *creation operator*

$$\hat{a}^\dagger = \frac{1}{\sqrt{2m\hbar\omega}}\hat{p} - i\sqrt{\frac{m\omega}{2\hbar}}\,\hat{q}, \tag{C.4}$$

the meanings of these terms well be clarified soon. Using the commutation relation C.2, we have

$$\hat{H} = \hbar\omega\hat{a}^\dagger\hat{a} + \frac{1}{2}\,\hbar\omega, \tag{C.5}$$

and the commutation relation

$$[\hat{a}, \hat{a}^\dagger] \equiv \hat{a}\hat{a}^\dagger - \hat{a}^\dagger\hat{a} = 1. \tag{C.6}$$

To find the eigenstates and energy levels of the Hamiltonian C.1,

$$\hat{H}|n\rangle = E_n|n\rangle, \tag{C.7}$$

it is sufficient to find the eigenstates and eigenvalues of the operator $\hat{a}^\dagger\hat{a}$,

$$\hat{a}^\dagger\hat{a}|n\rangle = u_n|n\rangle, \tag{C.8}$$

and $E_n = (u_n + \frac{1}{2})\hbar\omega$. As a consequence of Eq. C.6, if $|n\rangle$ is an eigenstate with eigenvalue u_n, then $\hat{a}|n\rangle$ is also an eigenstate

$$\hat{a}^\dagger\hat{a}\,\hat{a}|n\rangle = (u_n - 1)\,\hat{a}|n\rangle \tag{C.9}$$

with eigenvalue $u_n - 1$. Because $\langle n|\hat{a}^\dagger\hat{a}|n\rangle$ must not be negative, there must be an eigenstate with minimum value, $u_n = 0$. For such a state,

$$\hat{a}^\dagger\hat{a}|0\rangle = 0. \tag{C.10}$$

On the other hand, if $|n\rangle$ is an eigenstate with eigenvalue u_n, then $\hat{a}^\dagger|n\rangle$ is also an eigenstate

$$\hat{a}^\dagger\hat{a}(\hat{a}^\dagger|n\rangle) = (u_n + 1)(\hat{a}^\dagger|n\rangle) \tag{C.11}$$

with eigenvalue $u_n + 1$. Starting with $|0\rangle$, by applying $\hat{a}^\dagger$ many times, we have

$$\hat{a}^\dagger\hat{a}\left(\hat{a}^\dagger\right)^n|0\rangle = n\left(\hat{a}^\dagger\right)^n|0\rangle. \tag{C.12}$$

Therefore, we conclude that the eigenvalues of the operator $\hat{a}^\dagger\hat{a}$ are non-negative integers, and the eigenstates can be constructed from the state with a zero eigenvalue,

$$|n\rangle = C_n\left(\hat{a}^\dagger\right)^n|0\rangle. \tag{C.13}$$

The final step is to find the normalization constant C_n. The zeroth-order state is by definition normalized,

$$\langle 0|0\rangle = 1. \tag{C.14}$$

By applying $\hat{a}^\dagger$ many times, we have

$$\langle 0|\,(\hat{a})^n\,\left(\hat{a}^\dagger\right)^n|0\rangle = n!. \tag{C.15}$$

Therefore, $C_n = (n!)^{-1/2}$, and finally

$$|n\rangle = \frac{1}{\sqrt{n!}}\,\left(\hat{a}^\dagger\right)^n|0\rangle. \tag{C.16}$$

Those are also eigenstates of the harmonic oscillator problem,

$$\hat{H}|n\rangle = \left(n + \frac{1}{2}\right)\hbar\omega\,|n\rangle. \tag{C.17}$$

The energy level of the harmonic oscillator is thus quantized, with energy quanta $\hbar\omega$. The operator $\hat{a}^\dagger$ adds an energy quanta to the oscillator, thus the name *creation operator*; the operator $\hat{a}$ removes an energy quanta from the oscillator, thus the name *annihilation operator*. These operators play essential roles in quantum electrodynamics, or the *bona fide* quantum theroy of radiation.

C.2 Angular Momentum

The definition of angular momentum in quantum mechanics is similar to that in classical mechanics, except that the order of coordinate $\mathbf{r}$ and momentum $\mathbf{p}$ is fixed,

$$\hat{\mathbf{m}} = \hat{\mathbf{r}} \times \hat{\mathbf{p}}, \tag{C.18}$$

or in tensor notation,

$$\hat{m}_i = \epsilon_{ijk}\,\hat{x}_j\,\hat{p}_k, \tag{C.19}$$

where the *unit axial tensor* ϵ_{ijk} is a tensor antisymmetric to all three suffixes, with $\epsilon_{123} = 1$, and changes sign by exchanging two identical indices, where the value is zero. A sum over j and k is implied.

To simplify notation, a dimensionless version of the angular momentum is defined,

$$\hat{l}_i = \hbar^{-1}\hat{m}_i = \hbar^{-1}\,\epsilon_{ijk}\,\hat{x}_j\,\hat{p}_k. \tag{C.20}$$

The commutation relations can be obtained from the commutation relations of momentum and coordinate,

$$[\hat{p}_j, \hat{x}_k] = i\hbar\delta_{jk}, \tag{C.21}$$

which are

$$\begin{aligned} \hat{l}_x\hat{l}_y - \hat{l}_y\hat{l}_x &= i\hat{l}_z, \\ \hat{l}_y\hat{l}_z - \hat{l}_z\hat{l}_y &= i\hat{l}_x, \\ \hat{l}_z\hat{l}_x - \hat{l}_x\hat{l}_z &= i\hat{l}_y, \end{aligned} \tag{C.22}$$

or in tensor form,

$$[\hat{l}_i, \hat{l}_j] = i\epsilon_{ijk}\hat{l}_k. \tag{C.23}$$

From the components we can form an operator as the square of the modulus of the angular momentum vector,

$$\hat{\mathbf{l}}^2 = \hat{l}_x^2 + \hat{l}_y^2 + \hat{l}_z^2. \tag{C.24}$$

As a result of commutation relations C.22, $\hat{\mathbf{l}}^2$ commutes with a component, for example,

$$[\hat{\mathbf{l}}^2, \hat{l}_z] = 0. \tag{C.25}$$

Therefore, we can find states $|l, m\rangle$ which are simultaneously eigenstate of $\hat{\mathbf{l}}^2$ and $\hat{l}_z$,

$$\begin{aligned} \hat{\mathbf{l}}^2|l, m\rangle &= \lambda|l, m\rangle, \\ \hat{l}_z|l, m\rangle &= \mu|l, m\rangle. \end{aligned} \tag{C.26}$$

In the following, we introduce a pair of operators which are similar to the creation and annihilation operators in the problem of harmonic oscillators,

$$\begin{aligned} \hat{l}_+ &= \hat{l}_x + i\hat{l}_y, \\ \hat{l}_- &= \hat{l}_x - i\hat{l}_y, \end{aligned} \tag{C.27}$$

which have the commutation relations

$$[\hat{l}_z, \hat{l}_+] = \hat{l}_+, \tag{C.28}$$

$$[\hat{l}_z, \hat{l}_-] = -\hat{l}_- \tag{C.29}$$

and the identity

$$\hat{\mathbf{l}}^2 = \hat{l}_-\hat{l}_+ + \hat{l}_z^2 + \hat{l}_z. \tag{C.30}$$

Using arguments similar to those in the harmonic oscillator problem, we find that $\hat{l}_+|l,m\rangle$ is also an eigenstate of $\hat{l}_z$ with eigenvalue $\mu+1$ and $\hat{l}_-|l,m\rangle$ is an eigenstate of $\hat{l}_z$ with eigenvalue $\mu-1$. Applying those operators many times, we have

$$\hat{l}_z \left(\hat{l}_+\right)^n |l,m\rangle = (\mu+n)\left(\hat{l}_+\right)^n |l,m\rangle \tag{C.31}$$

and

$$\hat{l}_z \left(\hat{l}_-\right)^n |l,m\rangle = (\mu-n)\left(\hat{l}_-\right)^n |l,m\rangle. \tag{C.32}$$

However, because of Eq. C.24, the eigenvalue of $\hat{l}_z$ cannot grow indefinitely. Its absolute value must have a maximum. Because of the symmetry, the absolute value of the positive maximum and the absolute value of the negative maximum must beequal. Also, because the difference between the positive maximum and the negative maximum must be an integer, both must be one-half of an integer. Assigning this number as l, the possible eigenvalues of $\hat{l}_z$, often assigned as m, must be

$$m = -l, \quad -l+1, \quad -l+2, \quad ...l-2, \quad l-1, \quad l. \tag{C.33}$$

with

$$l = 0, \quad \tfrac{1}{2}, \quad 1, \quad \tfrac{3}{2}, \quad 2, \quad \tfrac{5}{2}, \quad 3\ldots \tag{C.34}$$

It is obvious that l, the maximum absolute value of $\hat{l}_z$, is also a quantum number for the total angular momentum $\hat{\mathbf{l}}^2$. In the following, we will find the eigenvalue of the operator $\hat{\mathbf{l}}^2$. Because $m = l$ is the maximum eigenvalue of $\hat{l}_z$, one must have

$$\hat{l}_+|l,l\rangle = 0. \tag{C.35}$$

In view of Eq. C.30,

$$\hat{\mathbf{l}}^2|l,l\rangle = \left(\hat{l}_-\hat{l}_+ + \hat{l}_z^2 + \hat{l}_z\right)|l,l\rangle = l(l+1)|l,l\rangle. \tag{C.36}$$

Because l and m are independent, finally we have the eigenvalues and eigenstates for the angular momentum operator,

$$\begin{aligned} \hat{\mathbf{m}}^2\,|l,m\rangle &= l(l+1)\hbar\,|l,m\rangle, \\ \hat{m}_z\,|l,m\rangle &= m\hbar\,|l,m\rangle. \end{aligned} \tag{C.37}$$

C.3 Hydrogen Atom

In classical physics, if the mass of the proton is large, the Hamiltonian of the hydrogen atom is

$$H = \frac{\mathbf{p}^2}{2m_e} - \frac{\kappa}{r}, \tag{C.38}$$

where $\kappa = e^2/4\pi\epsilon_0$; see Chapter 2. Because of the spherical symmetry of the problem, the angular momentum is conserved. The vector of angular momentum is always perpendicular to the plane of motion,

$$\mathbf{L} = \mathbf{r} \times \mathbf{p} = \text{const.} \tag{C.39}$$

In addition to angular momentum, there is another conserved vector related to the fixed orientation of the long axis of the orbital, which is a result of the Coulomb interaction. It is called the Runge–Lenz vector after its discoverers,

$$\mathbf{A} = \frac{\kappa\mathbf{r}}{r} - \frac{1}{m_e}(\mathbf{p} \times \mathbf{L}) = \text{const.} \tag{C.40}$$

Because the angular momentum vector $\mathbf{L}$ is perpendicular to the orbital plane and the Runge–Lenz vector is in the plane, the two vectors are perpendicular,

$$\mathbf{L} \cdot \mathbf{A} = 0. \tag{C.41}$$

In 1926, a year before Schrödinger discovered his differential equation and solved the hydrogen atom problem, Wolfgang Pauli solved the eigenvalue problem by using the algebraic method of Werner Heisenberg based on the two constants of motion. Pauli's treatment is as follows:

In quantum mechanics, the Hamiltonian is an operator,

$$\hat{H} = \frac{\hat{\mathbf{p}}^2}{2m_e} - \frac{\kappa}{r}. \tag{C.42}$$

The angular momentum, also an operator,

$$\hat{\mathbf{L}} = \hat{\mathbf{r}} \times \hat{\mathbf{p}}, \tag{C.43}$$

satisfies the commutation relation

$$[\hat{L}_i, \hat{L}_j] = i\hbar\epsilon_{ijk}\hat{L}_k. \tag{C.44}$$

It also commutes with the Hamiltonian C.42, thus a constant of motion.

Because $\hat{\mathbf{p}} \times \hat{\mathbf{L}}$ is not Hermitian, Pauli defined a Hermitian operator equivalent to the classical Runge–Lenz vector

$$\hat{\mathbf{A}} = \frac{\kappa \hat{\mathbf{r}}}{r} - \frac{1}{2m_e}\left(\hat{\mathbf{p}} \times \hat{\mathbf{L}} - \hat{\mathbf{L}} \times \hat{\mathbf{p}}\right). \tag{C.45}$$

With rather tedious but straightforward algebra, this operator is shown to commute with the quantum-mechanical Hamiltonian, equation C.42. Similar algebra results in the commutation relations,

$$[\hat{L}_i, \hat{A}_j] = i\hbar\epsilon_{ijk}\hat{A}_k, \tag{C.46}$$

$$[\hat{A}_i, \hat{A}_j] = i\hbar\left(\frac{-2\hat{H}}{m_e}\right)\epsilon_{ijk}\hat{L}_k, \tag{C.47}$$

and the relation

$$\hat{\mathbf{A}}^2 = \left(\frac{2\hat{H}}{m_e}\right)\left(\hat{\mathbf{L}}^2 + \hbar^2\right) + \kappa^2. \tag{C.48}$$

Because $\hat{\mathbf{A}}$ commutes with $\hat{H}$, and there are common eigenstates, a reduced vector can be defined,

$$\hat{\mathbf{B}} = \left(-\frac{m_e}{2\hat{H}}\right)^{1/2}\hat{\mathbf{A}}. \tag{C.49}$$

The commutation relations Eqs C.46 and C.47 are reduced to

$$[\hat{L}_i, \hat{B}_j] = i\hbar\epsilon_{ijk}\hat{B}_k, \tag{C.50}$$

$$[\hat{B}_i, \hat{B}_j] = i\hbar\epsilon_{ijk}\hat{L}_k. \tag{C.51}$$

We introduce a pair of operators

$$\begin{aligned}\hat{\mathbf{J}} &= \tfrac{1}{2}\left(\hat{\mathbf{L}} + \hat{\mathbf{B}}\right),\\ \hat{\mathbf{K}} &= \tfrac{1}{2}\left(\hat{\mathbf{L}} - \hat{\mathbf{B}}\right).\end{aligned} \tag{C.52}$$

The commutation relations are reduced to those of a pair of independent angular momenta

$$[\hat{J}_i, \hat{J}_j] = i\hbar\epsilon_{ijk}\hat{J}_k, \tag{C.53}$$

$$[\hat{K}_i, \hat{K}_j] = i\hbar\epsilon_{ijk}\hat{K}_k, \tag{C.54}$$

$$[\hat{J}_i, \hat{K}_j] = 0. \tag{C.55}$$

Using Eqs. C.49 and C.52, Eq. C.48 becomes

$$-\frac{1}{2}\mu\kappa^2\frac{1}{\hat{H}} = 2\left(\hat{\mathbf{J}}^2 + \hat{\mathbf{K}}^2\right) + \hbar^2. \tag{C.56}$$

From Eq. C.52, $\mathbf{J}^2 = \mathbf{K}^2$, and the eigenvalues of $\hat{\mathbf{J}}^2$ and $\hat{\mathbf{K}}^2$ are identical. According to the theory of angular momentum (Eq. C.37), both are $j(j+1)$. Therefore, the solution is

$$-\frac{1}{2}\mu\kappa^2\frac{1}{E_n} = 4j(j+1)\hbar^2 + \hbar^2 = (2j+1)^2\hbar^2. \tag{C.57}$$

From Eq. C.34, $n = 2j+1$ can be any positive integer. Finally, the energy eigenvalues of the hydrogen atom are

$$E_n = -\frac{\mu\kappa^2}{2\hbar^2 n^2}. \tag{C.58}$$

Appendix D

Statistics of Particles

In the microscopic treatment of matter, the particles, for example, molecules and electrons, are distributed in a system of *energy levels*. How the particles distribute among the energy levels determines the behavior of the system in a significant way. Regarding to applications with solar energy, an understanding of *Maxwell–Boltzmann statistics* and *Fermi–Dirac statistics* is essential. Maxwell–Boltzmann statistics is valid for a system of *distinguishable particles* in a system of energy levels allowing unlimited occupancy; whereas Fermi–Dirac statistics is valid for a system of *indistinguishable particles* in a system of energy levels allowing limited occupancy, that is, systems satisfying the Pauli exclusion principle.

The starting point of the derivation is the Boltzmann expression of entropy,

$$S = k_\mathrm{B} \ln W, \tag{D.1}$$

where k_B is Boltzmann's constant and W is the total number of configurations of the system.

Consider a situation of N particles in a system consisting of a series of energy levels. The occupancy number of the ith level is N_i. The energy of the ith level is E_i, and the total energy of the system is E. We have the conditions

$$\begin{aligned} N &= \textstyle\sum_i N_i, \\ E &= \textstyle\sum_i N_i\, E_i. \end{aligned} \tag{D.2}$$

The condition of equilibrium is that the entropy, Eq. D.1, reaches maximum under the two constraints in Eq. D.2.

Intuitively, the more uniform the distribution, the greater the randomness, or the greater the entropy. However, the condition of constant total energy adds another condition: It is preferable to have more particles in the levels of lower energy and less particles in the levels of higher energy. The problem can be resolved using *Fermat's theorem* together with the *Lagrange multiplier* method. Introducing two Lagrange multipliers α and β, the condition of equilibrium is

$$\frac{\partial}{\partial N_i}\left[S + \alpha\left(N - \sum_i N_i\right) + \beta\left(E - \sum_i N_i E_i\right)\right] = 0. \tag{D.3}$$

The distribution can be found by combining the solution of Eq. D.3 with the two constraints in Eqs. D.1 and D.2.

During the calculation, one needs an approximate value of the derivative of ln $N!$ for large N. This can be simply taken as

$$\frac{d}{dN}\ln N! \approx \ln N! - \ln(N-1)! \approx \ln N. \tag{D.4}$$

D.1 Maxwell–Boltzmann Statistics

In the case of Maxwell–Boltzmann statistics, applicable to classical atomic systems, the particles are distinguishable, and the number of particles per energy level is unlimited. The number of possible configurations W can be determined as follows.

Suppose that N particles are placed into containers with occupation numbers N_1, N_2, ... N_i, ... and so on. Consider now the number of ways to place N_1 particles into container 1. First, choose one of the N particles for the first position in container 1, that is, N ways. Next, choose one of the remaining $N-1$ particles for the second position in container 1. This generates $N-1$ ways. Altogether there are $N(N-1)$ ways. By continuing the process, the total number of ways is $N(N-1)(N-2)...(N-N_1-1)$. However, the placement of particles in the container is arbitrary. Therefore, there is a $N_1!$-fold redundancy. The net ways of placement are given as

$$W_1 = \frac{N(N-1)(N-2)...(N-N_1-1)}{N_1!} = \frac{N!}{(N-N_1)!\,N_1!}. \tag{D.5}$$

Similarly, the number of ways of placing the remaining $N-N_1$ particles into container 2 generates a factor

$$W_2 = \frac{(N-N_1)!}{(N-N_1-N_2)!\,N_2!}. \tag{D.6}$$

Similarly,

$$W_3 = \frac{(N-N_1-N_2)!}{(N-N_1-N_2-N_3)!\,N_3!}. \tag{D.7}$$

Continuing the process further, finally we find

$$W = W_1\,W_2\,...W_i... = \frac{N!}{\prod_i N_i!}. \tag{D.8}$$

The entropy is

$$S = k_{\mathrm{B}}\ln W = k_{\mathrm{B}}\left(\ln N! - \sum_i \ln N_i!\right). \tag{D.9}$$

The condition of thermal equilibrium gives

$$\frac{\partial}{\partial N_i}\left[S+\alpha\left(N-\sum_i N_i\right)+\beta\left(E-\sum_i N_i E_i\right)\right]=0. \tag{D.10}$$

The result is

$$k_{\mathrm{B}} \ln N_i = -\alpha - \beta E_i, \tag{D.11}$$

The meaning of the parameter β can be interpreted based on thermodynamics. Because the system is under the condition of constant temperature and constant volume, according to Eq. 6.28, we have

$$dE = T\,dS. \tag{D.12}$$

By treating S and E as variables, comparing Eq. D.3 and Eq. D.12, we find, heuristically,

$$\beta = \frac{1}{T}. \tag{D.13}$$

The constant α can be determined by the condition that the total number of particles is N. Equation D.11 can be rewritten as

$$n_i = \frac{N}{Z}\exp\left(\frac{-E_i}{k_{\mathrm{B}}T}\right), \tag{D.14}$$

where Z is a constant determined by the condition

$$\sum_i N_i = \sum_i \frac{N}{Z}\exp\left(\frac{-E_i}{k_{\mathrm{B}}T}\right) = N, \tag{D.15}$$

in other words,

$$Z = \sum_I \exp\left(\frac{-E_i}{k_{\mathrm{B}}T}\right). \tag{D.16}$$

By introducing a *probability* of the ith energy level, $p_i = N_i/N$, Eq. D.16 can be written as

$$p_i = \frac{1}{Z}\exp\left(\frac{-E_i}{k_{\mathrm{B}}T}\right). \tag{D.17}$$

Obviously, the sum of all probabilities is 1,

$$\sum_i p_i = 1. \tag{D.18}$$

D.2 Fermi–Dirac Statistics

Electrons are fermions obeying the Pauli exclusion principle. Each state can only be occupied by one electron. The electrons satisfies Fermi–Dirac statistics.

For each energy value, there can be multiple states. For example, each electron can have two spin states that have the same energy level. Let the *degeneracy*, that is, the number of states at energy E_i, be g_i. The number of electrons staying at that energy level, N_i, should not exceed g_i. The number of different ways of occupation in the g_i states is

$$W_i = \frac{g_i!}{(g_i - N_i)!\, N_i!}, \tag{D.19}$$

which is a special case of Eq. D.5. Following Eq. D.3, we obtain

$$k_{\rm B}\left[\ln(g_i - N_i) - \ln N_i\right] = \alpha + \beta E_i, \tag{D.20}$$

or

$$N_i = \frac{g_i}{\exp\left(\dfrac{\alpha + \beta E_i}{k_{\rm B}}\right) + 1}. \tag{D.21}$$

Using Eq. D.13 and introducing the probability $p_i = N_i/g_i$,

$$p_i = \frac{1}{\exp\left(\dfrac{E_i - E_F}{k_{\rm B}T}\right) + 1}. \tag{D.22}$$

Equation D.22 is called the *Fermi function*. At the *Fermi level* E_F, the probability is 1/2. Apparently,

$$\begin{aligned} p_i &\to 1, \quad E_i \ll E_F; \\ p_i &\to 0, \quad E_i \gg E_F. \end{aligned} \tag{D.23}$$

At low temperature, where $k_{\rm B}T \ll E_F$, the Fermi–Dirac statistics becomes

$$\begin{aligned} p_i &= 1, \quad E_i < E_F; \\ p_i &= 0, \quad E_i > E_F. \end{aligned} \tag{D.24}$$

At high temperatures, or at high energy levels $(E_i - E_F)/k_{\rm B}T \gg 1$, the Fermi–Dirac statistics reduces to Maxwell–Boltzmann statistics,

$$p_i \propto \exp\left(\frac{-E_i}{k_{\rm B}T}\right). \tag{D.25}$$

Appendix E

AM1.5 Reference Solar Spectrum

To facilitate testing of solar photovoltaic devices, American Society for Testing and Materials (ASTM) in conjenction with the photovoltaic (PV) industry and government research and development laboratories developed and defined two standard terrestrial solar spectral irradiance distributions: The standard direct normal spectral irradiance and the standard global spectral irradiance, incorporated into a single document, ASTM G-173-03; see Section 5.2.1.

The original data was presented in wavelength scale. The following table is presented in photon energy scale. The first column is photon energy in eV. The second column is the extraterrestrial solar radiation spectrum, AM0. The third column is the direct normal solar radiation spectrum. The fourth column is the global solar radiation spectrum, including radiation scattered from the sky. All spectral data are presented in watts per square meter per eV (W/m^2·eV).

Table E.1: AM1.5 Reference Solar Spectrum (W/m^2·eV)

ϵ (eV)	AM0	Direct	Global	ϵ (eV)	AM0	Direct	Global
0.32	116	94	94	0.47	232	0	0
0.33	128	105	105	0.48	240	0	0
0.34	135	104	104	0.49	251	11	11
0.35	141	110	109	0.50	265	104	105
0.36	146	75	74	0.51	274	163	164
0.37	156	37	37	0.52	283	199	201
0.38	163	27	26	0.53	292	247	250
0.39	168	52	52	0.54	301	268	271
0.40	178	25	25	0.55	313	294	299
0.41	184	32	32	0.56	322	294	299
0.42	192	16	16	0.57	328	310	315
0.43	200	0	0	0.58	338	321	326
0.44	208	0	0	0.59	345	286	290
0.45	216	0	0	0.60	358	252	256
0.46	223	0	0	0.61	370	154	156

AM1.5 Reference Solar Spectrum (continued)

ϵ (eV)	AM0	Direct	Global	ϵ (eV)	AM0	Direct	Global
0.62	380	214	218	1.02	577	487	510
0.63	385	46	47	1.03	579	485	508
0.64	399	2	2	1.04	577	461	483
0.65	409	0	0	1.05	582	467	489
0.66	406	0	0	1.06	577	398	418
0.67	427	0	0	1.07	584	245	257
0.68	433	31	31	1.08	579	204	213
0.69	447	214	219	1.09	583	141	148
0.70	456	367	376	1.10	578	129	135
0.71	460	396	406	1.11	581	298	314
0.72	477	448	460	1.12	578	445	468
0.73	482	459	471	1.13	571	506	534
0.74	494	470	483	1.14	576	523	553
0.75	497	472	485	1.15	578	532	563
0.76	507	489	503	1.16	584	539	571
0.77	513	479	493	1.17	586	546	578
0.78	516	474	487	1.18	586	550	582
0.79	523	507	522	1.19	589	553	587
0.80	528	505	520	1.20	591	555	589
0.81	529	472	486	1.21	589	552	586
0.82	530	392	404	1.22	597	558	593
0.83	533	212	218	1.23	592	553	587
0.84	535	147	151	1.24	599	559	595
0.85	536	116	119	1.25	600	546	581
0.86	538	52	53	1.26	599	484	514
0.87	541	20	21	1.27	596	454	482
0.88	544	1	1	1.28	595	360	382
0.89	547	0	0	1.29	593	268	284
0.90	547	0	0	1.30	601	233	247
0.91	549	4	4	1.31	593	229	243
0.92	551	204	211	1.32	607	149	158
0.93	553	327	340	1.33	603	351	374
0.94	558	429	447	1.34	596	471	503
0.95	564	494	516	1.35	602	408	435
0.96	562	528	552	1.36	599	422	450
0.97	575	488	510	1.37	596	430	460
0.98	573	516	540	1.38	598	471	504
0.99	574	547	573	1.39	596	545	585
1.00	576	545	571	1.40	591	544	584
1.01	579	522	547	1.41	590	544	584

AM1.5 Reference Solar Spectrum (continued)

ϵ (eV)	AM0	Direct	Global	ϵ (eV)	AM0	Direct	Global
1.42	591	544	586	1.82	556	471	520
1.43	579	532	572	1.83	554	467	517
1.44	596	544	586	1.84	553	464	513
1.45	554	504	543	1.85	555	464	512
1.46	589	534	577	1.86	556	456	504
1.47	593	530	573	1.87	552	443	489
1.48	589	507	548	1.88	519	422	467
1.49	590	471	508	1.89	519	420	465
1.50	586	453	488	1.90	538	432	477
1.51	587	439	474	1.91	536	425	471
1.52	594	481	521	1.92	542	439	486
1.53	587	521	566	1.93	533	430	474
1.54	585	514	558	1.94	539	430	475
1.55	584	514	557	1.95	535	422	467
1.56	581	509	555	1.96	526	403	446
1.57	592	525	572	1.97	537	403	446
1.58	586	526	572	1.98	519	401	442
1.59	587	526	572	1.99	528	406	450
1.60	583	519	567	2.00	521	404	448
1.61	576	452	493	2.01	511	396	439
1.62	587	212	229	2.02	519	400	443
1.63	584	414	452	2.03	518	397	441
1.64	583	517	564	2.04	522	396	440
1.65	578	511	560	2.05	513	386	430
1.66	577	508	557	2.06	510	383	427
1.67	560	488	535	2.07	513	379	422
1.68	575	485	532	2.08	511	373	413
1.69	570	447	490	2.09	507	369	411
1.70	565	414	454	2.10	481	347	386
1.71	570	427	469	2.11	504	376	420
1.72	559	384	421	2.12	509	380	424
1.73	562	477	526	2.13	499	369	411
1.74	566	485	535	2.14	492	356	398
1.75	567	479	527	2.15	487	351	393
1.76	561	462	509	2.16	496	358	400
1.77	561	467	514	2.17	473	342	383
1.78	564	444	491	2.18	484	354	396
1.79	565	420	462	2.19	466	342	383
1.80	555	396	434	2.20	468	343	385
1.81	556	471	519	2.21	458	336	377

AM1.5 Reference Solar Spectrum (continued)

ϵ (eV)	AM0	Direct	Global	ϵ (eV)	AM0	Direct	Global
2.22	458	337	379	2.62	365	241	284
2.23	466	342	385	2.63	349	229	271
2.24	460	337	381	2.64	356	233	277
2.25	455	333	376	2.65	339	221	263
2.26	448	326	368	2.66	349	228	271
2.27	451	328	371	2.67	355	230	272
2.28	443	322	365	2.68	355	229	273
2.29	415	302	342	2.69	341	218	261
2.30	434	316	358	2.70	342	218	261
2.31	447	323	366	2.71	347	220	265
2.32	415	299	339	2.72	333	212	253
2.33	443	318	361	2.73	324	204	245
2.34	433	310	354	2.74	344	215	260
2.35	392	282	320	2.75	338	210	254
2.36	429	309	351	2.76	325	201	244
2.37	409	295	335	2.77	320	198	240
2.38	394	283	323	2.78	298	182	221
2.39	356	254	291	2.79	304	184	226
2.40	403	288	330	2.80	302	182	225
2.41	392	279	320	2.81	284	171	209
2.42	414	294	337	2.82	248	148	183
2.43	406	286	328	2.83	274	163	201
2.44	396	277	319	2.84	286	170	210
2.45	410	286	328	2.85	248	146	181
2.46	386	268	309	2.86	255	149	185
2.47	377	263	304	2.87	219	127	158
2.48	385	269	311	2.88	180	104	130
2.49	394	274	316	2.89	234	135	168
2.50	395	274	318	2.90	241	137	172
2.51	383	266	308	2.91	252	143	179
2.52	371	255	297	2.92	247	139	175
2.53	372	256	296	2.93	251	141	178
2.54	357	244	283	2.94	258	144	182
2.55	340	231	269	2.95	237	131	166
2.56	378	256	298	2.96	238	130	166
2.57	385	258	302	2.97	248	135	172
2.58	381	254	297	2.98	249	135	173
2.59	378	250	293	2.99	236	127	163
2.60	367	243	285	3.00	244	130	168
2.61	371	246	289	3.01	234	123	159

AM1.5 Reference Solar Spectrum (continued)

ϵ (eV)	AM0	Direct	Global	ϵ (eV)	AM0	Direct	Global
3.02	208	109	142	3.42	106	39	59
3.03	236	123	160	3.43	101	36	55
3.04	217	112	146	3.44	110	39	60
3.05	221	114	148	3.45	82	29	45
3.06	229	117	153	3.46	84	29	46
3.07	227	115	151	3.47	92	32	50
3.08	235	119	157	3.48	108	37	58
3.09	226	113	150	3.49	115	39	62
3.10	213	106	140	3.50	111	37	59
3.11	161	79	106	3.51	95	32	50
3.12	94	46	61	3.52	99	32	52
3.13	156	75	101	3.53	105	34	55
3.14	114	54	73	3.54	95	30	49
3.15	98	46	63	3.55	90	28	46
3.16	160	75	102	3.56	90	28	46
3.17	157	73	100	3.57	93	29	47
3.18	139	64	88	3.58	92	28	46
3.19	121	56	77	3.59	82	25	41
3.20	124	56	78	3.60	87	26	43
3.21	121	54	75	3.61	98	29	48
3.22	114	51	71	3.62	95	28	47
3.23	85	38	53	3.63	90	26	44
3.24	104	46	64	3.64	98	27	46
3.25	140	61	85	3.65	90	25	43
3.26	130	56	79	3.66	86	23	41
3.27	150	64	91	3.67	77	20	35
3.28	149	63	90	3.68	75	20	34
3.29	126	53	76	3.69	84	22	39
3.30	117	49	70	3.70	90	23	41
3.31	105	43	62	3.71	85	21	36
3.32	119	48	70	3.72	86	21	38
3.33	131	53	77	3.73	88	21	38
3.34	130	52	76	3.74	87	20	36
3.35	141	56	82	3.75	94	22	39
3.36	127	50	73	3.76	91	21	38
3.37	135	52	77	3.77	81	18	32
3.38	139	53	79	3.78	84	17	31
3.39	123	47	70	3.79	86	18	34
3.40	113	43	64	3.80	81	16	30
3.41	112	42	63	3.81	70	13	23

AM1.5 Reference Solar Spectrum (continued)

ϵ (eV)	AM0	Direct	Global	ϵ (eV)	AM0	Direct	Global
3.82	64	12	23	4.12	31	0	0
3.83	56	10	19	4.13	33	0	0
3.84	59	10	18	4.14	35	0	0
3.85	60	10	19	4.15	33	0	0
3.86	65	11	20	4.16	37	0	0
3.87	64	9	16	4.17	34	0	0
3.88	59	8	16	4.18	38	0	0
3.89	57	7	14	4.19	39	0	0
3.90	65	8	14	4.20	36	0	0
3.91	56	7	13	4.21	36	0	0
3.92	47	5	9	4.22	38	0	0
3.93	52	5	10	4.23	36	0	0
3.94	55	5	9	4.24	38	0	0
3.95	57	4	8	4.25	40	0	0
3.96	54	4	8	4.26	41	0	0
3.97	55	3	7	4.27	38	0	0
3.98	58	3	6	4.28	32	0	0
3.99	50	2	5	4.29	24	0	0
4.00	39	1	3	4.30	20	0	0
4.01	46	1	3	4.31	22	0	0
4.02	49	1	2	4.32	23	0	0
4.03	47	1	2	4.33	15	0	0
4.04	43	0	1	4.34	10	0	0
4.05	44	0	1	4.35	13	0	0
4.06	48	0	1	4.36	19	0	0
4.07	46	0	0	4.37	20	0	0
4.08	47	0	0	4.38	20	0	0
4.09	41	0	0	4.39	17	0	0
4.10	33	0	0	4.40	13	0	0
4.11	34	0	0	4.41	9	0	0

List of Symbols

Foundamental Constants

c	speed of light in free space	2.998×10^8 m/s
h	Planck's constant	6.626×10^{-34} J·s
$\hbar$	Dirac's constant	1.055×10^{-34} J·s
k_B	Boltzmann's constant	1.38×10^{-23} J K^{-1}
q	elementary charge	1.60×10^{-19} C
σ	Stefan-Boltzmann constant	5.67×10^{-8} W m^{-2} K^{-4}
ε_0	electric constant	8.85×10^{-12} F/m
μ_0	magnetic constant	$4\pi \times 10^{17}$ H/m
m_e	electron mass	9.11×10^{-31} kg
R	gas constant	8.31×10^3 J kmole^{-1} K^{-1}

General Physics

ϵ	photon energy	J or eV
$\mathbf{E}$	electric field intensity	V/m
$\mathbf{B}$	magnetic field intensity	Tesla
$\mathbf{A}$	vector potential	V·s/m
α	absorption coefficient	m^{-1}
$A(\lambda)$	absorptivity	dimensionless
$R(\lambda)$	reflectivity	dimensionless
$T(\lambda)$	transmittivity	dimensionless

Semiconductor Physics

m^*	effective mass
E_C	bottom of the conduction band
E_V	top of the valance band
E_g	energy gap
E_F	Fermi level or Fermi energy
α_r	recombination rate of holes and electrons
μ_n	mobility of electrons
μ_p	mobility of holes
N_C	effective density of states in conduction band
N_V	effective density of states in valance band
N_A	concentration of acceptors
N_D	concentration of donors
n_i	intrinsic carrier concentration
n_n	concentration of electrons in an n-region

n_p	concentration of electrons in a p-region
n_{p0}	equilibrium concentration of electrons in a p-region
p_n	concentration of holes in an n-region
p_p	concentration of holes in a p-region
p_{n0}	equilibrium concentration of holes in an n-region
τ_n	lifetime of excess electrons in a p-region
τ_p	lifetime of excess holes in an n-region
D_n	diffusion coefficient of electrons
D_p	diffusion coefficient of holes
L_n	diffusion length of electrons
L_p	diffusion length of holes

Astromomy

S	solar constant	1366 W/m^2
$A_\odot$	distance between the Sun and Earth	1.50×10^{11} m
$r_\odot$	radius of the Sun	6.96×10^8 m
$L_\odot$	luminosity of the Sun	3.84×10^{26} W
$T_\odot$	surface temperature of the Sun	5800 K
$r_\oplus$	radius of the Earth	6378 m
ϕ	latitude	geographical coordinate
λ	longitude	geographical coordinate
h	height	also called altitude or elevation
A	azimuth	horizontal direction or bearing
δ	declination	angular distance to the equator
ω	hour angle	in radians, westward
ω_s	sunset hour angle	in radians, always positive
$\omega_{\rm ew}$	east-west hour angle	in radians, always positive
α	right ascension	absolute celestial coordinate
l	mean ecliptic longitude	on ecliptic plane
θ	true ecliptic longitude	on ecliptic plane
e	eccentricity of orbit	currently ≈ 0.0167
ε	obliquity of ecliptic	currently $\approx 23.44°$

Thermodynamics

T	temperature	K
$T_{\rm L}$	temperature of cold reservoir	
$T_{\rm H}$	temperature of hot reservoir	
η	efficeincy	

Bibliography

[1] Renewables Global Status Report 2007. *REN21 Publications*, 2007.

[2] Mineral Commodity Summaries. *U.S. Geological Survey*, 195:95, 2009.

[3] W. Adams. Cooking by Solar Heat. *Scientific American*, 1878:376, 1878.

[4] E. A. Alsema, M. J. de Wild-Scholten, and V. M. Fthenakis. Environmental impacts of PV electricity generation - a critical comparison of energy supply options. *Proceedings of 21st European Photovoltaic Solar Energy Conference*, 50:97–147, 2006.

[5] D. Banks. *An Introduction to Thermogeology: Ground Source Heating and Cooling.* Blackwell Publishing, Oxford, UK, 2008.

[6] A. J. Bard and M. A. Fox. Artificial photosynthesis solar splitting of water to hydrogen and oxygen. *Acc. Chem. Res.*, 28:141–145, 1995.

[7] J. K. Beatty, C. C. Peterson, and A. Chaokin. *The Solar System.* Cambridge University Press, Cambridge, UK, 1999.

[8] H. A. Bethe. Energy Production in Stars. *Physical Review*, 55:434–456, 1939.

[9] H. A. Bethe and C. L. Critchfield. The Formation of Deuterions by Proton Combination. *Physical Review*, 54:248–254, 1938.

[10] S. Blanco. Solar parking lot with electric car charging stations opens in Tennessee. *AutoblogGreen*, Aug 23, 2011.

[11] R. E. Blankenship. *Molecular Mechanisms of Photosynthesis.* Blackwell Science, Oxford, UK, 2002.

[12] J. B. Bolton and D. O. Hall. Photochemical conversion and storage of solar energy. *Ann. Rev. Energy*, 4:353–401, 1979.

[13] M. Born and E. Wolf. *Principles of Optics.* Seventh Edition, Cambridge University Press, Cambridge, 1999.

[14] G. Boyle. *Renewable Energy.* Second Edition, Oxford University Press, Oxford, UK, 2004.

[15] C. J. Brabec, N. S. Sariciftci, and J. C. Hummelen. Plastic solar cells. *Advanced Functional Materials*, 11:15–26, 2001.

[16] J. Britt and O. Ferekides. Thin-film CdS/CdTe solar cell with 15.8% efficiency. *Applied Physics Letters*, 62:2851–2852, 1993.

[17] K. Butti and J. Perlin. *A Golden Thread.* Marion Boyers, London Boston, 1980.

[18] C. J. Chen. *Introduction to Scanning Tunneling Microscopy.* Oxford University Press, Oxford, UK, 2007.

[19] K. L. Chopra, P. D. Paulson, and V. Dutta. Thin-film solar cells: an overview. *Progress in Photovoltaics*, 12:69–92, 2004.

[20] C. Darwin. *On the Origin of Spicies.* Oxford University Press, Oxford, 1859.

[21] K. S. Deffeyes. *Beyond Oil.* Hill and Wang, New York, 2005.

[22] P. A. M. Dirac. The Quantum Theory of the Emission and Absorption of Radiation. *Proc. R. Soc. London*, A114:243–265, 1927.

[23] A. Duffie and W. A. Beckman. *Solar Energy Thermal Processes.* John Wiley and Sons, New York, 1974.

[24] A. Duffie and W. A. Beckman. *Solar Engineering of Thermal Processes.* Third edition, John Wiley and Sons, Hoboken, NJ, 2006.

[25] US DoE EERE. *National Algal Biofuels Technology Roadmap.* US Department of Energy, 2010.

[26] A. Einstein. Ist die Trägheit eines Körpers von seinem Energieinhalt abhängig? *Annalen der Physik*, 18:639–641, 1905.

[27] A. Einstein. Über einen die Erzeugung und Verwandung des Lichts betreffenden heuristischen Gesichtspunkt. *Annalen der Physik*, 17:132–148, 1905.

[28] M. M. Farid, A. M. Khudhair, and S. A. K. Razack S. Al-Hallaj.

[29] E. Fermi. *Nuclear Physics.* U. of Chicago Press, Chicago, 1950.

[30] K. Garber. Steven Chu, Obama's Point Man on Energy, Says Conservation Is 'Sexy'. *U.S. News and World Report*, March 2009, 2009.

[31] H. P. Garg, S. C. Mullik, and A. K. Bhargava. *Solar Thermal Energy Storage.* D. Reider Publishing Company, Dordricht, 1985.

[32] M. Grätzel. Photoelectrochemical cells. *Nature*, 414:338–344, 2001.

[33] M. Grätzel. Dye-sensitized solar cells. *Journal of Photochemistry and Photobiology C: Photochemistry Reviews*, 4:145–153, 2003.

[34] M. A. Green. *Solar cells.* Prentice-Hal, Englewood Cliffs, NJ, 1982.

[35] M. A. Green. Limits on the open-circuit voltage and efficiency of silicon solar cells imposed by intrinsic auger processes. *IEEE Transactions on Electron Devices*, 31:671–678, 1996.

[36] M. A. Green, J. Zhao, A. Wang, and S. R. Wenham. Progress and outlook for high-efficiency crystalline silicon solar cells. *Solar Energy Materials and Solar Cells*, 65:9–16, 2001.

[37] W. Heitler. *The Quantum Theory of Radiation.* Clarendon Press, Oxford, 1954.

[38] P. Heremans, D. Cheyns, and B. P. Rand. Strategies for increasing the efficieny of heterojunction organic photovoltaic cells: Material selection and device architecture. *Accounts of Chemical research*, 42:1740–1747, 2009.

[39] H. C. Hottel and A. Whillier. Evaluation of flat-plate solar collector performance. *Transcaction of Conference on the Use of Solar Energy*, II:74–104, 1958.

[40] M. K. Hubbert. Nuclear Energy and the Fossil Fuels. *Shell Development Company Publications*, 95:1–40, 1956.

[41] C. E. Kennedy. Reiew of mid- to high-temperature absorber materials. *National Renewable Energy Laboratory report*, 520:31267, 2002.

[42] C. E. Kennedy and H. Price. Progress in development of high-temperature solar-selective coating. *Proceedings of ISEC 2005*, 520:36997, 2005.

[43] R. R. King, D. C. Law, K. M. Edmondson, C. M. Fetzer, G. S. Kinsey, H. Yoon, R. A. Sherif, and N. H. Karam. 40% efficient metamorphic GaInP/GaInAs/Ge multijunction solar cells. *Appl. Phys. Lett.*, 90:183516, 2007.

[44] S. A. Klein. Calculation of flat-plate collector loss coefficients. *Solar Energy*, 17:79–80, 1975.

[45] S. A. Klein. Why wind power works for Denmark. *Proc. ICE Civil Engineering.*, 2005.

[46] S. A. Klein, W. A. Beckman, and A. Duffie. A design procedure for solar heating systems. *Solar Energy*, 18:113–127, 1975.

[47] M. M. Koltun. *Selective optical surfaces for solar energy converters.* Allerton Press, Inc. New York, 1981.

[48] Sandia National Laboratory. Sandia, Stirling Energy Systems set new world record for solar-to-grid conversion efficiency. *News Release*, February 12, 2008.

[49] A. D. Leite. *Energy in Brazil.* Earthscan, London, 2009.

[50] P. Lenard. Über die lichtelektrische Wirkung. *Annalen der Physik*, 8:149–170, 1902.

[51] D. Linden and T. B. Reddy. *Handbook of Batteries.* Third Edition, McGraw-Hill, New York, 2002.

[52] B. Y. H. Liu and R. C. Jordan. The interrelationship and characteristic distribution of direct, diffuse, and total solar radiation. *Solar Energy*, 4(3):1–19, 1960.

[53] J. Lund, B. Sanner, L. Rybach, R. Curtis, and G. Hellstrom. Geothermal (ground-source) heat pumps: a world overview. *GHC Bullitin*, September:31267, 2004.

[54] P. J. Lunde. *Solar Thermal Engineering.* John Wiley and Sons, New York, 1980.

[55] C. Lyell. *Elements of Geology.* C. H. Key & Co., Pittzburgh, 1839.

[56] D. J. C. MacKay. *Sustainable Energy – Without Hot Air.* UIT, Cambridge, England, 2008.

[57] H. A. Macleod. *Thin Film Optical Filters.* American Elsevier Publishing Company, Inc., New York, 2005.

[58] J. C. Maxwell. *A Dynamic Theory of the Electromagnetic Field.* Reprinted by Wipf and Stock Publishers, 1996, 1864.

[59] McKinsey&Company. China's Green Revolution. *McKinsey Global Energy and Materials*, July 2009:1–136.

[60] McKinsey&Company. Unlocking Energy Efficiency in the U.S. Economy. *McKinsey Global Energy and Materials*, July 2009:1–144.

[61] R. A. Millikan. A Direct Photoelectric Determination of Planck's h. *Physical Review*, 7:355–388, 1916.

[62] R. A. Millikan and I. B. Cohen. *Autobiography of Robert A. Millikan.* Arno Press, New York, 1980.

[63] D. Moché. *Astronomy.* 7th Ed, John Wiley and Sons, Hoboken, 2009.

[64] B. O'Regan and M. Grätzel. A low-cost, high-efficiency solar cell based on dye-sensitized colloidal TiO_2 films. *Nature*, 353:737–740, 1991.

[65] J. I. Pankove. *Optical processes in semiconductors.* Cover Publications, Inc. New York, 1971.

[66] J. P. Peixoto and A. H. Oort. *Physics of Climate.* Third Edition, John Wiley and Sons, New York, 1992.

[67] J. Perlin. *From Space to Earth.* Aatec Publications, Ann Arbor, Michigan, 1999.

[68] J. H. Phillips. *Guide to the Sun.* Cambridge University Press, Cambridge, UK, 1992.

[69] M. Powalla and D. Bonnet. Thin-film solar cells based on polycrystalline compound semiconductors CIS and CdTe. *Advances in Optoelectronics*, 2007:97545, 2007.

[70] Jr. R. W. Bliss. The derivations of several plate efficiency factors in the design of flat-plate solar collectors. *Solar Energy*, 3:55–64, 1959.

[71] J. Risen. U.S. Identifies Vast Riches of Minerals in Afghanistan. *New York Times*, June 13, 2010.

[72] H. H. Rogner. *World Energy Assessment: Energy and the Challenge of Sustainability*. United Nations Development Programme, 2000.

[73] M. Romero, R. Buck, and J. E. Pacheco. Water-in-glass evacualte tube solar water heaters. *Journal of Solar Energy Engineering*, 124:98–108, 2002.

[74] J. J. Romm. *The Hype about Hydrogen: Fact and Fiction in the Race to Save the Climate*. Island Press, New York, 2005.

[75] H. J. Sauer and R. H. Howell. *Heat Pump Systems*. John Wiley and Sons, New York, 1983.

[76] J. L. Shay and S. Wagner. Efficient $CuInSe_2/CdS$ solar cells. *App. Phys. Lett.*, 27:89–90, 2007.

[77] W. Shockley and H. J. Queisser. Detailed balance limit of efficiency of p-n junction solar cells. *Journal of Applied Physics*, 32:510–519, 1961.

[78] B. J. Stanbery. Copper indium selenides and related materials for photovoltaic devices. *Critical Reviews in Solid State and Materials Sciences*, 27:73–117, 2002.

[79] M. Stix. *The Sun, an Introduction*. Second Edition, Springer, New York, 2002.

[80] C. W. Tang. Two-layer organic photovoltaic cell. *Appl. Phys. Lett.*, 48:183–185, 2004.

[81] W. Thompson. On the Age of Sun's Heat. *Macmillan's Magazine*, 5:388–393, 1862.

[82] W. Thompson. Nineteenth Century Clouds over the Dynamical Theory of Heat and Light. *Royal Institution Proceedings*, 16:363–397, 1900.

[83] T. Tiedje, E. Yablonovitch, G. D. Cody, and B. G. Brooks. Limiting efficiency of silicon solar cells. *IEEE Transactions on Electron Devices*, 31:711–716, 1996.

[84] D. Ugolini, J. Zachary, and J. Park. Options for hybrid solar and conventional fossil plants. *Bechtel Technology Journal*, 2:1–11, 2009.

[85] P.-F. Verhulst. Notice sur la loi que la population poursuit dans son accroissement. *Correspondance mathematique et physique*, 10:113–121, 1838.

[86] D. Voet and J. D. Voet. *Biochemistry*. John Wiley and Sons, Hoboken, NJ, 2004.

[87] S. Wagner, L. J. Shay, P. Migliorato, and H. M. Kasper. $CuImSe_2/CdS$ heterojunction photovoltaic detectors. *Applied Physics Letters*, 25:434–435, 1974.

[88] Renewable Energy World. Global concentrated solar power industry to reach 25 gw in 2020. *REW*, 2009.

[89] Q. C. Zhang. Recent progress in high-temperature solar-selective coatings. *Solar Energy Materials and Solar Cells*, 62:63–74, 2000.

[90] Q. C. Zhang, Y. Yin, and D. R. Mills. High-efficiency M0-Al_2O_3 cermet selective surfaces for high-temperature application. *Solar Energy Materials and Solar Cells*, 40:43–53, 1996.

[91] J. Zhao, A. Wang, P. P. Altermatt, S. R. Wenham, and M. A. Green. 24% efficient perl silicon solar cell: Recent improvements in high-efficiency silicon solar cell rezsearch. *Solar Energy Materials and Solar Cells*, 41–42:87–99, 1996.

Index